Student Solutions Guide for

CALCULUS:
AN APPLIED APPROACH

FIFTH EDITION

Larson / Edwards

Bruce H. Edwards

Department of Mathematics
University of Florida

Houghton Mifflin Company   Boston   New York

Sponsoring Editor: Jack Shira
Managing Editor: Cathy Cantin
Senior Associate Editor: Maureen Ross
Associate Editor: Michael J. Richards
Assistant Editor: Carolyn Johnson
Supervising Editor: Karen Carter
Project Editor: Patty Bergin
Editorial Assistant: Christine E. Lee
Art Supervisor: Gary Crespo
Marketing Manager: Ros Kane
Marketing Assistant: Erin Dionne
Design: Henry Rachlin
Composition and Art: Meridian Creative Group

Calculator Key font used with permission of Texas Instruments Incorporated. Copyright 1990, 1993, 1996.

Printed in the U.S.A.

ISBN: 0-395-93344-7

123456789-EB-02 01 00 99 98

Preface

The *Student Solutions Guide for Calculus: An Applied Approach,* Fifth Edition, is a supplement to the text by Ron Larson and Bruce H. Edwards. Solutions to all odd-numbered exercises in the textbook are given with all essential algebraic steps included.

If you have any corrections or suggestions for improving this solutions guide, I would appreciate hearing from you.

I would like to thank the staff at Larson Texts, Inc. for their help in the production of this guide.

Bruce H. Edwards
Department of Mathematics
University of Florida
Gainesville, Florida 32611
(be@math.ufl.edu)

CONTENTS

CHAPTER 0
A Precalculus Review

CHAPTER 0
A Precalculus Review

Section 0.1   The Real Line and Order

Solutions to Odd-Numbered Exercises

1. Since $0.7 = \dfrac{7}{10}$, it is rational.

3. $\dfrac{3\pi}{2}$ is irrational because π is irrational.

5. $4.\overline{3451}$ is rational because it has a repeating decimal expansion.

7. Since $\sqrt[3]{64} = 4$, it is rational.

9. $\sqrt[3]{60}$ is irrational, since 60 is not the cube of a rational number.

11. (a) Yes, if $x = 3$, then $5(3) - 12 = 3 > 0$.

    (b) No, if $x = -3$, then $5(-3) - 12 = -27 < 0$.

    (c) Yes, if $x = \frac{5}{2}$, then $5\left(\frac{5}{2}\right) - 12 = \frac{1}{2} > 0$.

    (d) No, if $x = \frac{3}{2}$, then $5\left(\frac{3}{2}\right) - 12 = -\frac{9}{2} < 0$.

13. $0 < \dfrac{x - 2}{4} < 2$

$0 < x - 2 < 8$

$2 < x < 10$

    (a) Yes, if $x = 4$, then $2 < x < 10$.

    (b) No, if $x = 10$, then x is not less than 10.

    (c) No, if $x = 0$, then x is not greater than 2.

    (d) Yes, if $x = \frac{7}{2}$, then $2 < x < 10$.

15. $\quad x - 5 \geq 7$

$x - 5 + 5 \geq 7 + 5$

$\qquad x \geq 12$

17. $4x + 1 < 2x$

$\quad 2x < -1$

$\qquad x < -\frac{1}{2}$

19. $4 - 2x < 3x - 1$

$\quad 4 - 5x < -1$

$\qquad -5x < -5$

$\qquad\quad x > 1$

21. $\quad -4 < 2x - 3 < 4$

$-4 + 3 < 2x - 3 + 3 < 4 + 3$

$\qquad -1 < 2x < 7$

$\qquad -\frac{1}{2} < x < \frac{7}{2}$

23. $\frac{3}{4} > x + 1 > \frac{1}{4}$

$-\frac{1}{4} > x > -\frac{3}{4}$

$-\frac{3}{4} < x < -\frac{1}{4}$

25. $\dfrac{x}{2} + \dfrac{x}{3} > 5$

$3x + 2x > 30$

$\quad 5x > 30$

$\qquad x > 6$

27. $x^2 - x \leq 0$

$x(x - 1) \leq 0$

Testing the intervals $(-\infty, 0)$, $(0, 1)$, and $(1, \infty)$, you see that the solution is $0 \leq x \leq 1$.

29.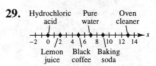

31. $R = 115.95x$ and $C = 95x + 750$ and we have:

$$R > C$$

$$115.95x > 95x + 750$$

$$20.95x > 750$$

$$x > \frac{750}{20.95} = 35.7995\ldots$$

Therefore, $x \geq 36$ units.

33. Let $x =$ length of the side of the square. Then, the area of the square is x^2, and we have

$$x^2 \geq 500$$

$$x \geq \sqrt{500}$$

$$x \geq 10\sqrt{5} \approx 22.36 \text{ meters.}$$

35. (a) True. Since $a < b$, $a - 4 < b - 4$.

   (c) True. Since $a < b$, $-3a > -3b$.

   (b) False. Since $a < b$, $-a > -b$ and $4 - a > 4 - b$.

   (d) True. Since $a < b$, $\frac{1}{4}a < \frac{1}{4}b$.

Section 0.2   Absolute Value and Distance on the Real Line

1. (a) The directed distance from a to b is $75 - 126 = -51$.

   (b) The directed distance from b to a is $126 - 75 = 51$.

   (c) The distance between a and b is $|75 - 126| = 51$.

3. (a) The directed distance from a to b is $-5.65 - 9.34 = -14.99$.

   (b) The directed distance from b to a is $9.34 - (-5.65) = 14.99$.

   (c) The distance between a and b is $|-5.65 - 9.34| = 14.99$.

5. Midpoint $= \dfrac{7 + 21}{2} = 14$

7. Midpoint $= \dfrac{-6.85 + 9.35}{2} = 1.25$

9. $-5 < x < 5$

11. $\dfrac{x}{2} < -3$ or $\dfrac{x}{2} > 3$

   $x < -6$ $x > 6$

13. $-5 < x + 2 < 5$

   $-7 < x < 3$

15. $\dfrac{x - 3}{2} \leq -5$ or $\dfrac{x - 3}{2} \geq 5$

   $\dfrac{x - 3}{2}(2) \leq -5(2)$ $\dfrac{x - 3}{2}(2) \geq 5(2)$

   $x - 3 \leq -10$ $x - 3 \geq 10$

   $x - 3 + 3 \leq -10 + 3$ $x - 3 + 3 \geq 10 + 3$

   $x \leq -7$ $x \geq 13$

17. $10 - x < -4$ or $10 - x > 4$

   $-x < -14$ $-x > -6$

   $x > 14$ $x < 6$

19. $-1 < 9 - 2x < 1$

   $-10 < -2x < -8$

   $5 > x > 4$

   $4 < x < 5$

21. $-b \leq x - a \leq b$

        $a - b \leq x \leq a + b$

23. $|x| \leq 2$

25. $|x| > 2$

27. $|x - 4| \leq 2$

29. $|x - 2| > 2$

31. $|x - 4| < 2$

33. $|y - a| \leq 2$

35. $\left|\dfrac{h - 68.5}{2.7}\right| \leq 1$

      $-1 \leq \dfrac{h - 68.5}{2.7} \leq 1$

      $-2.7 \leq h - 68.5 \leq 2.7$

      $65.8 \leq h \leq 71.2$

37. $|x - 200{,}000| \leq 25{,}000$

    $-25{,}000 \leq x - 200{,}000 \leq 25{,}000$

    $175{,}000 \leq x \leq 225{,}000$

39. (a) $|E - 4750| \leq 500 \Rightarrow 4250 \leq E \leq 5250$

    $0.05(4750) = 237.50$

    $|E - 4750| \leq 237.50 \Rightarrow 4512.50 \leq E \leq 4987.50$

  (b) \$5116.37 is not within 5% of the specified budgeted amount; at variance.

41. (a) $|E - 20{,}000| \leq 500 \Rightarrow 19{,}500 \leq E \leq 20{,}500$

    $0.05(20{,}000) = 1000$

    $|E - 20{,}000| \leq 1000 \Rightarrow 19{,}000 \leq E \leq 21{,}000$

  (b) \$22,718.35 is at variance with both budget restrictions.

Section 0.3   Exponents and Radicals

1. $-3(2)^3 = -3(8) = -24$

3. $4(2)^{-3} = 4\left(\frac{1}{8}\right) = \frac{1}{2}$

5. $\dfrac{1 + (2)^{-1}}{(2)^{-1}} = \dfrac{1 + (1/2)}{1/2} = \dfrac{3/2}{1/2} = 3$

7. $3(-2)^2 - 4(-2)^3 = 3(4) - 4(-8) = 12 + 32 = 44$

9. $6(10)^0 - [6(10)]^0 = 6(1) - 1 = 5$

11. $\sqrt[3]{27^2} = \left(\sqrt[3]{27}\right)^2 = 3^2 = 9$

13. $4^{-1/2} = \dfrac{1}{\sqrt{4}} = \dfrac{1}{2}$

15. $(-32)^{-2/5} = \dfrac{1}{\left(\sqrt[5]{-32}\right)^2} = \dfrac{1}{(-2)^2} = \dfrac{1}{4}$

17. $500(1.01)^{60} \approx 908.3483$

19. $\sqrt[3]{-154} \approx -5.3601$

21. $5x^4(x^2) = 5x^6$

23. $6y^2(2y^4)^2 = 6y^2(4y^8) = 24y^{10}$

25. $10(x^2)^2 = 10x^4$

27. $\dfrac{7x^2}{x^{-3}} = 7x^5$

29. $\dfrac{12(x + y)^3}{9(x + y)} = \dfrac{4}{3}(x + y)^2$

31. $\dfrac{3x\sqrt{x}}{x^{1/2}} = \dfrac{3x\sqrt{x}}{\sqrt{x}} = 3x$

33. $\left(\dfrac{\sqrt{2}\sqrt{x^3}}{\sqrt{x}}\right)^4 = \left(\dfrac{\sqrt{2}(x\sqrt{x})}{\sqrt{x}}\right)^4$

$\qquad = \left(\sqrt{2}\,x\right)^4$

$\qquad = \left(\sqrt{2}\right)^4 x^4$

$\qquad = 4x^4$

35. (a) $\sqrt{8} = \sqrt{4\cdot 2} = \sqrt{4}\sqrt{2} = 2\sqrt{2}$

$\qquad$ (b) $\sqrt{18} = \sqrt{9\cdot 2} = \sqrt{9}\sqrt{2} = 3\sqrt{2}$

37. (a) $\sqrt[3]{16x^5} = \sqrt[3]{(8x^3)(2x^2)} = \sqrt[3]{8x^3}\sqrt[3]{2x^2} = 2x\sqrt[3]{2x^2}$

$\qquad$ (b) $\sqrt[4]{32x^4z^5} = \sqrt[4]{16x^4z^4 2z} = \sqrt[4]{16x^4z^4}\sqrt[4]{2z} = 2|x|z\sqrt[4]{2z}$

[*Note:* Since x^4 is under the radical, x could be positive or negative. For z^5 to be under the radical, z must be positive.]

39. (a) $\sqrt{75x^2y^{-4}} = \sqrt{\dfrac{25x^2}{y^4}\cdot 3} = \dfrac{5\sqrt{3}\,|x|}{y^2}$

$\qquad$ (b) $\sqrt{5(x-y)^3} = \sqrt{(x-y)^2 5(x-y)}$

$\qquad\qquad = (x-y)\sqrt{5(x-y)}$

41. $\sqrt{x-1}$ is defined when $x \geq 1$. Therefore, the domain is $[1, \infty)$.

43. $\sqrt{x^2+3}$ is defined for all real numbers. Therefore, the domain is $(-\infty, \infty)$.

45. $\dfrac{1}{\sqrt[3]{x-1}}$ is defined for all real numbers except $x = 1$. Therefore, the domain is $(-\infty, 1)$ and $(1, \infty)$.

47. $\dfrac{1}{\sqrt[4]{2x-6}}$ is defined when $x > 3$. Therefore, the domain is $(3, \infty)$.

49. $\sqrt{x-1}$ is defined when $x \geq 1$, and $\sqrt{5-x}$ is defined when $x \leq 5$. Therefore, the domain of $\sqrt{x-1} + \sqrt{5-x}$ is $1 \leq x \leq 5$.

51. $A = 10{,}000\left(1 + \dfrac{0.065}{12}\right)^{120} \approx \$19{,}121.84$

53. $T = 2\pi\sqrt{\dfrac{L}{32}}$

$\qquad = 2\pi\sqrt{\dfrac{4}{32}}$

$\qquad = 2\pi\sqrt{\dfrac{1}{8}}$

$\qquad = 2\pi\dfrac{1}{2\sqrt{2}}$

$\qquad = \dfrac{\pi}{\sqrt{2}} = \dfrac{\pi\sqrt{2}}{2} \approx 2.22$ seconds

Section 0.4   Factoring Polynomials

1. Since $a = 6$, $b = -1$, and $c = -1$, we have

$$x = \dfrac{1 \pm \sqrt{1 - (-24)}}{12} = \dfrac{1 \pm 5}{12}.$$

Thus, $x = \dfrac{1+5}{12} = \dfrac{1}{2}$ or $x = \dfrac{1-5}{12} = -\dfrac{1}{3}$.

3. Since $a = 4$, $b = -12$, and $c = 9$, we have

$$x = \dfrac{12 \pm \sqrt{144 - 144}}{8} = \dfrac{12}{8} = \dfrac{3}{2}.$$

5. Since $a = 1$, $b = 4$, and $c = 1$, we have

$$y = \dfrac{-4 \pm \sqrt{16 - 4}}{2} = \dfrac{-4 \pm 2\sqrt{3}}{2} = -2 \pm \sqrt{3}.$$

7. $x^2 - 4x + 4 = (x - 2)^2$

9. $4x^2 + 4x + 1 = (2x + 1)^2$

11. $x^2 + x - 2 = (x + 2)(x - 1)$

13. $3x^2 - 5x + 2 = (3x - 2)(x - 1)$

15. $x^2 - 4xy + 4y^2 = (x - 2y)^2$

17. $81 - y^4 = (9 + y^2)(9 - y^2)$
$$= (9 + y^2)(3 + y)(3 - y)$$

19. $x^3 - 8 = x^3 - 2^3$
$$= (x - 2)(x^2 + 2x + 4)$$

21. $y^3 + 64 = y^3 + 4^3$
$$= (y + 4)(y^2 - 4y + 16)$$

23. $x^3 - 27 = x^3 - 3^3$
$$= (x - 3)(x^2 + 3x + 9)$$

25. $x^3 - 4x^2 - x + 4 = x^2(x - 4) - (x - 4)$
$$= (x - 4)(x^2 - 1)$$
$$= (x - 4)(x + 1)(x - 1)$$

27. $2x^3 - 3x^2 + 4x - 6 = x^2(2x - 3) + 2(2x - 3)$
$$= (2x - 3)(x^2 + 2)$$

29. $2x^3 - 4x^2 - x + 2 = 2x^2(x - 2) - (x - 2)$
$$= (x - 2)(2x^2 - 1)$$

31. $x^2 - 5x = 0$
$$x(x - 5) = 0$$
$$x = 0, 5$$

33. $x^2 - 9 = 0$
$$(x + 3)(x - 3) = 0$$
$$x = -3, 3$$

35. $x^2 - 3 = 0$
$$\left(x + \sqrt{3}\right)\left(x - \sqrt{3}\right) = 0$$
$$x = \pm\sqrt{3}$$

37. $(x - 3)^2 - 9 = 0$
$$x^2 - 6x + 9 - 9 = 0$$
$$x(x - 6) = 0$$
$$x = 0, 6$$

39. $x^2 + x - 2 = 0$
$$(x + 2)(x - 1) = 0$$
$$x = -2, 1$$

41. $x^2 - 5x + 6 = 0$
$$(x - 2)(x - 3) = 0$$
$$x = 2, 3$$

43. $x^3 + 64 = 0$
$$x^3 = -64$$
$$x = \sqrt[3]{-64} = -4$$

45. $x^4 - 16 = 0$
$$x^4 = 16$$
$$x = \pm\sqrt[4]{16} = \pm 2$$

47. $x^3 - x^2 - 4x + 4 = 0$
$$x^2(x - 1) - 4(x - 1) = 0$$
$$(x - 1)(x^2 - 4) = 0$$
$$(x - 1)(x - 2)(x + 2) = 0$$
$$x = 1, \pm 2$$

49. Since $\sqrt{x^2 - 7x + 12} = \sqrt{(x - 3)(x - 4)}$, the roots are $x = 3$ and $x = 4$. By testing points inside and outside the interval $[3, 4]$, we find that the expression is defined when $x \leq 3$ or $x \geq 4$. Thus, the domain is $(-\infty, 3] \cup [4, \infty)$.

51. Since $\sqrt{4 - x^2} = \sqrt{(2 + x)(2 - x)}$, the roots are $x = \pm 2$. By testing points inside and outside the interval $[-2, 2]$, we find that the expression is defined when $-2 \leq x \leq 2$. Thus, the domain is $[-2, 2]$.

53.

$$
\begin{array}{r|rrrr}
-2 & 1 & 0 & 0 & 8 \\
   &   & -2 & 4 & -8 \\
\hline
   & 1 & -2 & 4 & 0 \\
\end{array}
$$

Therefore, the factorization is
$$x^3 + 8 = (x + 2)(x^2 - 2x + 4).$$

55.

$$
\begin{array}{r|rrrr}
1 & 2 & -1 & -2 & 1 \\
  &   & 2 & 1 & -1 \\
\hline
  & 2 & 1 & -1 & 0 \\
\end{array}
$$

Therefore, the factorization is
$$2x^3 - x^2 - 2x + 1 = (x - 1)(2x^2 + x - 1).$$

57. Possible rational roots: ± 1
Using synthetic division for $x = 1$, we have the following.

$$
\begin{array}{r|rrrr}
1 & 1 & -1 & -1 & 1 \\
  &   & 1 & 0 & -1 \\
\hline
  & 1 & 0 & -1 & 0
\end{array}
$$

Therefore, we have

$$x^3 - x^2 - x + 1 = 0$$
$$(x - 1)(x^2 - 1) = 0$$
$$(x - 1)(x - 1)(x + 1) = 0$$
$$x = \pm 1.$$

59. Possible rational roots: $\pm 1, \pm 2, \pm 3, \pm 6$
Using synthetic division for $x = 1$, we have the following.

$$
\begin{array}{r|rrrr}
1 & 1 & -6 & 11 & -6 \\
  &   & 1 & -5 & 6 \\
\hline
  & 1 & -5 & 6 & 0
\end{array}
$$

Therefore, we have

$$x^3 - 6x^2 + 11x - 6 = 0$$
$$(x - 1)(x^2 - 5x + 6) = 0$$
$$(x - 1)(x - 2)(x - 3) = 0$$
$$x = 1, 2, 3.$$

61. Possible rational roots: $\pm 1, \pm \frac{1}{2}, \pm \frac{1}{4}$
Using synthetic division for $x = 1$, we have the following.

$$
\begin{array}{r|rrrr}
1 & 4 & -4 & -1 & 1 \\
  &   & 4 & 0 & -1 \\
\hline
  & 4 & 0 & -1 & 0
\end{array}
$$

Therefore, we have

$$4x^3 - 4x^2 - x + 1 = 0$$
$$(x - 1)(4x^2 - 1) = 0$$
$$(x - 1)(2x + 1)(2x - 1) = 0$$
$$x = 1, \pm \tfrac{1}{2}.$$

63. Possible rational roots: $\pm 1, \pm 2, \pm 4$
Using synthetic division for $x = 4$, we have the following.

$$
\begin{array}{r|rrrr}
4 & 1 & -3 & -3 & -4 \\
  &   & 4 & 4 & 4 \\
\hline
  & 1 & 1 & 1 & 0
\end{array}
$$

Therefore, we have

$$x^3 - 3x^2 - 3x - 4 = 0$$
$$(x - 4)(x^2 + x + 1) = 0.$$

Since $x^2 + x + 1$ has no real solutions, $x = 4$ is the only real solution.

65. $0.0003x^2 - 1200 = 0$

$$0.0003x^2 - 1200$$
$$x^2 = 4{,}000{,}000$$
$$x = 2000 \text{ units}$$

67.

$$1.8 \times 10^{-5} - \frac{x^2}{1.0 \times 10^{-4} - x}$$

$$1.8 \times 10^{-9} - 1.8 \times 10^{-5}x = x^2$$

$$x^2 + 1.8 \times 10^{-5}x - 1.8 \times 10^{-9} = 0$$

By the Quadratic Formula:

$$x = \frac{-1.8 \times 10^{-5} \pm \sqrt{(1.8 \times 10^{-5})^2 + 4 \times 1.8 \times 10^{-9}}}{2}$$

$$= \frac{-1.8 \times 10^{-5} \pm \sqrt{7.524 \times 10^{-9}}}{2}$$

$$\approx 3.437 \times 10^{-5} [H^+]$$

Section 0.5   Fractions and Rationalization

1. $\dfrac{5}{x - 1} + \dfrac{x}{x - 1} = \dfrac{5 + x}{x - 1} = \dfrac{x + 5}{x - 1}$

3. $\dfrac{2x}{x^2 + 2} - \dfrac{1 - 3x}{x^2 + 2} = \dfrac{2x - (1 - 3x)}{x^2 + 2} = \dfrac{5x - 1}{x^2 + 2}$

5. $\dfrac{4}{x} - \dfrac{3}{x^2} = \dfrac{4x}{x^2} - \dfrac{3}{x^2} = \dfrac{4x - 3}{x^2}$

7. $\dfrac{2}{x + 2} - \dfrac{1}{x - 2} = \dfrac{2(x - 2) - (x + 2)}{(x + 2)(x - 2)} = \dfrac{x - 6}{x^2 - 4}$

9. $\dfrac{5}{x-3} + \dfrac{3}{3-x} = \dfrac{5}{x-3} + \dfrac{-3}{x-3}$

$\qquad\qquad = \dfrac{2}{x-3}$

11. $\dfrac{A}{x-6} + \dfrac{B}{x+3} = \dfrac{A(x+3) + B(x-6)}{(x-6)(x+3)}$

$\qquad\qquad = \dfrac{Ax + 3A + Bx - 6B}{(x-6)(x+3)}$

$\qquad\qquad = \dfrac{(A+B)x + 3(A-2B)}{(x-6)(x+3)}$

13. $-\dfrac{1}{x} + \dfrac{2}{x^2+1} = \dfrac{-(x^2+1) + 2x}{x(x^2+1)}$

$\qquad\qquad = \dfrac{-x^2 + 2x - 1}{x(x^2+1)}$

$\qquad\qquad = \dfrac{-(x^2 - 2x + 1)}{x(x^2+1)}$

$\qquad\qquad = \dfrac{-(x-1)^2}{x(x^2+1)}$

15. $\dfrac{-x}{(x+1)^{3/2}} + \dfrac{2}{(x+1)^{1/2}} = \dfrac{-x + 2(x+1)}{(x+1)^{3/2}}$

$\qquad\qquad = \dfrac{x+2}{(x+1)^{3/2}}$

17. $\dfrac{2-t}{2\sqrt{1+t}} - \sqrt{1+t} = \dfrac{2-t}{2\sqrt{1+t}} - \dfrac{\sqrt{1+t}}{1} \cdot \dfrac{2\sqrt{1+t}}{2\sqrt{1+t}}$

$\qquad\qquad = \dfrac{(2-t) - 2(1+t)}{2\sqrt{1+t}}$

$\qquad\qquad = \dfrac{-3t}{2\sqrt{1+t}}$

19. $\dfrac{1}{x^2 - x - 2} - \dfrac{x}{x^2 - 5x + 6} = \dfrac{1}{(x+1)(x-2)} - \dfrac{x}{(x-2)(x-3)}$

$\qquad\qquad = \dfrac{(x-3) - x(x+1)}{(x+1)(x-2)(x-3)}$

$\qquad\qquad = \dfrac{-x^2 - 3}{(x+1)(x-2)(x-3)}$

$\qquad\qquad = -\dfrac{x^2 + 3}{(x+1)(x-2)(x-3)}$

21. $\dfrac{A}{x+1} + \dfrac{B}{(x+1)^2} + \dfrac{C}{x-2} = \dfrac{A(x+1)(x-2) + B(x-2) + C(x+1)^2}{(x+1)^2(x-2)}$

$\qquad\qquad = \dfrac{A(x^2 - x - 2) + B(x-2) + C(x^2 + 2x + 1)}{(x+1)^2(x-2)}$

$\qquad\qquad = \dfrac{Ax^2 - Ax - 2A + Bx - 2B + Cx^2 + 2Cx + C}{(x+1)^2(x-2)}$

$\qquad\qquad = \dfrac{(A+C)x^2 - (A - B - 2C)x - (2A + 2B - C)}{(x+1)^2(x-2)}$

23. $\left(2x\sqrt{x^2+1} - \dfrac{x^3}{\sqrt{x^2+1}}\right) \div (x^2+1) = \dfrac{2x(x^2+1) - x^3}{\sqrt{x^2+1}} \cdot \dfrac{1}{x^2+1}$

$\qquad\qquad = \dfrac{x^3 + 2x}{\sqrt{x^2+1}(x^2+1)}$

$\qquad\qquad = \dfrac{x(x^2+2)}{(x^2+1)^{3/2}}$

25. $\dfrac{(x^2 + 2)^{1/2} - x^2(x^2 + 2)^{-1/2}}{x^2} = \dfrac{(x^2 + 2)^{-1/2}[(x^2 + 2) - x^2]}{x^2}$

$$= \dfrac{2}{x^2\sqrt{x^2 + 2}}$$

27. $\dfrac{\dfrac{\sqrt{x + 1}}{\sqrt{x}} - \dfrac{\sqrt{x}}{\sqrt{x + 1}}}{2(x + 1)} = \dfrac{(x + 1) - x}{\sqrt{x}\sqrt{x + 1}} \cdot \dfrac{1}{2(x + 1)}$

$$= \dfrac{1}{2\sqrt{x}(x + 1)^{3/2}}$$

29. $\dfrac{3}{\sqrt{27}} = \dfrac{3}{3\sqrt{3}} = \dfrac{1}{\sqrt{3}} \cdot \dfrac{\sqrt{3}}{\sqrt{3}} = \dfrac{\sqrt{3}}{3}$

31. $\dfrac{\sqrt{2}}{3} = \dfrac{\sqrt{2}}{3} \cdot \dfrac{\sqrt{2}}{\sqrt{2}} = \dfrac{2}{3\sqrt{2}}$

33. $\dfrac{x}{\sqrt{x - 4}} = \dfrac{x}{\sqrt{x - 4}} \cdot \dfrac{\sqrt{x - 4}}{\sqrt{x - 4}} = \dfrac{x\sqrt{x - 4}}{x - 4}$

35. $\dfrac{\sqrt{y^3}}{6y} = \dfrac{y\sqrt{y}}{6y} = \dfrac{\sqrt{y}}{6} = \dfrac{y}{6\sqrt{y}}$

37. $\dfrac{49(x - 3)}{\sqrt{x^2 - 9}} = \dfrac{49(x - 3)}{\sqrt{x^2 - 9}} \cdot \dfrac{\sqrt{x^2 - 9}}{\sqrt{x^2 - 9}}$

$$= \dfrac{49(x - 3)\sqrt{x^2 - 9}}{(x + 3)(x - 3)}$$

$$= \dfrac{49\sqrt{x^2 - 9}}{x + 3}$$

39. $\dfrac{5}{\sqrt{14} - 2} = \dfrac{5}{\sqrt{14} - 2} \cdot \dfrac{\sqrt{14} + 2}{\sqrt{14} + 2}$

$$= \dfrac{5(\sqrt{14} + 2)}{14 - 4}$$

$$= \dfrac{\sqrt{14} + 2}{2}$$

41. $\dfrac{2x}{5 - \sqrt{3}} = \dfrac{2x}{5 - \sqrt{3}} \cdot \dfrac{5 + \sqrt{3}}{5 + \sqrt{3}}$

$$= \dfrac{2x(5 + \sqrt{3})}{25 - 3}$$

$$= \dfrac{x(5 + \sqrt{3})}{11}$$

43. $\dfrac{1}{\sqrt{6} + \sqrt{5}} = \dfrac{1}{\sqrt{6} + \sqrt{5}} \cdot \dfrac{\sqrt{6} - \sqrt{5}}{\sqrt{6} - \sqrt{5}}$

$$= \dfrac{\sqrt{6} - \sqrt{5}}{6 - 5}$$

$$= \sqrt{6} - \sqrt{5}$$

45. $\dfrac{\sqrt{3} - \sqrt{2}}{x} = \dfrac{\sqrt{3} - \sqrt{2}}{x} \cdot \dfrac{\sqrt{3} + \sqrt{2}}{\sqrt{3} + \sqrt{2}}$

$$= \dfrac{3 - 2}{x(\sqrt{3} + \sqrt{2})}$$

$$= \dfrac{1}{x(\sqrt{3} + \sqrt{2})}$$

47. $\dfrac{2x - \sqrt{4x - 1}}{2x - 1} = \dfrac{2x - \sqrt{4x - 1}}{2x - 1} \cdot \dfrac{2x + \sqrt{4x - 1}}{2x + \sqrt{4x - 1}}$

$$= \dfrac{4x^2 - (4x - 1)}{(2x - 1)(2x + \sqrt{4x - 1})}$$

$$= \dfrac{(2x - 1)^2}{(2x - 1)(2x + \sqrt{4x - 1})}$$

$$= \dfrac{2x - 1}{2x + \sqrt{4x - 1}}$$

49. $P = 10{,}000$, $r = 0.14$, $N = 60$

$$M = 10{,}000\left[\dfrac{0.14/12}{1 - \left(\dfrac{1}{(0.14/12) + 1}\right)^{60}}\right] \approx \$232.68$$

Practice Test for Chapter 0

1. Determine whether $\sqrt[4]{81}$ is rational or irrational.

2. Determine whether the given value of x satisfies the inequality $3x + 4 \le x/2$.

   (a) $x = -2$ (b) $x = 0$ (c) $x = -\frac{8}{5}$ (d) $x = -6$

3. Solve the inequality $3x + 4 \ge 13$.

4. Solve the inequality $x^2 < 6x + 7$.

5. Determine which of the two given real numbers is greater, $\sqrt{19}$ or $\frac{13}{3}$.

6. Given the interval $[-3, 7]$, find (a) the distance between -3 and 7 and (b) the midpoint of the interval.

7. Solve the inequality $|3x + 1| \le 10$.

8. Solve the inequality $|4 - 5x| > 29$.

9. Solve the inequality $\left| 3 - \dfrac{2x}{5} \right| < 8$.

10. Use absolute values to describe the interval $[-3, 5]$.

11. Simplify $\dfrac{12x^3}{4x^{-2}}$.

12. Simplify $\left(\dfrac{\sqrt{3}\sqrt{x^3}}{x} \right)^0$, $x \ne 0$.

13. Remove all possible factors from the radical $\sqrt[3]{32x^4 y^3}$.

14. Complete the factorization: $\frac{3}{2}(x + 1)^{-1/3} + \frac{1}{4}(x + 1)^{2/3} = \frac{1}{4}(x + 1)^{-1/3}(\quad)$

15. Find the domain: $\dfrac{1}{\sqrt{5 - x}}$

16. Factor completely: $3x^2 - 19x - 14$

17. Factor completely: $25x^2 - 81$

18. Factor completely: $x^3 + 8$

19. Use the Quadratic Formula to find all real roots of $x^2 + 6x - 2 = 0$.

20. Use the Rational Zero Theorem to find all real roots of $x^3 - 4x^2 + x + 6 = 0$.

21. Combine terms and simplify: $\dfrac{x}{x^2 + 2x - 3} - \dfrac{1}{x - 1}$

22. Combine terms and simplify: $\dfrac{3 - x}{2\sqrt{x + 5}} + \sqrt{x + 5}$

23. Combine terms and simplify: $\dfrac{\dfrac{\sqrt{x + 2}}{\sqrt{x}} - \dfrac{\sqrt{x}}{\sqrt{x + 2}}}{2(x + 2)}$

24. Rationalize the denominator: $\dfrac{3y}{\sqrt{y^2 + 9}}$

25. Rationalize the numerator: $\dfrac{\sqrt{x} + \sqrt{x + 7}}{14}$

Graphing Calculator Required

26. Use a graphing calculator to find the real solutions of $x^3 - 5x^2 + 2x + 8 = 0$ by graphing $y = x^3 - 5x^2 + 2x + 8$ and finding the x-intercepts.

C H A P T E R 1
Functions, Graphs, and Limits

C H A P T E R 1
Functions, Graphs, and Limits

Section 1.1 The Cartesian Plane and the Distance Formula

Solutions to Odd-Numbered Exercises

1. (a) $a = 4$

$b = 3$

$c = \sqrt{(4 - 0)^2 + (3 - 0)^2} = 5$

(b) $a^2 + b^2 = 16 + 9 = 25 = c^2$

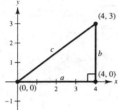

3. (a) $a = 10$

$b = 3$

$c = \sqrt{(7 + 3)^2 + (4 - 1)^2} = \sqrt{109}$

(b) $a^2 + b^2 = 100 + 9 = 109 = c^2$

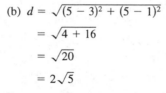

5. (a)

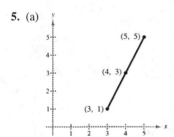

(b) $d = \sqrt{(5 - 3)^2 + (5 - 1)^2}$

$= \sqrt{4 + 16}$

$= \sqrt{20}$

$= 2\sqrt{5}$

(c) Midpoint $= \left(\dfrac{3 + 5}{2}, \dfrac{1 + 5}{2} \right) = (4, 3)$

7. (a)

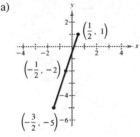

(b) $d = \sqrt{[-(3/2) - (1/2)]^2 + (-5 - 1)^2}$

$= \sqrt{4 + 36}$

$= \sqrt{4(1 + 9)}$

$= 2\sqrt{10}$

(c) Midpoint $= \left(\dfrac{(1/2) + (-3/2)}{2}, \dfrac{1 + (-5)}{2} \right) = \left(-\dfrac{1}{2}, -2 \right)$

9. (a)

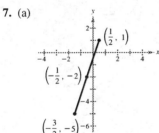

(b) $d = \sqrt{(4 - 2)^2 + (14 - 2)^2}$

$= \sqrt{4 + 144}$

$= \sqrt{4(1 + 36)}$

$= 2\sqrt{37}$

(c) Midpoint $= \left(\dfrac{2 + 4}{2}, \dfrac{2 + 14}{2} \right) = (3, 8)$

11. (a)

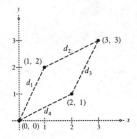

(b) $d = \sqrt{(-1 - 1)^2 + \left(1 - \sqrt{3}\right)^2}$

$= \sqrt{8 - 2\sqrt{3}}$

(c) Midpoint $= \left(\dfrac{1 + (-1)}{2}, \dfrac{\sqrt{3} + 1}{2}\right) = \left(0, \dfrac{\sqrt{3} + 1}{2}\right)$

13. $d_1 = \sqrt{(3 - 0)^2 + (7 - 1)^2} = \sqrt{45} = 3\sqrt{5}$

$d_2 = \sqrt{(4 - 0)^2 + (-1 - 1)^2} = \sqrt{20} = 2\sqrt{5}$

$d_3 = \sqrt{(3 - 4)^2 + [7 - (-1)]^2} = \sqrt{65}$

Since $d_1{}^2 + d_2{}^2 = d_3{}^2$, the triangle is a right triangle.

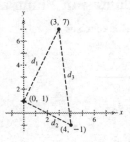

15. $d_1 = \sqrt{(1 - 0)^2 + (2 - 0)^2} = \sqrt{5}$

$d_2 = \sqrt{(3 - 1)^2 + (3 - 2)^2} = \sqrt{5}$

$d_3 = \sqrt{(2 - 3)^2 + (1 - 3)^2} = \sqrt{5}$

$d_4 = \sqrt{(0 - 2)^2 + (0 - 1)^2} = \sqrt{5}$

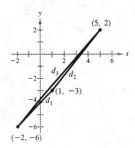

17. $d_1 = \sqrt{(2 - 0)^2 + (0 + 4)^2} = \sqrt{20} = 2\sqrt{5}$

$d_2 = \sqrt{(3 - 2)^2 + (2 - 0)^2} = \sqrt{5}$

$d_3 = \sqrt{(3 - 0)^2 + (2 + 4)^2} = \sqrt{45} = 3\sqrt{5}$

Since $d_1 + d_2 = d_3$, the points are collinear.

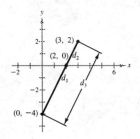

19. $d_1 = \sqrt{[1 - (-2)]^2 + [-3 - (-6)]^2} = \sqrt{18} = 3\sqrt{2}$

$d_2 = \sqrt{(5 - 1)^2 + [2 - (-3)]^2} = \sqrt{41}$

$d_3 = \sqrt{[5 - (-2)]^2 + [2 - (-6)]^2} = \sqrt{113}$

Since $d_1 + d_2 \neq d_3$, the points are not collinear.

21. $d = \sqrt{(x - 1)^2 + (-4 - 0)^2} = 5$

$\sqrt{x^2 - 2x + 17} = 5$

$x^2 - 2x + 17 = 25$

$x^2 - 2x - 8 = 0$

$(x - 4)(x + 2) = 0$

$x = 4, -2$

23. $d = \sqrt{(3 - 0)^2 + (y - 0)^2} = 8$

$\sqrt{9 + y^2} = 8$

$9 + y^2 = 64$

$y^2 = 55$

$y = \pm\sqrt{55}$

25. Midpoint $= \left(\dfrac{x_1 + x_2}{2}, \dfrac{y_1 + y_2}{2}\right)$

The point one-fourth of the way between (x_1, y_1) and (x_2, y_2) is the midpoint of the line segment from (x_1, y_1) to

$$\left(\frac{x_1 + x_2}{2}, \frac{y_1 + y_2}{2}\right),$$

which is

$$\left(\frac{x_1 + \dfrac{x_1 + x_2}{2}}{2}, \frac{y_1 + \dfrac{y_1 + y_2}{2}}{2}\right) = \left(\frac{3x_1 + x_2}{4}, \frac{3y_1 + y_2}{4}\right).$$

The point three-fourths of the way between (x_1, y_1) and (x_2, y_2) is the midpoint of the line segment from

$$\left(\frac{x_1 + x_2}{2}, \frac{y_1 + y_2}{2}\right)$$

to (x_2, y_2), which is

$$\left(\frac{\dfrac{x_1 + x_2}{2} + x_2}{2}, \frac{\dfrac{y_1 + y_2}{2} + y_2}{2}\right) = \left(\frac{x_1 + 3x_2}{4}, \frac{y_1 + 3y_2}{4}\right).$$

Thus,

$$\left(\frac{3x_1 + x_2}{4}, \frac{3y_1 + y_2}{4}\right), \left(\frac{x_1 + x_2}{2}, \frac{y_1 + y_2}{2}\right), \text{ and } \left(\frac{x_1 + 3x_2}{4}, \frac{y_1 + 3y_2}{4}\right)$$

are the three points that divide the line segment joining (x_1, y_1) and (x_2, y_2) into four equal parts.

27. (a) $\left(\dfrac{3(1) + 4}{4}, \dfrac{3(-2) - 1}{4}\right) = \left(\dfrac{7}{4}, -\dfrac{7}{4}\right)$

(b) $\left(\dfrac{3(-2) + 0}{4}, \dfrac{3(-3) + 0}{4}\right) = \left(-\dfrac{3}{2}, -\dfrac{9}{4}\right)$

$\left(\dfrac{1 + 4}{2}, \dfrac{-2 - 1}{2}\right) = \left(\dfrac{5}{2}, -\dfrac{3}{2}\right)$

$\left(\dfrac{-2 + 0}{2}, \dfrac{-3 + 0}{2}\right) = \left(-1, -\dfrac{3}{2}\right)$

$\left(\dfrac{1 + 3(4)}{4}, \dfrac{-2 + 3(-1)}{4}\right) = \left(\dfrac{13}{4}, -\dfrac{5}{4}\right)$

$\left(\dfrac{-2 + 3(0)}{4}, \dfrac{-3 + 3(0)}{4}\right) = \left(-\dfrac{1}{2}, -\dfrac{3}{4}\right)$

29. (a) $x^2 = 16^2 + 5^2, \quad x > 0$

$x^2 = 281$

$x = \sqrt{281} \approx 16.76$ feet

(b) $A = 2(40)\left(\sqrt{281}\right) = 80\sqrt{281} \approx 1341.04$ square feet

31.

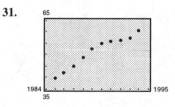

33. (a) 3800   (b) 3700

(c) 4800   (d) 5625

35. (a) 55 billion   (b) 115 billion

(c) 110 billion   (d) 128 billion

37. (a) Revenue midpoint $= \left(\dfrac{1992 + 1996}{2}, \dfrac{13{,}074 + 18{,}546}{2}\right) = (1994, 15{,}810)$

Profit midpoint $= \left(\dfrac{1992 + 1996}{2}, \dfrac{1883.8 + 3492}{2}\right) = (1994, 2687.9)$

(b) Actual revenue in 1994: 16,172 million

Actual profit in 1994: 2554 million

(c) Yes, the pattern seems linear.

(d) Expenses for 1992: \$11,190.2 million; 1994: \$13,618 million; 1996: \$15,054 million

(e) Answers vary.

39. (a) $(0, 0)$ is translated to $(0 + 2, 0 + 3) = (2, 3)$.

$(-3, -1)$ is translated to $(-3 + 2, -1 + 3) = (-1, 2)$.

$(-1, -2)$ is translated to $(-1 + 2, -2 + 3) = (1, 1)$.

(b)

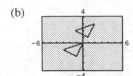

41. (a)

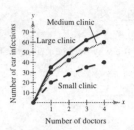

(b) The larger the clinic, the more patients a doctor can handle.

Section 1.2   Graphs of Equations

1. (a) This is not a solution point since $2x - y - 3 = 2(1) - 2 - 3 = -3 \neq 0$.

(b) This is a solution point since $2x - y - 3 = 2(1) - (-1) - 3 = 0$.

(c) This is a solution point since $2x - y - 3 = 2(4) - 5 - 3 = 0$.

3. (a) This is not a solution point since $x^2 y + x^2 - 5y = 0^2\left(\frac{1}{5}\right) + 0^2 - 5\left(\frac{1}{5}\right) = -1 \neq 0$.

(b) This is a solution point since $x^2 y + x^2 - 5y = 2^2(4) + 2^2 - 5(4) = 0$.

(c) This is not a solution point since $x^2 y + x^2 - 5y = (-2)^2(-4) + (-2)^2 - 5(-4) = 8 \neq 0$.

5. The graph of $y = x - 2$ is a straight line with y-intercept at $(0, -2)$. Thus, it matches (e).

7. The graph of $y = x^2 + 2x$ is a parabola opening up with vertex at $(-1, -1)$. Thus, it matches (c).

9. The graph of $y = |x| - 2$ has a y-intercept at $(0, -2)$ and has x-intercepts at $(-2, 0)$ and $(2, 0)$. Thus, it matches (a).

11. To find the y-intercept, let $x = 0$ to obtain

$$2(0) - y - 3 = 0$$
$$y = -3.$$

Thus, the y-intercept is $(0, -3)$. To find the x-intercept, let $y = 0$ to obtain

$$2x - (0) - 3 = 0$$
$$x = \frac{3}{2}.$$

Thus, the x-intercept is $\left(\frac{3}{2}, 0\right)$.

13. The y-intercept occurs at $(0, -2)$. To find the x-intercepts, let $y = 0$ to obtain

$$x^2 + x - 2 = 0$$
$$(x + 2)(x - 1) = 0$$
$$x = -2, 1.$$

Thus, the x-intercepts are $(-2, 0)$ and $(1, 0)$.

15. The y-intercept occurs at $(0, 0)$. To find the x-intercepts, let $y = 0$ to obtain

$$x^2\sqrt{9 - x^2} = 0$$
$$x = 0, \pm 3.$$

Thus, the x-intercepts are $(0, 0)$, $(-3, 0)$, and $(3, 0)$.

17. The y-intercept occurs at $(0, 2)$. The x-intercept occurs when the numerator equals zero and the denominator does not equal zero. Thus, $(-2, 0)$ is the only x-intercept.

19. Let $x = 0$. Then $4y = 0 \Longrightarrow y = 0$.

Let $y = 0$. Then $-x^2 = 0 \Longrightarrow x = 0$.

The x-intercept and y-intercept both occur at $(0, 0)$.

21. The graph of $y = x + 3$ is a straight line with intercepts at $(0, 3)$ and $(-3, 0)$.

x	0	1	2	3
y	3	4	5	6

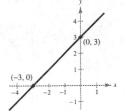

23. The graph of $y = 2x - 3$ is a straight line with intercepts at $\left(\frac{3}{2}, 0\right)$ and $(0, -3)$.

x	0	$\frac{3}{2}$	1	2
y	-3	0	-1	1

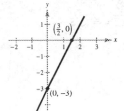

25.

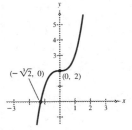

Intercepts: $\left(-\sqrt[3]{2}, 0\right)$ and $(0, 2)$

x	-2	-1	0	1	2
y	-6	1	2	3	10

27.

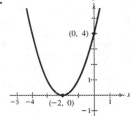

The graph is a parabola with vertex at $(-2, 0)$ and intercepts at $(-2, 0)$ and $(0, 4)$.

x	-3	-2	-1	0	1
y	1	0	1	4	9

29.

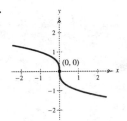

Intercept: $(0, 0)$

x	-8	-1	0	1	8
y	2	1	0	-1	-2

31.

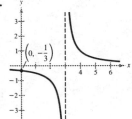

x	-1	0	1	2	2.5	3.5	4	5	6
y	$-\frac{1}{4}$	$-\frac{1}{3}$	$-\frac{1}{2}$	-1	-2	2	1	$\frac{1}{2}$	$\frac{1}{3}$

33.

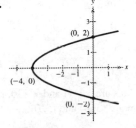

x	5	0	-3	-4
y	±3	±2	±1	0

The graph is a parabola with vertex at $(-4, 0)$ and intercepts at $(-4, 0)$, $(0, 2)$, and $(0, -2)$.

35. $(x - 0)^2 + (y - 0)^2 = 3^2$

$$x^2 + y^2 = 9$$

$$x^2 + y^2 - 9 = 0$$

37. $(x - 2)^2 + (y + 1)^2 = 4^2$

$$x^2 - 4x + 4 + y^2 + 2y + 1 = 16$$

$$x^2 + y^2 - 4x + 2y - 11 = 0$$

39. Since the point $(0, 0)$ lies on the circle, the radius must be the distance between $(0, 0)$ and $(-1, 2)$.

Radius $= \sqrt{(0 + 1)^2 + (0 - 2)^2} = \sqrt{5}$

$$(x + 1)^2 + (y - 2)^2 = 5$$

$$x^2 + y^2 + 2x - 4y = 0$$

41. Center $=$ midpoint $= (0, 3)$

Radius $=$ distance from center to an endpoint

$$= \sqrt{(0 - 3)^2 + (3 - 3)^2} = 3$$

$$(x - 0)^2 + (y - 3)^2 = 3^2$$

$$x^2 + y^2 - 6y = 0$$

43. $(x^2 - 2x + 1) + (y^2 + 6y + 9) = -6 + 1 + 9$

$$(x - 1)^2 + (y + 3)^2 = 4$$

Center: $(1, -3)$

Radius: 2

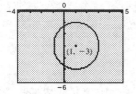

45. $(x^2 + 4x + 4) + (y^2 + 6y + 9) = 3 + 4 + 9$

$$(x + 2)^2 + (y + 3)^2 = 16$$

Center: $(-2, -3)$

Radius: 4

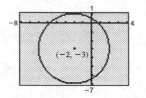

47. $x^2 + y^2 - x - y = \frac{3}{2}$

$$\left(x^2 - x + \tfrac{1}{4}\right) + \left(y^2 - y + \tfrac{1}{4}\right) = \tfrac{3}{2} + \tfrac{1}{4} + \tfrac{1}{4}$$

$$\left(x - \tfrac{1}{2}\right)^2 + \left(y - \tfrac{1}{2}\right)^2 = 2$$

Center: $\left(\tfrac{1}{2}, \tfrac{1}{2}\right)$

Radius: $\sqrt{2}$

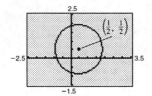

49. $x^2 + y^2 + x + \frac{5}{2}y = \frac{7}{16}$

$$\left(x^2 + x + \tfrac{1}{4}\right) + \left(y^2 + \tfrac{5}{2}y + \tfrac{25}{16}\right) = \tfrac{7}{16} + \tfrac{1}{4} + \tfrac{25}{16}$$

$$\left(x + \tfrac{1}{2}\right)^2 + \left(y + \tfrac{5}{4}\right)^2 = \tfrac{9}{4}$$

Center: $\left(-\tfrac{1}{2}, -\tfrac{5}{4}\right)$

Radius: $\tfrac{3}{2}$

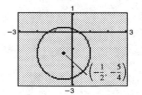

51. Solving for y in the equation $x + y = 2$ yields $y = 2 - x$, and solving for y in the equation $2x - y = 1$ yields $y = 2x - 1$. Then setting these two y-values equal to each other, we have

$$2 - x = 2x - 1$$

$$3 = 3x$$

$$x = 1.$$

The corresponding y-value is $y = 2 - 1 = 1$, so the point of intersection is $(1, 1)$.

53. Solving for y in the second equation yields $y = 10 - 2x$ and substituting this into the first equation gives

$$x^2 + (10 - 2x)^2 = 25$$

$$x^2 + 100 + 4x^2 - 40x = 25$$

$$5x^2 - 40x + 75 = 0$$

$$x^2 - 8x + 15 = 0$$

$$(x - 3)(x - 5) = 0$$

$$x = 3, 5.$$

The corresponding y-values are $y = 4$ and $y = 0$, so the points of intersection are $(3, 4)$ and $(5, 0)$.

55. By equating the *y*-values for the two equations, we have

$$x^3 = 2x$$
$$x^3 - 2x = 0$$
$$x(x^2 - 2) = 0$$
$$x = 0, \pm\sqrt{2}.$$

The corresponding *y*-values are $y = 0$, $y = -2\sqrt{2}$, and $y = 2\sqrt{2}$, so the points of intersection are $(0, 0)$, $\left(-\sqrt{2}, -2\sqrt{2}\right)$, and $\left(\sqrt{2}, 2\sqrt{2}\right)$.

57. By equating the *y*-values for the two equations, we have

$$x^4 - 2x^2 + 1 = 1 - x^2$$
$$x^4 - x^2 = 0$$
$$x^2(x + 1)(x - 1) = 0$$
$$x = 0, \pm 1.$$

The corresponding *y*-values are $y = 1$, 0, and 0, so the points of intersection are $(-1, 0)$, $(0, 1)$, and $(1, 0)$.

59. (a) $C = 11.8x + 5000$

$$R = 19.3x$$

(b) By equating *R* and *C*, we have

$$R = C$$
$$19.3x = 11.8x + 5000$$
$$7.5x = 5000$$
$$x = \frac{5000}{7.5} = 666\tfrac{2}{3} \approx 667 \text{ units.}$$

(c) Profit = Revenue − Cost

$$100 = 19.3x - (11.8x + 5000)$$
$$= 7.5x - 5000$$
$$5100 = 7.5x$$
$$x = \frac{5100}{7.5} = 680 \text{ units}$$

61. $R = C$

$$1.55x = 0.85x + 35,000$$
$$0.7x = 35,000$$
$$x = \frac{35,000}{0.7} = 50,000 \text{ units}$$

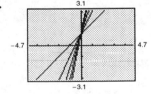

63. $R = C$

$$9950x = 8650x + 250,000$$
$$1300x = 250,000$$
$$x = \frac{250,000}{1300} \approx 193 \text{ units}$$

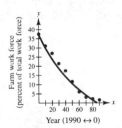

65.

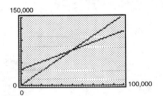

The greater the value of *c*, the steeper the line.

67. (a)

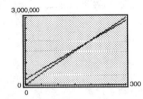

(b) The model predicts -2.91% for the year 2000 $(t - 100)$.

(c) The model is not valid.

69. If *C* and *R* represent the cost and revenue for a business, the break-even point is that value of *x* for which $C = R$. For example, if $C = 100,000 + 10x$ and $R = 20x$, then the break-even point is $x = 10,000$ units.

71. Intercept: $(0, 5.36)$

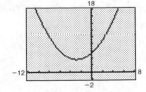

73. Intercepts: $(1.4780, 0)$, $(12.8553, 0)$, $(0, 2.3875)$

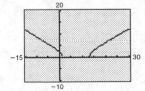

Section 1.3   Lines in the Plane and Slope

1. The slope is $m = 1$ since the line rises one unit vertically for each unit of horizontal change from left to right.

3. The slope is $m = 0$ since the line is horizontal.

5. The points are plotted in the accompanying graph and the slope is

$$m = \frac{2 - (-4)}{5 - 3} = 3.$$

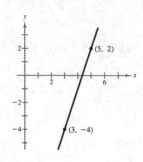

7. The points are plotted in the accompanying graph and the slope is

$$m = \frac{2 - 2}{6 - (1/2)} = 0.$$

Thus, the line is horizontal.

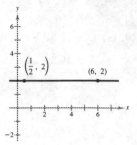

9. The points are plotted in the accompanying graph. The slope is undefined since

$$m = \frac{-5 - (-3)}{-8 - 8}. \quad \text{(undefined slope)}$$

Thus, the line is vertical.

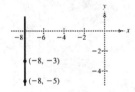

11. The points are plotted in the accompanying graph and the slope is

$$m = \frac{2 - 2}{-2 - 1} = 0.$$

Thus, the line is horizontal.

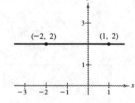

13. The equation of this horizontal line is $y = 1$. Therefore, three additional points are $(0, 1)$, $(1, 1)$, and $(3, 1)$.

15. The equation of this line is

$$y + 4 = -2(x - 6)$$
$$y = -2x + 8.$$

Therefore, three additional points are $(0, 8)$, $(1, 6)$, and $(2, 4)$.

17. The equation of the line is

$$y - 7 = -3(x - 1)$$
$$y = -3x + 10.$$

Therefore, three additional points are $(0, 10)$, $(2, 4)$, and $(3, 1)$.

19. The equation of this vertical line is $x = -8$. Therefore, three additional points are $(-8, 0)$, $(-8, 2)$, and $(-8, 3)$.

21. $x + 5y = 20$

$$y = -\frac{1}{5}x + 4$$

Therefore, the slope is $m = -\frac{1}{5}$ and the y-intercept is $(0, 4)$.

23. $7x - 5y = 15$

$$y = \frac{7}{5}x - 3$$

Therefore, the slope is $m = \frac{7}{5}$ and the y-intercept is $(0, -3)$.

25. Since the line is vertical, the slope is undefined and there is no y-intercept.

27. The slope of the line is

$$m = \frac{3 - (-5)}{4 - 0} = 2.$$

Using the point-slope form, we have

$$y + 5 = 2(x - 0)$$

$$y = 2x - 5.$$

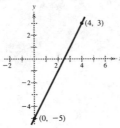

29. The slope of the line is

$$m = \frac{3 - 0}{-1 - 0} = -3.$$

Using the point-slope form, we have

$$y = -3x$$

$$3x + y = 0.$$

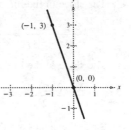

31. The slope of the line is undefined, so the line is vertical and its equation is

$$x = 2$$

$$x - 2 = 0.$$

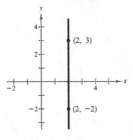

33. The slope of the line is $m = 0$, so the line is horizontal and its equation is

$$y = -2$$

$$y + 2 = 0.$$

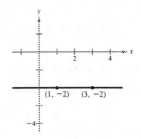

35. Using the slope-intercept form, we have

$$y = \tfrac{3}{4}x + 3$$

$$4y = 3x + 12$$

$$3x - 4y + 12 = 0.$$

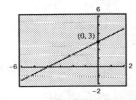

37. Using the slope-intercept form, we have

$$y = \tfrac{2}{3}x + 0$$

$$2x - 3y = 0.$$

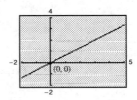

39. Using the slope-intercept form, we have

$$y - 7 = -3(x + 2)$$

$$y = -3x + 1.$$

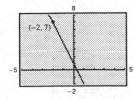

41. Using the slope-intercept form, we have

$$y = 4x + 2$$

$$4x - y + 2 = 0.$$

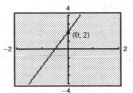

43. Using the slope-intercept form, we have

$$y = \tfrac{3}{4}x + \tfrac{2}{3}$$

$$12y = 9x + 8$$

$$9x - 12y + 8 = 0.$$

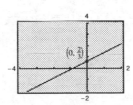

45. The slope of the line joining $(-2, 1)$ and $(-1, 0)$ is

$$\frac{1 - 0}{-2 - (-1)} = \frac{1}{-1} = -1.$$

The slope of the line joining $(-1, 0)$ and $(2, -2)$ is

$$\frac{0 - (-2)}{-1 - 2} = \frac{2}{-3} = -\frac{2}{3}.$$

Since the slopes are different, the points are not collinear.

$$d_1 = \sqrt{[-2 - (-1)]^2 + (1 - 0)^2} = \sqrt{1 + 1} = \sqrt{2} \approx 1.41421$$

$$d_2 = \sqrt{(-1 - 2)^2 + [0 - (-0)]^2} = \sqrt{9 + 4} = \sqrt{13} \approx 3.60555$$

$$d_3 = \sqrt{(-2 - 2)^2 + [1 - (-2)]^2} = \sqrt{6 + 9} = 5$$

Since $d_1 + d_2 \neq d_3$, the points are collinear.

47. Since the line is vertical, it has an undefined slope and its equation is

$$x = 3$$

$$x - 3 = 0.$$

49. Given line: $y = -x + 7$, $m_1 = -1$

(a) Parallel; $m_1 = -1$

$$y - 2 = -1(x + 3)$$

$$x + y + 1 = 0$$

(b) Perpendicular; $m_2 = 1$

$$y - 2 = -1(x + 3)$$

$$x - y + 5 = 0$$

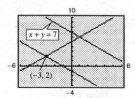

51. Given line: $y = -\frac{3}{4}x + \frac{7}{4}$, $m_1 = -\frac{3}{4}$

(a) Parallel; $m_1 = -\frac{3}{4}$

$$y - 4 = -\frac{3}{4}(x + 6)$$

$$4y - 16 = -3x - 18$$

$$3x + 4y + 2 = 0$$

(b) Perpendicular; $m_2 = \frac{4}{3}$

$$y - 4 = \frac{4}{3}(x + 6)$$

$$3y - 12 = 4x + 24$$

$$4x - 3y + 36 = 0$$

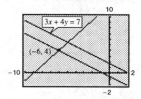

53. Given line: $y = -3$ is horizontal and $m_1 = 0$

(a) Parallel: $y = 0$ or the x-axis

(b) Perpendicular: $x = -1$ or $x + 1 = 0$

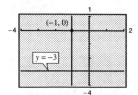

55. Given line: $y = \frac{2}{3}x - 1$

(a) Parallel; $m_1 = \frac{2}{3}$

$$y - 1 = \frac{2}{3}(x - 1)$$

$$y = \frac{2}{3}x + \frac{1}{3}$$

$$3y - 2x - 1 = 0$$

(b) Perpendicular; $m_2 = -\frac{3}{2}$

$$y - 1 = -\frac{3}{2}(x - 1)$$

$$y = -\frac{3}{2}x + \frac{5}{2}$$

$$2y + 3x - 5 = 0$$

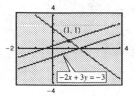

57. $y = -2$ is a horizontal line.

x	-2	-1	0	1
y	-2	-2	-2	-2

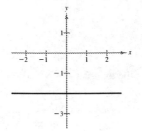

59. $y - 2x - 3$

x	-1	0	1	2
y	-5	-3	-1	1

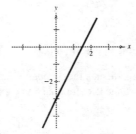

61.

x	-1	0	1	2
y	3	1	-1	-3

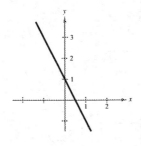

63. $y = -4x - 6$ has slope -4 and y-intercept $(0, -6)$.

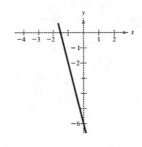

65. (a) $y - 3146 = \dfrac{3278 - 3146}{5 - 0}(t - 0)$

$$y = 26.4t + 3146$$

The slope $m = 26.4$ tells you that the population increases 26.4 thousand per year.

(b) If $t = 2$, $y = 26.4(2) + 3146 = 3198.8$ thousand.

(c) If $t = 6$, $y = 26.4(6) + 3146 = 3304.4$ thousand.

(d) 1992: 3207.2 thousand
1996: 3300.9 thousand
The estimates are good.

(e) The model could possibly be used to predict the population in 2000 ($t = 10$) if the population continues to grow at the same linear rate.

67. Using $F = \frac{9}{5}C + 32$, you have $C = \frac{5}{9}(F - 32)$.

(a) If $F = 102.5$, then $C = \frac{5}{9}(102.5 - 32) \approx 39.17°$.

(b) If $F = 68$, then $C = \frac{5}{9}(68 - 32) = 20°$. Since $20° < 29.8°$, gallium is solid.

69. (a) $W = 0.80x + 8.75$ (union plan)

     $W = 1.15x + 6.35$ (corporation plan)

(b) $0.80x + 8.75 = 1.15x + 6.35$

           $2.40 = 0.35x$

                $x = \dfrac{2.40}{0.35} \approx 6.857$

              $W \approx 14.236 \Longrightarrow (6.857, 14.236)$

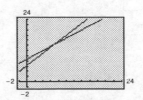

(c) The point of intersection indicates the number of units (6.857) a worker needs to produce for the two plans to be equivalent. If a worker produces more than 6 units, the corporation plan pays better.

71. Use the points $(0, 825,000)$ and $(25, 75,000)$.

    $y - 825,000 = \dfrac{825,000 - 75,000}{0 - 25}(t - 0)$

    $y - 825,000 = -30,000t$

            $y = -30,000t + 825,000, \quad 0 \le t \le 25$

73. (a) The slope is $\dfrac{47 - 50}{425 - 380} = \dfrac{-3}{45} = -\dfrac{1}{15}$. Hence,

        $x - 50 = -\dfrac{1}{15}(p - 380)$

          $x = -\dfrac{1}{15}p + \dfrac{226}{3}.$

(b) If $p = 455$, $x = -\dfrac{1}{15}(455) + \dfrac{226}{3} = 45$ units.

(c) If $p = 395$, $x = -\dfrac{1}{15}(395) + \dfrac{226}{3} = 49$ units.

75. (a) $Y - 4792 = \dfrac{6102 - 4792}{5 - 0}(t - 0)$

        $Y = 262t + 4792$

(b) If $t = 2$, $Y = 262(2) + 4792 = 5316$ billion.

(c) If $t = 6$, $Y = 262(6) + 4792 = 6364$ billion.

(d) 1992: 5264.2 billion
    1996: 6449.5 billion

77. $23,500 + 3100x \le 100,000$

        $3100x \le 76,500$

            $x \le 24.677$

            $x \le 24$ units or 24.67 units if fractional units are allowed.

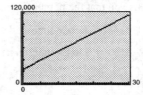

79. $C = 18,375 + 1150x \le 100,000$

        $1150x \le 81,625$

           $x \le 70.978$

           $x \le 70$ units

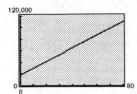

81. $C = 75,500 + 89x \le 100,000$

        $89x \le 24,500$

          $x \le 275.28$

          $x \le 275$ units

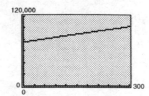

Section 1.4   Functions

1. (a) $f(0) = 2(0) - 3 = -3$

    (c) $f(x - 1) = 2(x - 1) - 3 = 2x - 5$

    (b) $f(-3) = 2(-3) - 3 = -9$

    (d) $f(1 + \Delta x) = 2(1 + \Delta x) - 3 = 2\Delta x - 1$

3. (a) $g(2) = \dfrac{1}{2}$

    (c) $g(x + \Delta x) = \dfrac{1}{x + \Delta x}$

    (b) $g\left(\dfrac{1}{4}\right) = \dfrac{1}{1/4} = 4$

    (d) $g(x + \Delta x) - g(x) = \dfrac{1}{x + \Delta x} - \dfrac{1}{x}$

$$= \dfrac{x - (x + \Delta x)}{(x + \Delta x)x} = \dfrac{-\Delta x}{x(x + \Delta x)}$$

5. $\dfrac{f(x + \Delta x) - f(x)}{\Delta x} = \dfrac{[3(x + \Delta x) - 1] - (3x - 1)}{\Delta x}$

$$= \dfrac{3x + 3\Delta x - 1 - 3x + 1}{\Delta x}$$

$$= \dfrac{3\Delta x}{\Delta x} = 3, \quad \Delta x \neq 0$$

7. $\dfrac{f(x + \Delta x) - f(x)}{\Delta x} = \dfrac{(x + \Delta x)^3 - (x + \Delta x) - (x^3 - x)}{\Delta x}$

$$= \dfrac{x^3 + 3x^2(\Delta x) + 3x(\Delta x)^2 + (\Delta x)^3 - x - \Delta x - x^3 + x}{\Delta x}$$

$$= \dfrac{3x^2(\Delta x) + 3x(\Delta x)^2 + (\Delta x)^3 - \Delta x}{\Delta x}$$

$$= 3x^2 + 3x(\Delta x) + (\Delta x)^2 - 1, \quad \Delta x \neq 0$$

9. $y = \pm\sqrt{4 - x^2}$

    y is *not* a function of x since there are two values of y for some x.

11. $\frac{1}{2}x - 6y = -3$

$$y = \tfrac{1}{12}x + \tfrac{1}{2}$$

    y *is* a function of x since there is only one value of y for each x.

13. $y = 4 - x^2$

    y *is* a function of x since there is only one value of y for each x.

15. $y = \pm\sqrt{x^2 - 1}$

    y is *not* a function of x since there are two values of y for some x.

17. Domain: $[-3, 3]$

    Range: $[0, 3]$

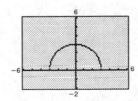

19. Domain: $(-\infty, 0) \cup (0, \infty)$

    Range: $\{-1, 1\}$

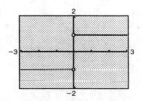

21. Domain: $(-\infty, -4) \cup (-4, \infty)$

    Range: $(-\infty, 1) \cup (1, \infty)$

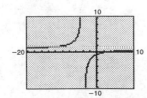

23. Domain: $(-\infty, \infty)$

Range: $(-\infty, \infty)$

25. Domain: $(-\infty, \infty)$

Range: $(-\infty, 4]$

27. y is *not* a function of x.

29. y *is* a function of x.

31. (a) $f(x) + g(x) = (x^2 + 1) + (x - 1) = x^2 + x$

(b) $f(x) \cdot g(x) = (x^2 + 1)(x - 1) = x^3 - x^2 + x - 1$

(c) $\dfrac{f(x)}{g(x)} = \dfrac{x^2 + 1}{x - 1}$

(d) $f(g(x)) = f(x - 1) = (x - 1)^2 + 1 = x^2 - 2x + 2$

(e) $g(f(x)) = g(x^2 + 1) = (x^2 + 1) - 1 = x^2$

33. (a) $f(x) + g(x) = x^2 + 5 + \sqrt{1 - x}, \quad x \le 1$

(b) $f(x) \cdot g(x) = (x^2 + 5)\sqrt{1 - x}, \quad x \le 1$

(c) $\dfrac{f(x)}{g(x)} = \dfrac{x^2 + 5}{\sqrt{1 - x}}, \quad x < 1$

(d) $f(g(x)) = f\left(\sqrt{1 - x}\right)$

$\qquad = \left(\sqrt{1 - x}\right)^2 + 5$

$\qquad = 6 - x, \quad x \le 1$

(e) $g(f(x))$ is not defined since the domain of g is $(-\infty, 1]$ and the range of f is $[5, \infty)$. The range of f is not in the domain of g.

35. $f(g(x)) = f\left(\dfrac{x - 1}{5}\right) = 5\left(\dfrac{x - 1}{5}\right) + 1 = x$

$g(f(x)) = g(5x + 1) = \dfrac{(5x + 1) - 1}{5} = x$

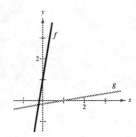

37. $f(g(x)) = f\left(\sqrt{9 - x}\right) = 9 - \left(\sqrt{9 - x}\right)^2 = x$

$g(f(x)) = g(9 - x^2)$

$\qquad = \sqrt{9 - (9 - x^2)}$

$\qquad = \sqrt{x^2}$

$\qquad = x, \quad x \ge 0$

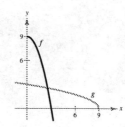

39.　$f(x) = 2x - 3 = y$

$2y - 3 = x$

$y = \dfrac{x + 3}{2}$

$f^{-1}(x) = \dfrac{x + 3}{2}$

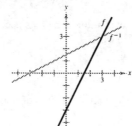

41.　$f(x) = x^5 = y$

$x = y^5$

$y = \sqrt[5]{x}$

$f^{-1}(x) = \sqrt[5]{x}$

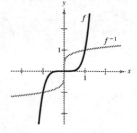

43.　$f(x) = \sqrt{9 - x^2} = y, \quad 0 \le x \le 3$

$x = \sqrt{9 - y^2}$

$x^2 = 9 - y^2$

$y^2 = 9 - x^2$

$y = \sqrt{9 - x^2}$

$f^{-1}(x) = \sqrt{9 - x^2}, \quad 0 \le x \le 3$

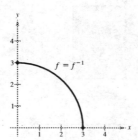

45. $f(x) = x^{2/3} = y, \quad x \geq 0$

$x = y^{2/3}$

$y = x^{3/2}$

$f^{-1}(x) = x^{3/2}$

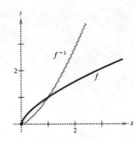

47. $f(x) = 3 \quad 7x$ is one-to-one.

$y = 3 - 7x$

$x = 3 - 7y$

$y = \dfrac{3 - x}{7}$

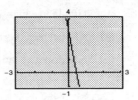

49. $f(x) = x^2 = y$

f is *not* one-to-one since $f(1) = 1 = f(-1)$.

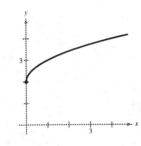

51. $f(x) = |x - 2| = y$

f is *not* one-to-one since $f(0) = 2 = f(4)$.

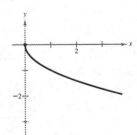

53. $f(x) = \sqrt{x}, \quad g(x) = x^2 - 1$

(a) $f(g(1)) = f(1^2 - 1) = f(0) = 0$

(b) $g(f(1)) = g(\sqrt{1}) = g(1) = 0$

(c) $g(f(0)) = g(0) = -1$

(d) $f(g(-4)) = f(15) = \sqrt{15}$

(e) $f(g(x)) = f(x^2 - 1) = \sqrt{x^2 - 1}$

(f) $g(f(x)) = g(\sqrt{x}) = x - 1, \quad x \geq 0$

55. The data fits the function (b) $g(x) = -2x^2$ with $c = -2$.

57. The data fits the function (d) $r(x) = \dfrac{32}{x}$ with $c = 32$.

59. (a) $y = \sqrt{x} + 2$

(b) $y = -\sqrt{x}$

(c) $y = \sqrt{x - 2}$

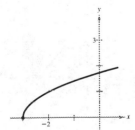

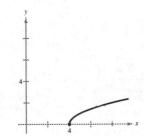

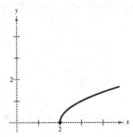

(d) $y = \sqrt{x + 3}$

(e) $y = \sqrt{x - 4}$

(f) $y = 2\sqrt{x}$

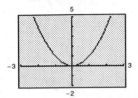

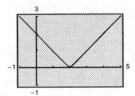

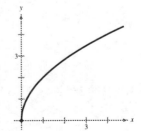

61. (a) Shifted three units to the left: $y = (x + 3)^2$

(b) Shifted three units upward: $y = x^2 + 3$

(c) Shifted three units to the right, six units upward, and reflected: $y = -(x - 3)^2 + 6$

(d) Shifted six units to the left, three units downward, and reflected: $y = -(x + 6)^2 - 3$

63. (a) $C = 0.95x + 6000$

(b) $\overline{C} = \dfrac{C}{x} = \dfrac{0.95x + 6000}{x} = 0.95 + \dfrac{6000}{x}$

(c) $0.95 + \dfrac{6000}{x} < 1.69$

$\dfrac{6000}{x} < 0.74$

$\dfrac{6000}{0.74} < x$ since $x > 0$.

$8108.108 < x$

Must sell 8109 units before the average cost per unit falls below the selling price.

65. Cost = Cost on land + Cost underwater

$= 10(5280)(3 - x) + 15(5280)\sqrt{x^2 + \frac{1}{4}}$

$= 5(5280)\left[2(3 - x) + 3\sqrt{x^2 + \frac{1}{4}}\right]$

67. (a) If $0 \leq x \leq 100$, then $p = 90$. If $100 < x \leq 1600$, then $p = 90 - 0.01(x - 100) = 91 - 0.01x$. If $x > 1600$, then $p = 75$. Thus,

$$p = \begin{cases} 90, & 0 \leq x \leq 100 \\ 91 - 0.01x, & 100 < x \leq 1600 \\ 75, & x > 1600. \end{cases}$$

(b) $P = px - 60x$

$$P = \begin{cases} 90x - 60x, & 0 \leq x \leq 100 \\ (91 - 0.01x)x - 60x, & 100 < x \leq 1600 \\ 75x - 60x, & x > 1600 \end{cases}$$

$$= \begin{cases} 30x, & 0 \leq x \leq 100 \\ 31x - 0.01x^2, & 100 < x \leq 1600 \\ 15x, & x > 1600 \end{cases}$$

69. $r = 8 - 0.05(n - 80), \quad n \geq 80$

(a) Revenue $= R = rn = [8 - 0.05(n - 80)]n$

(b)

n	90	100	110	120	130	140	150
R	675	700	715	720	715	700	675

(c) The revenue begins to decrease for $n > 120$.

71. $f(x) = x\sqrt{9 - x^2}$

Zeros: $x = 0, \pm 3$

The function is *not* one-to-one.

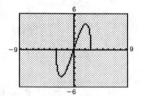

73. $g(t) = \dfrac{t + 3}{1 - t}$

Zero: $t = -3$

The function *is* one-to-one.

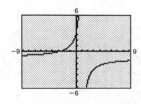

75. $f(x) = \dfrac{4 - x^2}{x}$

Zeros: $x = \pm 2$

The function is *not* one-to-one.

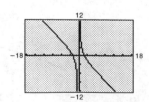

77. Answers will vary.

Section 1.5   Limits

1.

x	1.9	1.99	1.999	2	2.001	2.01	2.1
$f(x)$	13.5	13.95	13.995	14	14.005	14.05	14.5

$$\lim_{x \to 2} (5x + 4) = 14$$

3.

x	1.9	1.99	1.999	2	2.001	2.01	2.1
$f(x)$	0.2564	0.2506	0.2501	undefined	0.2499	0.2494	0.2439

$$\lim_{x \to 2} \frac{x - 2}{x^2 - 4} = \frac{1}{4}$$

5.

x	-0.1	-0.01	-0.001	0	0.001	0.01	0.1
$f(x)$	0.2911	0.2889	0.2887	undefined	0.2887	0.2884	0.2863

$$\lim_{x \to 0} \frac{\sqrt{x + 3} - \sqrt{3}}{x} = \frac{1}{2\sqrt{3}} \approx 0.289$$

7.

x	1.5	1.9	1.99	1.999	2
$f(x)$	0.3780	0.1601	0.0501	0.0158	undefined

$$\lim_{x \to 2^-} \frac{2 - x}{\sqrt{4 - x^2}} = 0$$

9. (a) $\lim_{x \to 0} f(x) = 1$ (b) $\lim_{x \to -1} f(x) = 3$

11. (a) $\lim_{x \to 0} g(x) = 1$ (b) $\lim_{x \to -1} g(x) = 3$

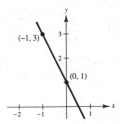

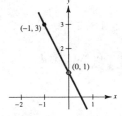

13. (a) $\lim_{x \to c} [f(x) + g(x)] = \lim_{x \to c} f(x) + \lim_{x \to c} g(x)$

$$= 3 + 9 = 12$$

(b) $\lim_{x \to c} [f(x)g(x)] = \left[\lim_{x \to c} f(x)\right]\left[\lim_{x \to c} g(x)\right] = 3 \cdot 9 = 27$

(c) $\lim_{x \to c} \dfrac{f(x)}{g(x)} = \dfrac{\lim_{x \to c} f(x)}{\lim_{x \to c} g(x)} = \dfrac{3}{9} = \dfrac{1}{3}$

15. (a) $\lim_{x \to 3^+} f(x) = 1$

(b) $\lim_{x \to 3^-} f(x) = 1$

(c) $\lim_{x \to 3} f(x) = 1$

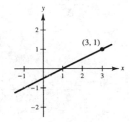

17. (a) $\lim\limits_{x \to 3^+} f(x) = 0$

(b) $\lim\limits_{x \to 3^-} f(x) = 0$

(c) $\lim\limits_{x \to 3} f(x) = 0$

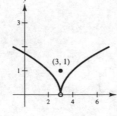

19. (a) $\lim\limits_{x \to 3^+} f(x) = 3$

(b) $\lim\limits_{x \to 3^-} f(x) = -3$

(c) $\lim\limits_{x \to 3} f(x)$ does not exist.

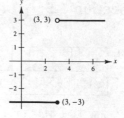

21. $\lim\limits_{x \to -3} (3x + 2) = 3(-3) + 2 = -7$

23. $\lim\limits_{x \to 1} (1 - x^2) = 1 - 1^2 = 0$

25. $\lim\limits_{x \to 3} \sqrt{x + 1} = \sqrt{3 + 1} = 2$

27. $\lim\limits_{x \to -3} \dfrac{2}{x + 2} = \dfrac{2}{-3 + 2} = -2$

29. $\lim\limits_{x \to -2} \dfrac{x^2 - 1}{2x} = \dfrac{(-2)^2 - 1}{2(-2)} = \dfrac{3}{-4} = -\dfrac{3}{4}$

31. $\lim\limits_{x \to -1} \dfrac{x^2 - 1}{x + 1} = \lim\limits_{x \to -1} \dfrac{(x + 1)(x - 1)}{x + 1}$

$= \lim\limits_{x \to -1} (x - 1) = -2$

33. $\lim\limits_{x \to 2} \dfrac{x - 2}{x^2 - 4x + 4} = \lim\limits_{x \to 2} \dfrac{x - 2}{(x - 2)(x - 2)}$

$= \lim\limits_{x \to 2} \dfrac{1}{x - 2}$ does not exist.

35. $\lim\limits_{t \to 5} \dfrac{t - 5}{t^2 - 25} = \lim\limits_{t \to 5} \dfrac{t - 5}{(t + 5)(t - 5)}$

$= \lim\limits_{t \to 5} \dfrac{1}{t + 5} = \dfrac{1}{10}$

37. $\lim\limits_{x \to -2} \dfrac{x^3 + 8}{x + 2} = \lim\limits_{x \to -2} \dfrac{(x + 2)(x^2 - 2x + 4)}{x + 2}$

$= \lim\limits_{x \to -2} (x^2 - 2x + 4) = 12$

39. $\lim\limits_{x \to 0^-} \dfrac{|x|}{x} = -1$

$\lim\limits_{x \to 0^+} \dfrac{|x|}{x} = 1$

Therefore, $\lim\limits_{x \to 0} \dfrac{|x|}{x}$ does not exist.

41. $\lim\limits_{x \to 3^-} f(x) = \lim\limits_{x \to 3^-} \left(\dfrac{1}{3}x - 2\right) = 1 - 2 = -1$

$\lim\limits_{x \to 3^+} f(x) = \lim\limits_{x \to 3^+} (-2x + 5) = 6 + 5 = -1$

Therefore, $\lim\limits_{x \to 3} = -1$.

43. $\lim\limits_{\Delta x \to 0} \dfrac{2(x + \Delta x) - 2x}{\Delta x} = \lim\limits_{\Delta x \to 0} \dfrac{2x + 2\Delta x - 2x}{\Delta x}$

$= \lim\limits_{\Delta x \to 0} 2 = 2$

45. $\lim\limits_{\Delta x \to 0} \dfrac{(x + \Delta x)^3 - x^3}{\Delta x} = \lim\limits_{\Delta x \to 0} \dfrac{1 + 3x^2 \Delta x + 3x(\Delta x)^2 + (\Delta x)^3 - x^3}{\Delta x}$

$= \lim\limits_{\Delta x \to 0} [3x^2 + 3x\Delta x + (\Delta x)^2] = 3x^2$

47. $\lim\limits_{\Delta t \to 0} \dfrac{(t + \Delta t)^2 - 5(t + \Delta t) - (t^2 - 5t)}{\Delta t} = \lim\limits_{\Delta t \to 0} \dfrac{t^2 + 2t(\Delta t) + (\Delta t)^2 - 5t - 5(\Delta t) - t^2 + 5t}{\Delta t}$

$= \lim\limits_{\Delta t \to 0} \dfrac{2t(\Delta t) + (\Delta t)^2 - 5(\Delta t)}{\Delta t}$

$= \lim\limits_{\Delta t \to 0} 2t + (\Delta t) - 5$

$= 2t - 5$

49.

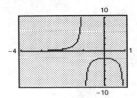

$$\lim_{x \to 1^-} \frac{2}{x^2 - 1} = -\infty$$

x	0	0.5	0.9	0.99	0.999	0.9999	1
$f(x)$	-2	-2.67	-10.53	-100.5	-1000.5	$-10{,}000.5$	undefined

Because $f(x) = \dfrac{2}{x^2 - 1}$ decreases without bound as x tends to 1 from the left, the limit does not exist.

$$\lim_{x \to 1^-} \frac{2}{x^2 - 1} = -\infty$$

51.

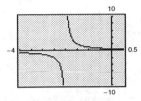

$$\lim_{x \to -2^-} \frac{1}{x + 2} = -\infty$$

x	-3	-2.5	-2.1	-2.01	-2.001	-2.0001	-2
$f(x)$	-1	-2	-10	-100	-1000	$-10{,}000$	undefined

Because $f(x) = \dfrac{1}{x + 2}$ decreases without bound as x tends to -2 from the left, the limit does not exist.

$$\lim_{x \to -2^-} \frac{1}{x + 2} = -\infty$$

53. $C = \dfrac{25{,}000p}{100 - p}, \quad 0 \le p < 100$

(a) If $p = 50$, $C = \dfrac{25{,}000(50)}{100 - 50} = \$25{,}000$.

(b) If $C = 100{,}000 = \dfrac{25{,}000p}{100 - p}$, then

$$4(100 - p) = p$$
$$400 = 5p$$
$$p = 80 \Rightarrow 80\%.$$

(c) $\displaystyle \lim_{p \to 100^-} C = \infty$

The cost function increases without bound as p approaches 100%.

55. (a)

x	-0.01	-0.001	-0.0001	0	0.0001	0.001	0.01
$f(x)$	2.732	2.720	2.718	undefined	2.718	2.717	2.705

$$\lim_{x \to 0} (x + x)^{1/x} \approx 2.718$$

(b)

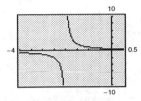

(c) Domain: $(-1, 0) \cup (0, \infty)$

Range: $(1, e) \cup (e, \infty)$

57. $f(x) = \dfrac{x^2 - 5x + 6}{x^2 - 4x + 4} = \dfrac{(x-2)(x-3)}{(x-2)(x-2)}$

$\displaystyle\lim_{x \to 2} f(x) = \lim_{x \to 2} \dfrac{x-3}{x-2}$ does not exist.

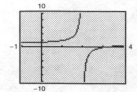

59. $\displaystyle\lim_{x \to -4} \dfrac{x^3 + 4x^2 + x + 4}{2x^2 + 7x - 4} = \lim_{x \to -4} \dfrac{(x+4)(x^2+1)}{(x+4)(2x-1)}$

$\qquad\qquad = \dfrac{17}{9} \approx -1.889$

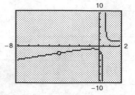

61. (a)

1400

1200

0   1

(b) For $x = 0.25$, $A = 1342.53$.

For $x = \frac{1}{365}$, $A = 1358.95$.

(c) $\displaystyle\lim_{x \to 0^+} 500(1 + 0.10x)^{10/x} = 500e \approx \$1359.14$

Continuous compounding

Section 1.6   Continuity

1. The polynomial $f(x) = 5x^3 - x^2 + 2$ is continuous on the entire real line.

3. The rational function $f(x) = \dfrac{1}{x^2 - 4}$ is not continuous on the entire real line. It is continuous at all $x \neq \pm 2$.

5. $f(x) = -\dfrac{x^3}{2}$ is continuous on $(-\infty, \infty)$.

7. $f(x) = \dfrac{x^2 - 1}{x + 1}$ is continuous on $(-\infty, -1) \cup (-1, \infty)$.

9. $f(x) = x^2 - 2x + 1$ is continuous on $(-\infty, \infty)$.

11. $f(x) = \dfrac{1}{x - 1}$ is continuous on $(-\infty, 1) \cup (1, \infty)$.

13. $f(x) = \dfrac{x}{x^2 + 1}$ is continuous on $(-\infty, \infty)$.

15. $f(x) = \dfrac{x - 5}{x^2 - 9 + 20} = \dfrac{x - 5}{(x - 5)(x - 4)}$ is continuous on $(-\infty, 4) \cup (4, 5) \cup (5, \infty)$.

17. $f(x)$ is continuous on $(-\infty, 1) \cup (1, \infty)$.

19. $\displaystyle\lim_{x \to 1^-} f(x) = \lim_{x \to 1^-} (-2x + 3) = 1$

$\displaystyle\lim_{x \to 1^+} f(x) = \lim_{x \to 1^+} x^2 = 1$

f is continuous on $(-\infty, \infty)$.

21. $\displaystyle\lim_{x \to 2^-} f(x) = \lim_{x \to 2^-} (3 + x) = 5$

$\displaystyle\lim_{x \to 2^+} f(x) = \lim_{x \to 2^+} (x^2 + 1) = 5$

Since $f(2) = 5$, f is continuous on the entire real line.

23. $\displaystyle\lim_{x \to -1^-} \dfrac{|x + 1|}{x + 1} = \lim_{x \to -1^-} \dfrac{-(x + 1)}{x + 1} = -1$

$\displaystyle\lim_{x \to -1^+} \dfrac{|x + 1|}{x + 1} = \lim_{x \to -1^+} \dfrac{x + 1}{x + 1} = 1$

Since $\displaystyle\lim_{x \to -1} f(x)$ does not exist, f is continuous on $(-\infty, -1) \cup (-1, \infty)$.

25. $\lim_{x \to c^-} [\![x - 1]\!] = c - 2,$ c is any integer.

$\lim_{x \to c^+} [\![x - 1]\!] = c - 1,$ c is any integer.

Since $\lim_{x \to c} [\![x - 1]\!]$ does not exist, f is continuous on all intervals $(c, c + 1)$.

27. $h(x) = f(g(x)) = f(x - 1) = \dfrac{1}{\sqrt{x - 1}},$ $x > 1$

Thus, h is continuous on its entire domain $(1, \infty)$.

29. Since $f(x) = x^2 - 4x - 5$ is a polynomial, it is continuous on $[-1, 5]$.

31. Since $\lim_{x \to 2^+} \dfrac{1}{x - 2} = \infty$, f has a nonremovable discontinuity at $x = 2$ on the closed interval $[1, 4]$.

33. $f(x) = \dfrac{x^2 - 16}{x - 4} = \dfrac{(x + 4)(x - 4)}{x - 4} = x + 4,$ $x \neq 4$

Removable discontinuity at $x = 4$

Continuous on $(-\infty, 4) \cup (4, \infty)$

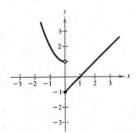

35. $f(x) = \dfrac{x^3 + x}{x} = \dfrac{x(x^2 + 1)}{x} = x^2 + 1,$ $x \neq 0$

Removable discontinuity at $x = 0$

Continuous on $(-\infty, 0) \cup (0, \infty)$

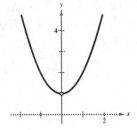

37. $f(x) = \begin{cases} x^2 + 1, & x < 0 \\ x - 1, & x \geq 0 \end{cases}$

Nonremovable discontinuity at $x = 0$

Continuous on $(-\infty, 0), (0, \infty)$

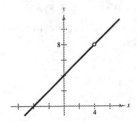

39. $\lim_{x \to 2^-} f(x) = \lim_{x \to 2^-} x^3 = 8$

$\lim_{x \to 2^+} f(x) = \lim_{x \to 2^+} ax^2 = 4a$

Therefore, $8 = 4a$ and $a = 2$.

41. $h(x) = \dfrac{1}{x^2 - x - 2} = \dfrac{1}{(x - 2)(x + 1)}$

h is not continuous at $x = 2$ and $x = -1$.

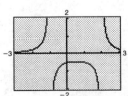
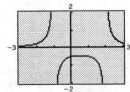

43. $f(x) = \dfrac{x}{x^2 + 1}$ is continuous on $(-\infty, \infty)$.

45. $f(x) = \dfrac{1}{2}[\![2x]\!]$ is continuous on all intervals of the form $\left(\dfrac{c}{2}, \dfrac{c + 1}{2}\right)$, where c is an integer.

47. $f(x) = \dfrac{x^2 + x}{x} = \dfrac{x(x+1)}{x}$ appears to be continuous on $[-4, 4]$. But it is not continuous at $x = 0$ (removable discontinuity).

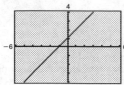

49. $A = 7500(1.015)^{[\![4t]\!]}, \quad t \geq 0$

(a)

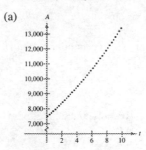

The graph has nonremovable discontinuities at $t = \frac{1}{4}, \frac{1}{2}, \frac{3}{4}, 1, \frac{5}{4}, \ldots$ (every 3 months).

(b) For $t = 7$, $A = 7500(1.015)^{[\![4\cdot7]\!]} = \$11,379.17$.

51. (a) $C(t) = \begin{cases} 1.04, & 0 < t \leq 2 \\ 1.04 + 0.36[\![t - 1]\!], & t > 2t, t \text{ is not an integer.} \\ 1.04 + 0.36(t - 2), & t > 2, t \text{ is an integer.} \end{cases}$

C is not continuous at $t = 2, 3, 4, \ldots$

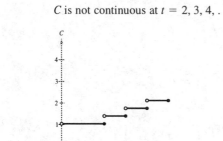

(b) $C(9) = 1.04 + 0.36[\![9 - 2]\!] = \$3.56$

53. Nonremovable discontinuities at $t = 2, 4, 6, 8, \ldots$
$N \to 0$ when $t \to 2^-, 4^-, 6^-, 8^-, \ldots$, so the inventory is replenished every two months.

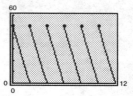

55. There are nonremovable discontinuities at $t = 1, 2, 3, 4, 5,$ and 6.

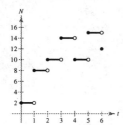

Review Exercises for Chapter 1

1. Population of Texas matches (a).

3. Number of U.S. Business Failures matches (b).

5. Distance $= \sqrt{(0 - 5)^2 + (0 - 2)^2}$
$= \sqrt{25 + 4} = \sqrt{29}$

7. Distance $= \sqrt{[-1 - (-4)]^2 + (3 - 6)^2}$
$= \sqrt{9 + 9} = \sqrt{18} = 3\sqrt{2}$

9. Midpoint $= \left(\dfrac{5 + 9}{2}, \dfrac{6 + 2}{2}\right) = (7, 4)$

11. Midpoint $= \left(\dfrac{-10 - 6}{2}, \dfrac{4 + 8}{2}\right) = (-8, 6)$

13. The taller bars in the back represent revenues. The middle bars represent costs. The smaller bars in front represent profits, since $P = R - C$.

15. The translated vertices are $(1 + 3, 3 + 4) = (4, 7)$, $(2 + 3, 4 + 4) = (5, 8)$, and $(5 + 3, 6 + 4) = (8, 10)$.

17.

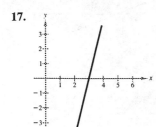

19.

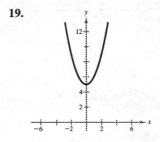

21.

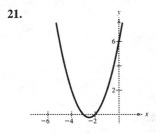

23.

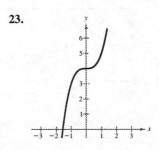

25. y-intercept: $x = 0 \Rightarrow y = -3 \Rightarrow (0, -3)$

x-intercept: $y = 0 \Rightarrow x = -\frac{3}{4} \Rightarrow \left(-\frac{3}{4}, 0\right)$

27. $(x - 0)^2 + (y - 0)^2 = r^2$

$$x^2 + y^2 = r^2$$
$$2^2 + \left(\sqrt{5}\right)^2 = r^2$$
$$9 = r^2$$
$$3 = r$$

Elevation: $x^2 + y^2 = 3^2$

29.

$$x^2 + y^2 - 6x + 8y = 0$$
$$(x^2 - 6x + 9) + (y^2 + 8y + 16) = 9 + 16$$
$$(x - 3)^2 + (y + 4)^2 = 25$$

Center: $(3, -4)$

Radius: 5

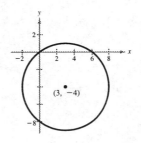

31. $x + y = 2 \Rightarrow y = 2 - x$ $2 - x = 2x - 1$

$2x - y = 1 \Rightarrow y = 2x - 1$ $3 = 3x$

$1 = x$

Point of intersection: $(1, 1)$

33. $y = x^3$ $x^3 = x$

$y = x$ $x^3 - x = 0$

$x(x - 1)(x + 1) = 0$

Points of intersection: $(0, 0), (1, 1), (-1, -1)$

35. (a) $C = 200 + 2x + 8x = 200 + 10x$

$R = 14x$

(b) $C = R$

$200 + 10x = 14x$

$200 = 4x$

$x = 50$ shirts

37. $3x + y = -2$

$y = -3x - 2$

Slope: -3

y-intercept: $(0, -2)$

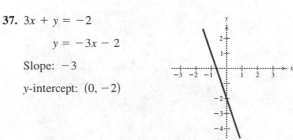

39. $y = -\frac{5}{3}$

Slope: 0 (horizontal line)

y-intercept: $\left(0, -\frac{5}{3}\right)$

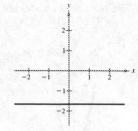

41. $-2x - 5y - 5 = 0$

$$5y = -2x - 5$$

$$y = -\frac{2}{5}x - 1$$

Slope: $-\frac{2}{5}$

y-intercept: $(0, -1)$

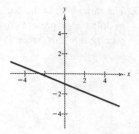

43. Slope $= \dfrac{6 - 0}{7 - 0} = \dfrac{6}{7}$

45. Slope $= \dfrac{17 - (-3)}{10 - (-11)} = \dfrac{20}{21}$

47. $y - (-1) = -2(x - 3)$

$$y = -2x + 5$$

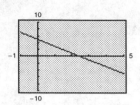

49. (a) $y - 6 = \frac{7}{8}[x - (-3)]$

$$y = \frac{7}{8}x + \frac{69}{8}$$

(b) $4x + 2y = 7 \Longrightarrow y = -2x + \frac{7}{2}$; slope $= -2$

$$y - 6 = -2[x - (-3)]$$

$$y = -2x$$

(c) The line through $(0, 0)$ and $(-3, 6)$ has slope

$$\frac{6}{-3} = -2 \Longrightarrow y = -2x.$$

(d) $3x - 2y = 2 \Longrightarrow y = \frac{3}{2}x - 1$

Slope of perpendicular is $-\frac{2}{3}$.

$$y - 6 = -\frac{2}{3}[x - (-3)]$$

$$y = -\frac{2}{3}x + 4$$

51. $(32, 750), (37, 700)$

$$m = \frac{750 - 700}{32 - 37} = \frac{50}{-5} = -10$$

(a) $x - 750 = -10(p - 32)$

$$x = -10p + 1070$$

(b) If $p = 34.50$, $x = -10(34.50) + 1070 = 725$.

(c) If $p = 42.00$, $x = -10(42.00) + 1070 = 650$.

53. Yes

55. No

57. $f(x) = 3x + 4$

(a) $f(1) = 3(1) + 4 = 7$

(b) $f(x + 1) = 3(x + 1) + 4 = 3x + 7$

(c) $f(2 + \Delta x) = 3(2 + \Delta x) + 4 = 10 + 3\Delta x$

59. $f(x) = x^2 + 3x + 2$

$$= \left(x + \frac{3}{2}\right)^2 - \frac{1}{4}$$

Domain: $(-\infty, \infty)$

Range: $\left[-\frac{1}{4}, \infty\right)$

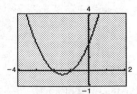

61. $f(x) = \sqrt{x + 1}$

Domain: $[-1, \infty]$

Range: $[0, \infty]$

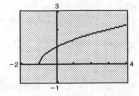

63. $f(x) = -|x| + 3$

Domain: $(-\infty, \infty)$

Range: $(-\infty, 3]$

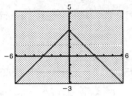

65. (a) $f(x) + g(x) = (1 + x^2) + (2x - 1) = x^2 + 2x$

(b) $f(x) - g(x) = (1 + x^2) - (2x - 1) = x^2 - 2x + 2$

(c) $f(x)g(x) = (1 + x^2)(2x - 1) = 2x^3 - x^2 + 2x - 1$

(d) $\dfrac{f(x)}{g(x)} = \dfrac{1 + x^2}{2x - 1}$

(e) $f(g(x)) = f(2x - 1) = 1 + (2x - 1)^2 = 4x^2 - 4x + 2$

(f) $g(f(x)) = g(1 + x^2) = 2(1 + x^2) - 1 = 2x^2 + 1$

67. $f(x) = \frac{3}{2}x$ has an inverse by the horizontal line test.

$$y = \tfrac{3}{2}x$$

$$x = \tfrac{3}{2}y$$

$$y = \tfrac{2}{3}x$$

$$f^{-1}(x) = \tfrac{2}{3}x$$

69. $f(x) = -x^2 + \frac{1}{2}$ does not have an inverse by the horizontal line test.

71. $\lim\limits_{x \to 2} (5x - 3) = 5(2) - 3 = 7$

73. $\lim\limits_{x \to 2} (5x - 3)(2x + 3) = [5(2) - 3][2(2) + 3] = 49$

75. $\lim\limits_{t \to 3} \dfrac{t^2 + 1}{t} = \dfrac{(3)^2 + 1}{3} = \dfrac{10}{3}$

77. $\lim\limits_{t \to 0^-} \dfrac{t^2 + 1}{t} = -\infty$

$\lim\limits_{t \to 0^+} \dfrac{t^2 + 1}{t} = \infty$

$\lim\limits_{t \to 0} \dfrac{t^2 + 1}{t}$ does not exist.

79. $\lim\limits_{x \to -2} \dfrac{x + 2}{x^2 - 4} = \lim\limits_{x \to -2} \dfrac{x + 2}{(x + 2)(x - 2)}$

$$= \lim\limits_{x \to -2} \dfrac{1}{x - 2} = -\dfrac{1}{4}$$

81. $\lim\limits_{x \to 0^+} \left(x - \dfrac{1}{x}\right) = \lim\limits_{x \to 0^+} \dfrac{x^2 - 1}{x} = -\infty$

83. $\lim\limits_{x \to 0} \dfrac{[1/(x - 2)] - 1}{x} = \lim\limits_{x \to 0} \dfrac{1 - (x - 2)}{x(x - 2)} = \lim\limits_{x \to 0} \dfrac{3 - x}{x(x - 2)}$ does not exist.

85. $\lim\limits_{\Delta x \to 0} \dfrac{(x + \Delta x)^3 - (x + \Delta x) - (x^3 - x)}{\Delta x} = \lim\limits_{\Delta x \to 0} \dfrac{x^3 + 3x^2\Delta x + 3x(\Delta x)^2 + (\Delta x)^3 - x - \Delta x - x^3 + x}{\Delta x}$

$$= \lim\limits_{\Delta x \to 0} \dfrac{3x^2\Delta x + 3x(\Delta x)^2 + (\Delta x)^3 - \Delta x}{\Delta x}$$

$$= \lim\limits_{\Delta x \to 0} [3x^2 + 3x\Delta x + (\Delta x)^2 - 1]$$

$$= 3x^2 - 1$$

87.

x	1.1	1.01	1.001	1.0001
$f(x)$	0.5680	0.5764	0.5773	0.5773

$\lim\limits_{x \to 1^+} \dfrac{\sqrt{2x + 1} - \sqrt{3}}{x - 1} = \dfrac{1}{\sqrt{3}} \approx 0.5774$

89. The statement is false since $\lim\limits_{x \to 0^-} \dfrac{|x|}{x} = -1$.

91. The statement is false since $\lim\limits_{x \to 0^-} \sqrt{x}$ is undefined.

93. The statement is false since $\lim\limits_{x \to 2^+} f(x) = \lim\limits_{x \to 2^+} 0 = 0$.

95. $f(x) = \dfrac{1}{(x + 4)^2}$ is continuous on the intervals $(-\infty, -4) \cup (-4, \infty)$.

97. $f(x) = \dfrac{3}{x + 1}$ is continuous on the intervals $(-\infty, -1) \cup (-1, \infty)$.

99. $f(x) = [\![x + 3]\!]$ is continuous on all intervals of the form $(c, c + 1)$, where c is an integer.

101. $f(x)$ is continuous on the intervals $(-\infty, 0) \cup (0, \infty)$.

103. $\lim\limits_{x \to 3^-} f(x) = \lim\limits_{x \to 3^-} (-x + 1) = -2$

$\lim\limits_{x \to 3^+} f(x) = \lim\limits_{x \to 3^+} (ax - 8) = 3a - 8$

Thus, $-2 = 3a - 8$ and $a = 2$.

105. (a)

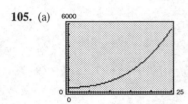

(b)

t	0	5	10	15	20	21	22	23
Debt	381	542	914	1817	3266	3599	4083	4436
Model	387.94	518.59	951.74	1837.39	3325.54	3708.67	4123.10	4570.03

(c) In 1999, $t = 29$ and $D \approx 8001.31$ billion.

107. Yellow sweet maize:

Intercepts: $(0, 45), (5, 0)$

Line: $y - 45 = \dfrac{45 - 0}{0 - 5}(x - 0)$

$\quad\quad y = -9x + 45$

White flint maize:

Intercepts: $(0, 30), (5.5, 0)$

Line: $y - 30 = \dfrac{30 - 0}{0 - 5.5}(x - 0)$

$\quad\quad y = -5.45x + 30$

Practice Test for Chapter 1

1. Find the distance between $(3, 7)$ and $(4, -2)$.

2. Find the midpoint of the line segment joining $(0, 5)$ and $(2, 1)$.

3. Determine whether the points $(0, -3)$, $(2, 5)$, and $(-3, -15)$ are collinear.

4. Find x so that the distance between $(0, 3)$ and $(x, 5)$ is 7.

5. Sketch the graph of $y = 4 - x^2$.

6. Sketch the graph of $y = \sqrt{x - 2}$.

7. Sketch the graph of $y = |x - 3|$.

8. Write the equation of the circle in standard form and sketch its graph.

$$x^2 + y^2 - 8x + 2y + 8 = 0$$

9. Find the points of intersection of the graphs of $x^2 + y^2 = 25$ and $x - 2y = 10$.

10. Find the general equation of the line passing through the points $(7, 4)$ and $(6, -2)$.

11. Find the general equation of the line passing through the point $(-2, -1)$ with a slope of $m = \frac{2}{3}$.

12. Find the general equation of the line passing through the point $(6, -8)$ with undefined slope.

13. Find the general equation of the line passing through the point $(0, 3)$ and perpendicular to the line given by $2x - 5y = 7$.

14. Given $f(x) = x^2 - 5$, find the following.

    (a) $f(3)$ (b) $f(-6)$ (c) $f(x - 5)$ (d) $f(x + \Delta x)$

15. Find the domain and range of $f(x) = \sqrt{3 - x}$.

16. Given $f(x) = 2x + 3$ and $g(x) = x^2 - 1$, find the following.

    (a) $f(g(x))$ (b) $g(f(x))$

17. Given $f(x) = x^3 + 6$, find $f^{-1}(x)$.

18. Find $\lim\limits_{x \to -4} (2 - 5x)$.

19. Find $\lim\limits_{x \to 6} \dfrac{x^2 - 36}{x - 6}$.

20. Find $\lim\limits_{x \to -1} \dfrac{|x + 1|}{x + 1}$.

21. Find $\lim\limits_{x \to 0} \dfrac{\sqrt{x + 5} - \sqrt{5}}{x}$.

22. Find $\lim\limits_{x \to 1} f(x)$, where $f(x) = \begin{cases} 2x + 3, & x \le 1 \\ x^2 + 4, & x > 1. \end{cases}$

23. Find the discontinuities of $f(x) = \dfrac{x - 8}{x^2 - 64}$. Which are removable?

24. Find the discontinuities of $f(x) = \dfrac{|x - 3|}{x - 3}$. Which are removable?

25. Sketch the graph of $f(x) = \dfrac{x^2 - 5x + 6}{x - 3}$.

Graphing Calculator Required

26. Solve the equation for y and graph the resulting two equations on the same set of coordinate axes.

$$x^2 + y^2 + 6x + 5 = 0$$

27. Use a graphing calculator to graph $f(x) = \dfrac{x^2 - 9}{x - 3}$ and find $\lim\limits_{x \to 3} f(x)$. Is the graph displayed correctly at $x = 3$?

CHAPTER 2
Differentiation

C H A P T E R 2
Differentiation

Section 2.1   The Derivative and the Slope of a Graph

Solutions to Odd-Numbered Exercises

1. The tangent line at (x_1, y_1) has a positive slope. The tangent line at (x_2, y_2) has a negative slope.

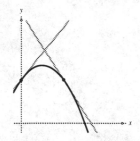

3. The tangent line at (x_1, y_1) has a positive slope. The tangent line at (x_2, y_2) has zero slope.

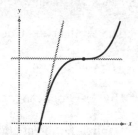

5. The slope is $m = 1$.

7. The slope is $m = 0$.

9. The slope is $m = -\frac{1}{3}$.

11. In 1991, $t = 1$ and the slope is about

$$\frac{4300 - 2500}{3 - 0} = 600 \text{ million/year.}$$

In 1995, $t = 5$ and the slope is about

$$\frac{6000 - 3600}{2 - 0} = 1200 \text{ million/year.}$$

13.
$$f(x + \Delta x) = 3$$
$$f(x + \Delta x) - f(x) = 0$$
$$\frac{f(x + \Delta x) - f(x)}{\Delta x} = 0$$
$$\lim_{\Delta x \to 0} \frac{f(x + \Delta x) - f(x)}{\Delta x} = 0$$

15.
$$f(x + \Delta x) = -5(x + \Delta x) + 3 = -5x - 5\Delta x + 3$$
$$f(x + \Delta x) - f(x) = -5\Delta x$$
$$\frac{f(x + \Delta x) - f(x)}{\Delta x} = -5$$
$$\lim_{\Delta x \to 0} \frac{f(x + \Delta x) - f(x)}{\Delta x} = -5$$

17.
$$f(x + \Delta x) = (x + \Delta x)^2 = x^2 + 2x\Delta x + (\Delta x)^2$$
$$f(x + \Delta x) - f(x) = 2x\Delta x + (\Delta x)^2$$
$$\frac{f(x + \Delta x) - f(x)}{\Delta x} = 2x + \Delta x$$
$$\lim_{\Delta x \to 0} \frac{f(x + \Delta x) - f(x)}{\Delta x} = 2x$$

19.
$$f(x + \Delta x) = 3(x + \Delta x)^2 - 5(x + \Delta x) - 2$$
$$= 3x^2 + 6x\Delta x + 3(\Delta x)^2 - 5x - 5(\Delta x) - 2$$
$$f(x + \Delta x) - f(x) = 6x\Delta x + 3(\Delta x)^2 - 5(\Delta x)$$
$$\frac{f(x + \Delta x) - f(x)}{\Delta x} = 6x + 3(\Delta x) - 5$$
$$\lim_{\Delta x \to 0} \frac{f(x + \Delta x)}{\Delta x} = 6x - 5$$

21.
$$h(t + \Delta t) = \sqrt{t + \Delta t - 1}$$
$$h(t + \Delta t) - h(t) = \sqrt{t + \Delta t - 1} - \sqrt{t - 1}$$
$$= \frac{\sqrt{t + \Delta t - 1} - \sqrt{t - 1}}{1} \cdot \frac{\sqrt{t + \Delta t - 1} + \sqrt{t - 1}}{\sqrt{t + \Delta t - 1} + \sqrt{t - 1}}$$
$$= \frac{\Delta t}{\sqrt{t + \Delta t - 1} + \sqrt{t - 1}}$$
$$\frac{h(t + \Delta t) - h(t)}{\Delta t} = \frac{1}{\sqrt{t + \Delta t - 1} + \sqrt{t - 1}}$$
$$\lim_{\Delta t \to 0} \frac{h(t + \Delta t) - h(t)}{\Delta t} = \frac{1}{2\sqrt{t - 1}}$$

23.
$$f(t + \Delta t) = (t + \Delta t)^3 - 12(t + \Delta t)$$
$$= t^3 + 3t^2\Delta t + 3t(\Delta t)^2 + (\Delta t)^3 - 12t - 12\Delta t$$
$$f(t + \Delta t) - f(t) = 3t^2\Delta t + 3t(\Delta t)^2 + (\Delta t)^3 - 12\Delta t$$
$$\frac{f(t + \Delta t) - f(t)}{\Delta t} = 3t^2 + 3t\Delta t + (\Delta t)^2 - 12$$
$$\lim_{\Delta t \to 0} \frac{f(t + \Delta t) - f(t)}{\Delta t} = 3t^2 - 12$$

25.
$$f(x + \Delta x) = \frac{1}{(x + \Delta x)^2}$$
$$f(x + \Delta x) - f(x) = \frac{1}{(x + \Delta x)^2} - \frac{1}{x^2}$$
$$= \frac{x^2 - [x^2 + 2x\Delta x + (\Delta x)^2]}{(x + \Delta x)^2 x^2}$$
$$= \frac{-2x\Delta x - (\Delta x)^2}{(x + \Delta x)^2 x^2}$$
$$\frac{f(x + \Delta x) - f(x)}{\Delta x} = \frac{-2x - \Delta x}{(x + \Delta x)^2 x^2}$$
$$\lim_{\Delta x \to 0} \frac{f(x + \Delta x) - f(x)}{\Delta x} = \frac{-2x}{x^4} = -\frac{2}{x^3}$$

27.
$$f(x + \Delta x) = 6 - 2(x + \Delta x)$$
$$= 6 - 2x - 2\Delta x$$
$$f(x + \Delta x) - f(x) = -2\Delta x$$
$$\frac{f(x + \Delta x) - f(x)}{\Delta x} = -2$$
$$\lim_{\Delta x \to 0} \frac{f(x + \Delta x) - f(x)}{\Delta x} = -2$$

At $(2, 2)$, the slope of the tangent line is $m = -2$. The figure shows the graph of f and the tangent line.

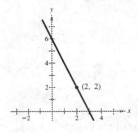

29.

$$f(x + \Delta x) = -(x + \Delta x)$$

$$f(x + \Delta x) - f(x) = -(x + \Delta x) - (-x) = -\Delta x$$

$$\frac{f(x + \Delta x) - f(x)}{\Delta x} = -1$$

$$\lim_{\Delta x \to 0} \frac{f(x + \Delta x) - f(x)}{\Delta x} = -1$$

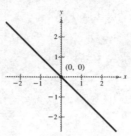

At $(0, 0)$, the slope of the tangent line is -1. The figure shows the graph of f and the tangent line.

31.

$$f(x + \Delta x) = (x + \Delta x)^2 - 2 = x^2 + 2x\Delta x + (\Delta x)^2 - 2$$

$$f(x + \Delta x) - f(x) = 2x\Delta x + (\Delta x)^2$$

$$\frac{f(x + \Delta x) - f(x)}{\Delta x} = 2x + \Delta x$$

$$\lim_{\Delta x \to 0} \frac{f(x + \Delta x) - f(x)}{\Delta x} = 2x$$

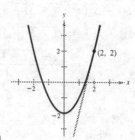

At $(2, 2)$, the slope of the tangent line is $m = 2(2) = 4$. The figure shows the graph of f and the tangent line.

33.

$$f(x + \Delta x) = x^3 + 3x^2\Delta x + 3x(\Delta x)^2 + (\Delta x)^3$$

$$f(x + \Delta x) - f(x) = 3x^2\Delta x + 3x(\Delta x)^2 + (\Delta x)^3$$

$$\frac{f(x + \Delta x) - f(x)}{\Delta x} = 3x^2 + 3x\Delta x + (\Delta x)^2$$

$$\lim_{\Delta x \to 0} \frac{f(x + \Delta x) - f(x)}{\Delta x} = 3x^2$$

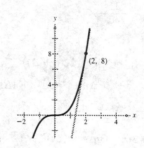

At $(2, 8)$, the slope of the tangent line is $m = 3(2)^2 = 12$. The figure shows the graph of f and the tangent line.

35.

$$f(x + \Delta x) = (x + \Delta x)^3 + 2(x + \Delta x)$$

$$= x^3 + 3x^2\Delta x + 3x(\Delta x)^2 + (\Delta x)^3 + 2x + 2\Delta x$$

$$f(x + \Delta x) - f(x) = 3x^2\Delta x + 3x(\Delta x)^2 + (\Delta x)^3 + 2\Delta x$$

$$\frac{f(x + \Delta x) - f(x)}{\Delta x} = 3x^2 + 3\Delta x + (\Delta x)^2 + 2$$

$$\lim_{\Delta x \to 0} \frac{f(x + \Delta x) - f(x)}{\Delta x} = 3x^2 + 2$$

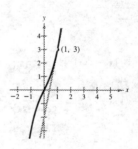

At $(1, 3)$, the slope of the tangent line is $m = 3(1)^2 + 2 = 5$. The figure shows the graph of f and the tangent line.

37.

$$f(x + \Delta x) = \frac{1}{2}(x + \Delta x)^2 = \frac{1}{2}x^2 + x\Delta x + \frac{1}{2}\Delta x^2$$

$$f(x + \Delta x) - f(x) = x\Delta x + \frac{1}{2}(\Delta x)^3$$

$$\frac{f(x + \Delta x) - f(x)}{\Delta x} = x + \frac{1}{2}(\Delta x)$$

$$\lim_{\Delta x \to 0} \frac{f(x + \Delta x) - f(x)}{\Delta x} = x$$

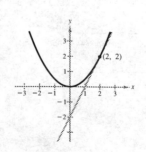

At the point $(2, 2)$, the slope of the tangent line is $m = 2$. The equation of the tangent line is

$$y - 2 = 2(x - 2)$$

$$y = 2x - 2.$$

39.
$$f(x + \Delta x) = [(x + \Delta x) - 1]^2$$
$$= x^2 + 2x\Delta x + \Delta x^2 - 2x - 2\Delta x + 1$$
$$f(x + \Delta x) - f(x) = 2x\Delta x + (\Delta x)^2 - 2\Delta x$$
$$\frac{f(x + \Delta x) - f(x)}{\Delta x} = 2x + \Delta x - 2$$
$$\lim_{\Delta x \to 0} \frac{f(x + \Delta x) - f(x)}{\Delta x} = 2x - 2$$

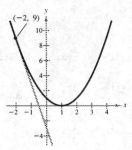

At the point $(-2, 9)$, the slope of the tangent line is $m = 2(-2) - 2 = -6$. The equation of the tangent line is
$$y - 9 = -6[x - (-2)]$$
$$y = -6x - 3.$$

41.
$$f(x + \Delta x) = \sqrt{x + \Delta x} + 1$$
$$f(x + \Delta x) - f(x) = \sqrt{x + \Delta x} - \sqrt{x}$$
$$= \frac{\sqrt{x + \Delta x} - \sqrt{x}}{1} \cdot \frac{\sqrt{x + \Delta x} + \sqrt{x}}{\sqrt{x + \Delta x} + \sqrt{x}}$$
$$= \frac{\Delta x}{\sqrt{x + \Delta x} + \sqrt{x}}$$
$$\frac{f(x + \Delta x) - f(x)}{\Delta x} = \frac{1}{\sqrt{x + \Delta x} + \sqrt{x}}$$
$$\lim_{\Delta x \to 0} \frac{f(x + \Delta x) - f(x)}{\Delta x} = \frac{1}{\sqrt{x} + \sqrt{x}} = \frac{1}{2\sqrt{x}}$$

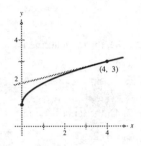

At the point $(4, 3)$, $f'(4) = \frac{1}{2\sqrt{4}} = \frac{1}{4}$. The equation of the tangent line is
$$y - 3 = \frac{1}{4}(x - 4)$$
$$y = \frac{1}{4}x + 2.$$

43.
$$f(x + \Delta x) = -\frac{1}{4}(x + \Delta x)^2 = -\frac{1}{4}[x^2 + 2x\Delta x + (\Delta x)^2]$$
$$f(x + \Delta x) - f(x) = -\frac{1}{4}[2x\Delta x + (\Delta x)^2]$$
$$\frac{f(x + \Delta x) - f(x)}{\Delta x} = -\frac{x}{2} - \frac{1}{4}\Delta x$$
$$\lim_{\Delta x \to 0} \frac{f(x + \Delta x) - f(x)}{\Delta x} = -\frac{x}{2} \quad \text{(slope of tangent line)}$$

Since the slope of the given line is -1, we have
$$-\frac{x}{2} = -1$$
$$x = 2.$$

Therefore, at the point $(2, -1)$, the tangent line parallel to $x + y = 0$ is
$$y - (-1) = -1(x - 2)$$
$$y = -x + 1.$$

45.
$$f(x + \Delta x) = -\frac{1}{2}(x + \Delta x)^3 = -\frac{1}{2}[x^3 + 3x^2\Delta x + 3x(\Delta x)^2 + (\Delta x)^3]$$

$$f(x + \Delta x) - f(x) = -\frac{1}{2}[3x^2\Delta x + 3x(\Delta x)^2 + (\Delta x)^3]$$

$$\frac{f(x + \Delta x) - f(x)}{\Delta x} = -\frac{1}{2}[3x^2 + 3x\Delta x + (\Delta x)^2]$$

$$\lim_{\Delta x \to 0}\frac{f(x + \Delta x) - f(x)}{\Delta x} = -\frac{1}{2}(3x^2) = -\frac{3}{2}x^2 \quad \text{(slope of tangent line)}$$

Since the slope of the given line is -6, we have

$$-\frac{3}{2}x^2 = -6$$

$$x^2 = 4$$

$$x = \pm 2.$$

At the point $(2, -4)$, the tangent line is

$$y + 4 = -6(x - 2)$$

$$y = -6x + 8.$$

At the point $(-2, 4)$, the tangent line is

$$y - 4 = -6(x + 2)$$

$$y = -6x - 8.$$

47. y is not differentiable when $x = -3$. At $(-3, 0)$, the graph has a node. y is differentiable for all $x \neq -3$.

49. y is differentiable everywhere except at $x = 3$. At $(3, 0)$, the graph has a node.

51. f is differentiable on the open interval $(1, \infty)$.

53. f is differentiable everywhere except at $x = 0$, which is a nonremovable discontinuity.

55. $f(x) = \frac{1}{4}x^3$

x	-2	$-\frac{3}{2}$	-1	$-\frac{1}{2}$	0	$\frac{1}{2}$	1	$\frac{3}{2}$	2
$f(x)$	-2	-0.8438	-0.25	-0.0313	0	0.0313	0.25	0.8438	2
$f'(x)$	3	1.6875	0.75	0.1875	0	0.1875	0.75	1.6875	3

Analytically, the slope of $f(x) = \frac{1}{4}x^3$ is

$$m = \lim_{\Delta x \to 0}\frac{f(x + \Delta x) - f(x)}{\Delta x}$$

$$= \lim_{\Delta x \to 0}\frac{\frac{1}{4}(x + \Delta x)^3 - \frac{1}{4}x^3}{\Delta x}$$

$$= \lim_{\Delta x \to 0}\frac{\frac{1}{4}[3x^2\Delta x + 3x(\Delta x)^2 + (\Delta x)^3]}{\Delta x}$$

$$= \lim_{\Delta x \to 0}\frac{1}{4}[3x^2 + 3x\Delta x + (\Delta x)^2]$$

$$= \frac{3}{4}x^2.$$

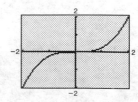

57. $f(x) = -\frac{1}{2}x^3$

x	-2	$-\frac{3}{2}$	-1	$-\frac{1}{2}$	0	$\frac{1}{2}$	1	$\frac{3}{2}$	2
$f(x)$	4	1.6875	0.5	0.0625	0	-0.0625	-0.5	-1.6875	-4
$f'(x)$	-6	-3.375	-1.5	-0.375	0	-0.375	-1.5	-3.375	-6

Analytically, the slope of $f(x) = -\frac{1}{2}x^3$ is

$$m = \lim_{\Delta x \to 0} \frac{f(x + \Delta x) - f(x)}{\Delta x}$$

$$= \lim_{\Delta x \to 0} \frac{-\frac{1}{2}(x + \Delta x)^3 + \frac{1}{2}x^3}{\Delta x}$$

$$= \lim_{\Delta x \to 0} \frac{-\frac{1}{2}[x^3 + 3x^2\Delta x + 3x(\Delta x)^2 + (\Delta x)^3] + \frac{1}{2}x^3}{\Delta x}$$

$$= \lim_{\Delta x \to 0} \frac{-\frac{1}{2}[3x^2\Delta x + 3x(\Delta x)^2 + (\Delta x)^3]}{\Delta x}$$

$$= \lim_{\Delta x \to 0} -\frac{1}{2}[3x^2 + 3x(\Delta x) + (\Delta x)^2]$$

$$= -\frac{3}{2}x^2.$$

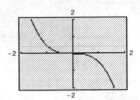

59. True

61. True. See page 89.

63. $f'(x) = \lim_{\Delta x \to 0} \frac{f(x + \Delta x) - f(x)}{\Delta x}$

$$= \lim_{\Delta x \to 0} \frac{(x + \Delta x)^2 - 4(x + \Delta x) - (x^2 - 4x)}{\Delta x}$$

$$= \lim_{\Delta x \to 0} \frac{2x\Delta x + (\Delta x)^2 - 4\Delta x}{\Delta x}$$

$$= \lim_{\Delta x \to 0} (2x + \Delta x - 4)$$

$$= 2x - 4$$

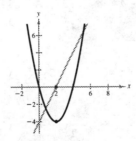

The x-intercept of the derivative indicates a point of horizontal tangency for f.

65. The graph of $f(x) = x^2 + 1$ is smooth at $(0, 1)$, but the graph of $g(x) = |x| + 1$ has a sharp point at $(0, 1)$. The function g is not differentiable at $(0, 1)$.

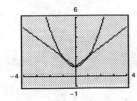

Section 2.2 Some Rules for Differentiation

1. (a) $y = x^2$

$y' = 2x$

At $(1, 1)$, $y' = 2$.

(b) $y = x^{1/2}$

$y' = \frac{1}{2}x^{-1/2} = \frac{1}{2\sqrt{x}}$

At $(1, 1)$, $y' = \frac{1}{2}$.

3. (a) $y = x^{-1}$

$y' = -x^{-2} = -\frac{1}{x^2}$

At $(1, 1)$, $y' = -1$.

(b) $y = x^{-1/3}$

$y' = -\frac{1}{3}x^{-4/3} = -\frac{1}{3x^{4/3}}$

At $(1, 1)$, $y' = -\frac{1}{3}$.

5. $y' = 0$

7. $f'(x) = 1$

9. $g'(x) = 2x$

11. $f'(t) = -6t + 2$

13. $s'(t) = 3t^2 - 2$

15. $y' = 4\left(\frac{4}{3}\right)t^{1/3} = \frac{16}{3}t^{1/3}$

17. $f(x) = 4\sqrt{x} = 4x^{1/2}$

$f'(x) = 4\left(\frac{1}{2}\right)x^{(1/2)-1} = 2x^{-1/2} = \frac{2}{\sqrt{x}}$

19. $y' = 4(-2)x^{-2-1} + 2(2)x^{2-1}$

$= -8x^{-3} + 4x^1 = -\frac{8}{x^3} + 4x$

	Function	Rewrite	Differentiate	Simplify
21.	$y = \frac{1}{4x^3}$	$y = \frac{1}{4}x^{-3}$	$y' = -\frac{3}{4}x^{-4}$	$y' = -\frac{3}{4x^4}$
23.	$y = \frac{1}{(4x)^3}$	$y = \frac{1}{64}x^{-3}$	$y' = -\frac{3}{64}x^{-4}$	$y' = -\frac{3}{64x^4}$
25.	$y = \frac{\sqrt{x}}{x}$	$y = x^{-(1/2)}$	$y' = -\frac{1}{2}x^{-(3/2)}$	$y' = -\frac{1}{2x^{3/2}}$

27. $f(x) = \frac{1}{x} = x^{-1}$

$f'(x) = -x^{-2} = -\frac{1}{x^2} \Rightarrow f'(1) = -1$

29. $f(t) = 4 - \frac{4}{3t} = 4 - \frac{4}{3}t^{-1}$

$f'(t) = \frac{4}{3}t^{-2} = \frac{4}{3t^2}$

$f'\left(\frac{1}{2}\right) = \frac{16}{3}$

31. $y = (2x + 1)^2 = 4x^2 + 4x + 1$

$y' = 8x + 4$

At $(0, 1)$, $y' = 4$.

33. $f(x) = x^2 - \frac{4}{x} = x^2 - 4x^{-1}$

$f'(x) = 2x + 4x^{-2} = 2x + \frac{4}{x^2} = \frac{2(x^3 + 2)}{x^2}$

35. $f(x) = x^2 - 2x - \frac{2}{x^4} = x^2 - 2 - 2x^{-4}$

$f'(x) = 2x - 2 + 8x^{-5} = 2x - 2 + \frac{8}{x^5}$

37. $f(x) = \frac{2x^3 - 4x^2 + 3}{x^2} = 2x - 4 + 3x^{-2}$

$f'(x) = 2 - 6x^{-3} = 2 - \frac{6}{x^3} = \frac{2x^3 - 6}{x^3}$

39. $f(x) = x(x^2 + 1) = x^3 + x$

$f'(x) = 3x^2 + 1$

41. $f'(x) = \dfrac{4}{5}x^{-1/5} = \dfrac{4}{5x^{1/5}}$

43. $y' = -4x^3 + 6x$

At $(1, 0)$, the slope is $m = y' = -4 + 6 = 2$. The equation of the tangent line is

$y - 0 = 2(x - 1)$

$y = 2x - 2.$

45. $f(x) = x^{1/3} + x^{1/5}$

$f'(x) = \dfrac{1}{3}x^{-2/3} + \dfrac{1}{5}x^{-4/5} = \dfrac{1}{3x^{2/3}} + \dfrac{1}{5x^{4/5}}$

$f'(1) = \dfrac{1}{3} + \dfrac{1}{5} = \dfrac{8}{15}$

$y - 2 = \dfrac{8}{15}(x - 1)$

$y = \dfrac{8}{15}x + \dfrac{22}{15}$

47. $y' = -4x^3 + 6x = 2x(3 - 2x^2) = 0$

$x = 0, \quad x = \pm\sqrt{\dfrac{3}{2}} = \pm\sqrt{\dfrac{6}{2}}$

If $x = \pm\sqrt{\dfrac{6}{2}}$, then

$y = -\left(\dfrac{\sqrt{6}}{2}\right)^4 + 3\left(\dfrac{\sqrt{6}}{2}\right)^2 - 1$

$= -\dfrac{9}{4} + 3\left(\dfrac{3}{2}\right) - 1$

$= \dfrac{5}{4}.$

The function has horizontal tangent lines at the points

$(0, 1), \left(-\dfrac{\sqrt{6}}{2}, \dfrac{5}{4}\right),$ and $\left(\dfrac{\sqrt{6}}{2}, \dfrac{5}{4}\right).$

49. $f'(x) = 3x^2 + 1$

$f'(x) \neq 0$ for any value of x. The graph of $f(x)$ has no horizontal tangent lines.

51. (a)

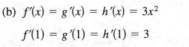

(b) $f'(x) = g'(x) = h'(x) = 3x^2$

$f'(1) = g'(1) = h'(1) = 3$

(c)

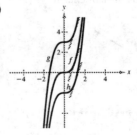

53. (a) $h(x) = f(x) - 2$

$h'(x) = f'(x) - 0 = f'(x)$

$h'(1) = f'(1) = 3$

(c) $h(x) = -f(x)$

$h'(x) = -f'(x)$

$h'(1) = -f'(1) = -3$

(b) $h(x) = 2f(x)$

$h'(x) = 2f'(x)$

$h'(1) = 2f'(1) = 2(3) = 6$

(d) $h(x) = -1 + 2f(x)$

$h'(x) = 0 + 2f'(x) = 2f'(x)$

$h'(1) = 2f'(1) = 2(3) = 6$

55. (a) $R'(t) = -382.5t^4 + 4354.8t^3 - 15,585.9t^2 + 17,561.6t - 1763.4$

$R'(2) = -265.4$

$R'(5) = 1684.6$

(b) The results of part (a) are more accurate.

(c) The units of the slope are millions of dollars per year.

57. $C = 0.60x + 250$

$R = 1.00x$

$P = R - C$

$\quad = 1.00x - (0.60x + 250)$

$\quad = 0.40x - 250$

$\dfrac{dP}{dx} = 0.40$

Therefore, the derivative is constant and is equal to the profit on each candy bar sold.

59.

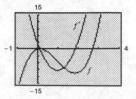

There are horizontal tangents at (0.11, 0.14) and (1.84, −10.49).

Section 2.3   Rates of Change: Velocity and Marginals

1. (a) $\dfrac{88 - 73}{1980 - 1975} = 3$ billion per year

(b) $\dfrac{121 - 88}{1985 - 1980} = 6.6$ billion per year

(c) $\dfrac{134.5 - 121}{1990 - 1985} = 2.7$ billion per year

(d) $\dfrac{132 - 73}{1995 - 1975} = 2.95$ billion per year

3. $f'(t) = 2$

Average rate of change: $\dfrac{\Delta y}{\Delta t} = \dfrac{f(2) - f(1)}{2 - 1} = \dfrac{11 - 9}{1} = 2$

Instantaneous rates of change: $f'(1) = 2$, $f'(2) = 2$

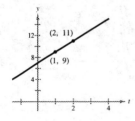

5. $h'(x) = 2x$

Average rate of change: $\dfrac{\Delta y}{\Delta x} = \dfrac{h(2) - h(-2)}{2 - (-2)} = \dfrac{0}{4} = 0$

Instantaneous rates of change: $h'(-2) = -4$, $h'(2) = 4$

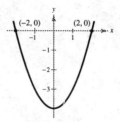

7. $f'(x) = -\dfrac{1}{x^2}$

Average rate of change: $\dfrac{\Delta y}{\Delta x} = \dfrac{f(4) - f(1)}{4 - 1} = \dfrac{(1/4) - 1}{3} = -\dfrac{1}{4}$

Instantaneous rates of change: $f'(1) = -1$, $f'(4) = -\dfrac{1}{16}$

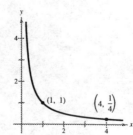

9. $g'(x) = 4x^3 - 2x$

Average rate of change: $\dfrac{g(3) - g(1)}{3 - 1} = \dfrac{74 - 2}{2} = 36$

Instantaneous rates of change: $g'(1) = 2, \quad g'(3) = 102$

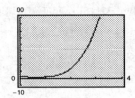

11. $E'(t) = \dfrac{1}{27}(9 + 6t - 3t^2) = \dfrac{1}{9}(3 + 2t - t^2)$

(a) $\dfrac{E(1) - E(0)}{1 - 0} = \dfrac{(11/27) - 0}{1 - 0} = \dfrac{11}{27}$

   $E'(0) = \dfrac{1}{3}$

   $E'(1) = \dfrac{4}{9}$

(b) $\dfrac{E(2) - E(1)}{2 - 1} = \dfrac{(22/27) - (11/27)}{2 - 1} = \dfrac{11}{27}$

   $E'(1) = \dfrac{4}{9}$

   $E'(2) = \dfrac{1}{3}$

(c) $\dfrac{E(3) - E(2)}{3 - 2} = \dfrac{1 - (22/27)}{3 - 2} = \dfrac{5}{27}$

   $E'(2) = \dfrac{1}{3}$

   $E'(3) = 0$

(d) $\dfrac{E(4) - E(3)}{4 - 3} = \dfrac{(20/27) - 1}{4 - 3} = -\dfrac{7}{27}$

   $E'(3) = 0$

   $E'(4) = -\dfrac{5}{9}$

13. $s = -16t^2 + 555$

(a) Average velocity $= \dfrac{5(3) - 5(2)}{3 - 2} = \dfrac{411 - 491}{1} = -80 \text{ ft/sec}$

(b) $v = s'(t) = -32t, \quad v(2) = -64 \text{ ft/sec}, \quad v(3) = -96 \text{ ft/sec}$

(c) $\quad s = -16t^2 + 555 = 0$

   $16t^2 = 555$

   $t^2 = \dfrac{555}{16}$

   $t \approx 5.89 \text{ seconds}$

(d) $v(5.89) \approx -188.5 \text{ ft/sec}$

15. $s'(t) = 15\sqrt{t} = v(t)$

(a) $v(0) = 0 \text{ ft/sec}$ (b) $v(1) = 15 \text{ ft/sec}$

(c) $v(4) = 30 \text{ ft/sec}$ (d) $v(9) = 45 \text{ ft/sec}$

17. $\dfrac{dC}{dx} = 1.47$

19. $\dfrac{dC}{dx} = 470 - 0.5x$

21. $\dfrac{dR}{dx} = 50 - x$

23. $\dfrac{dR}{dx} = -18x^2 + 16x + 200$

25. $\dfrac{dP}{dx} = -4x + 72$

27. $\dfrac{dP}{dx} = -0.0005x + 12.2$

29. (a) $R(15) - R(14) = 2(15)[900 + 32(15) - 15^2] - 2(14)[900 + 32(14) - 14^2]$

$$= 34,650 - 32,256 = 2394 \text{ dollars}$$

(b) $\quad R = 1800x + 64x^2 - 2x^3$

$R'(x) = 1800 + 128x - 6x^2$

$R'(14) = 2416 \text{ dollars}$

(c) The answers are nearly the same.

31. (a) $P(8) - P(7) = -90.4 - (-90.9) = 0.50 \text{ dollars}$ (c) The answers are nearly the same.

(b) $P'(x) = -0.2x + 2$

$P'(7) = 0.60 \text{ dollars}$

33. (a)

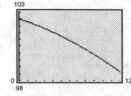

(c) $T(0) = 102.5 \text{ degrees}$

$T(4) = 101.54 \text{ degrees}$

$T(8) = 100.26 \text{ degrees}$

$T(12) = 98.66 \text{ degrees}$

(b) Slopes are negative which means that the patient's temperature is decreasing.

(d) $T'(t) = -0.02t - 0.2$ is the rate of change of the patient's temperature with respect to time.

(e) $T'(0) = -0.2 \text{ degree/hour}$

$T'(4) = -0.28 \text{ degree/hour}$

$T'(8) = -0.36 \text{ degree/hour}$

$T'(12) = -0.44 \text{ degree/hour}$

35. $p = 5 - 0.001x, \quad C = 35 + 1.5x$

(a) $R = xp = x(5 - 0.001x) = 5x - 0.001x^2$

(b) $P = R - C = (5x - 0.001x^2) - (35 + 1.5x)$

$$= 3.5x - 0.001x^2 - 35$$

(c) $R'(x) = 5 - 0.002x$

$P'(x) = 3.5 - 0.002x$

x	600	1200	1800	2400	3000
dR/dx	3.8	2.6	1.4	0.2	-1.0
dP/dx	2.3	1.1	-0.1	-1.3	-2.5
P	1705	2725	3025	2605	1465

37. $(36{,}000, 6), (33{,}000, 7)$

$$\text{Slope} = \frac{7 - 6}{33{,}000 - 36{,}000} = \frac{1}{-3000}$$

$$p - 6 = -\frac{1}{3000}(x - 36{,}000)$$

$$p = -\frac{1}{3000}x + 18 \quad \text{(demand function)}$$

(a) $P = R - C = xp - (0.20x + 85{,}000)$

$$= x\left(-\frac{1}{3000}x + 18\right) - 0.2x - 85{,}000$$

$$= \frac{x^2}{3000} + 17.8x - 85{,}000$$

(b)

The slope of P is positive at $x = 18{,}000$.

(c) $P'(x) = -\dfrac{x}{1500} + 17.8$

$P'(18{,}000) = 5.8 \text{ dollars}$

39. $C = v(x) + k$

Marginal cost: $C' = v'(x) + 0 = v'(x)$

Thus, the marginal cost is independent of the fixed cost.

41. $\dfrac{dC}{dQ} = -\dfrac{1,008,000}{Q^2} + \dfrac{63}{10}$

$C(351) - C(350) \approx -1.91$

$\dfrac{dC}{dQ} = -1.93$ when $Q = 350$.

43. $C(x) = \dfrac{15,000}{x}(1.10)$ dollars/mile

$C'(x) = \dfrac{-16,500}{x^2}$

x	10	15	20	25	30	35	40
C	1650	1100	825	660	550	471.43	412.50
dC/dx	-165	-73.33	-41.25	-26.40	-18.33	-13.47	-10.31

The car that gets 15 miles/gallon would benefit more.

45. $f(x) = \frac{1}{4}x^3$, $f'(x) = \frac{3}{4}x^2$

f has a horizontal tangent at $x = 0$.

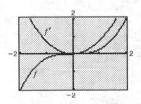

47. The population in each phase is increasing. During the acceleration phase the growth is the greatest. Therefore, the slopes of the tangent lines are greater than the slopes of the tangent lines during the lag phase and the deceleration phase. Possible reasons for the changing rates could be seasonal growth or food supplies.

Section 2.4   The Product and Quotient Rules

1. $f'(x) = x^2(9x^2) + 2x(3x^3 - 1) = 15x^4 - 2x$

$f'(1) = 13$

3. $f'(x) = \frac{1}{3}(6x^2) = 2x^2$

$f'(0) = 0$

5. $g'(x) = (x^2 - 4x + 3)(1) + (2x - 4)(x - 2)$

$\qquad = 3x^2 - 12x + 11$

$g'(4) = (16 - 16 + 3) + (8 - 4)(4 - 2) = 3 + 8 = 11$

7. $h'(x) = \dfrac{(x - 5)(1) - (x)(1)}{(x - 5)^2} = \dfrac{-5}{(x - 5)^2}$

$h'(6) = -5$

9. $f'(t) = \dfrac{3t(4t) - (2t^2 - 3)(3)}{(3t)^2} = \dfrac{6t^2 + 9}{9t^2} = \dfrac{2t^2 + 3}{3t^2}$

$f'(2) = \dfrac{2(2)^2 + 3}{3(2)^2} = \dfrac{11}{12}$

Function	Rewrite	Differentiate	Simplify
11. $y = \dfrac{x^2 + 2x}{x}$	$y = x + 2$	$y' = 1$	$y' = 1$
13. $y = \dfrac{7}{3x^3}$	$y = \dfrac{7}{3}x^{-3}$	$y' = -7x^{-4}$	$y' = -\dfrac{7}{x^4}$
15. $y = \dfrac{4x^2 - 3x}{8\sqrt{x}}$	$y = \dfrac{1}{2}x^{3/2} - \dfrac{3}{8}x^{1/2}$	$y' = \dfrac{3}{4}x^{1/2} - \dfrac{3}{16}x^{-1/2}$	$y' = \dfrac{3}{4}\sqrt{x} - \dfrac{3}{16\sqrt{x}}$
17. $y = \dfrac{x^2 - 4x + 3}{x - 1}$	$y = x - 3,\ x \neq 1$	$y' = 1,\ x \neq 1$	$y' = 1,\ x \neq 1$

19. $f'(x) = (x^3 - 3x)(4x + 3) + (3x^2 - 3)(2x^2 + 3x + 5)$

$\qquad = 4x^4 + 3x^3 - 12x^2 - 9x + 6x^4 + 9x^3 + 9x^2 - 9x - 15$

$\qquad = 10x^4 + 12x^3 - 3x^2 - 18x - 15$

21. $g'(t) = (2t^2 - 3)(-2t - 4t^3) + (4 - t^2 - t^4)(4t)$

$\qquad = -4t^3 + 6t - 8t^5 + 12t^3 + 16t - 4t^3 - 4t^5$

$\qquad = -12t^5 + 4t^3 + 22t$

23. $f(x) = \sqrt[3]{x}(\sqrt{x} + 3) = x^{1/3}(x^{1/2} + 3) = x^{5/6} + 3x^{1/3}$

$\qquad f'(x) = \dfrac{5}{6}x^{-1/6} + x^{-2/3} = \dfrac{5}{6x^{1/6}} + \dfrac{1}{x^{2/3}}$

25. $f(x) = \dfrac{3x - 2}{2x - 3}$

$\qquad f'(x) = \dfrac{(2x - 3)(3) - (3x - 2)(2)}{(2x - 3)^2} = \dfrac{-5}{(2x - 3)^2}$

27. $f(x) = \dfrac{3 - 2x - x^2}{x^2 - 1}$

$\qquad = \dfrac{(3 + x)(1 - x)}{(x + 1)(x - 1)} = \dfrac{-(3 + x)}{x + 1}, \quad x \neq -1$

$\qquad f'(x) = \dfrac{(x + 1)(-1) + (3 + x)(1)}{(x + 1)^2} = \dfrac{2}{(x + 1)^2}, \quad x \neq -1$

29. $f(x) = x\left(1 - \dfrac{2}{x + 1}\right) = x - \dfrac{2x}{x + 1}$

$\qquad f'(x) = 1 - \dfrac{(x + 1)(2) - (2x)(1)}{(x + 1)^2}$

$\qquad = 1 - \dfrac{2}{(x + 1)^2} = \dfrac{x^2 + 2x - 1}{(x + 1)^2}$

31. $g'(s) = \dfrac{(s^2 + 2s + 2)(0) - 5(2s + 2)}{(s^2 + 2s + 2)^2}$

$\qquad = \dfrac{-10(s + 1)}{(s^2 + 2s + 2)^2}$

33. $g(x) = \dfrac{x - 3}{x + 4}(x^2 + 2x + 1) = \dfrac{x^3 - x^2 - 5x - 3}{x + 4}$

$\qquad g'(x) = \dfrac{(x + 4)(3x^2 - 2x - 5) - (x^3 - x^2 - 5x - 3)(1)}{(x + 4)^2}$

$\qquad = \dfrac{3x^3 + 10x^2 - 13x - 20 - x^3 + x^2 + 5x + 3}{(x + 4)^2}$

$\qquad = \dfrac{2x^3 + 11x^2 - 8x - 17}{(x + 4)^2}$

35. $f(0) = (-1)(2) = -2$

$\qquad f'(x) = (x - 1)(2x - 3) + (x^2 - 3x + 2)(1)$

$\qquad f'(0) = (-1)(-3) + 2 = 5$

$\qquad y + 2 = 5(x - 0)$

$\qquad y = 5x - 2$

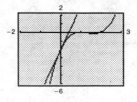

37. $f(x) = \dfrac{x^4 - 2x^3}{x + 1}$

$\qquad f(1) = -\dfrac{1}{2}$

$\qquad f'(x) = \dfrac{(x + 1)(4x^3 - 6x^2) - (x^4 - 2x^3)(1)}{(x + 1)^2}$

$\qquad f'(1) = \dfrac{2(-2) - (-1)}{2^2} = -\dfrac{3}{4}$

$\qquad y - \left(-\dfrac{1}{2}\right) = -\dfrac{3}{4}(x - 1)$

$\qquad y = -\dfrac{3}{4}x + \dfrac{1}{4}$

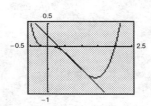

39. $f(0) = -3$

$$f(x) = \frac{x^4 + x^3 + x^2 - x^3 - x^2 - x - 3x^2 - 3x - 3}{x^2 + 1} = \frac{x^4 - 3x^2 - 4x - 3}{x^2 + 1}$$

$$f'(x) = \frac{(x^2 + 1)(4x^3 - 6x - 4) - (x^4 - 3x^2 - 4x - 3)(2x)}{(x^2 + 1)^2}$$

$$f'(0) = \frac{(1)(-4) - (-3)(0)}{1} = -4$$

$$y - (-3) = -4(x - 0)$$

$$y = -4x - 3$$

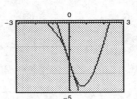

41. $f'(x) = \dfrac{(x - 1)(2x) - x^2(1)}{(x - 1)^2} = \dfrac{x^2 - 2x}{(x - 1)^2}$

$f'(x) = 0$ when $x^2 - 2x = x(x - 2) = 0$, which implies that $x = 0$ or $x = 2$. Thus, the horizontal tangent lines occur at $(0, 0)$ and $(2, 4)$.

43. $f'(x) = \dfrac{(x^3 + 1)(4x^3) - x^4(3x^2)}{(x^3 + 1)^2} = \dfrac{x^6 + 4x^3}{(x^3 + 1)^2}$

$f'(x) = 0$ when $x^6 + 4x^3 = x^3(x^3 + 4) = 0$. Thus, the horizontal tangent lines occur at $x = 0$ and $x = \sqrt[3]{-4}$.

45. $f(x) = x(x + 1) = x^2 + x$

$f'(x) = 2x + 1$

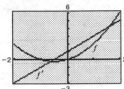

47. $f(x) = x(x + 1)(x - 1)$

$\quad\quad = x^3 - x$

$f'(x) = 3x^2 - 1$

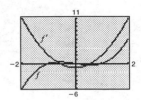

49. $x = 275\left(1 - \dfrac{3p}{5p + 1}\right)$

$$\frac{dx}{dp} = -275\left[\frac{(5p + 1)3 - (3p)(5)}{(5p + 1)^2}\right] = -275\left[\frac{3}{(5p + 1)^2}\right]$$

When $p = 4$,

$$\frac{dx}{dp} = -275\left[\frac{3}{(21)^2}\right] \approx -1.87.$$

51. $f'(t) = \dfrac{(t^2 + 1)(2t - 1) - (t^2 - t + 1)(2t)}{(t^2 + 1)^2} = \dfrac{t^2 - 1}{(t^2 + 1)^2}$

(a) $f'(0.5) = -0.48$

(b) $f'(2) = 0.12$

(c) $f'(8) \approx 0.015$

53. $P' = 500\left[\dfrac{(50 + t^2)(4) - (4t)(2t)}{(50 + t^2)^2}\right] = 500\left[\dfrac{200 - 4t^2}{(50 + t^2)^2}\right]$

When $t = 2$, $P' = 500\left[\dfrac{184}{(54)^2}\right] \approx 31.55$ bacteria/hour.

55. (a) $\quad\quad x = \dfrac{k}{p^2}, \quad x \ge 5$

$$16 = \frac{k}{1000^2}$$

$$16,000,000 = k$$

$$x = \frac{16,000,000}{p^2}$$

$$p^2 = \frac{16,000,000}{x}$$

$$p = \frac{4000}{\sqrt{x}}$$

(b) $C = 250x + 10,000$

(c) $P = R - C = xp - C$

$$= x\left(\frac{4000}{\sqrt{x}}\right) - (250x + 10,000)$$

$$= 4000\sqrt{x} - 250x - 10,000$$

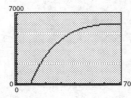

Let $x = 64$ and $p = \$500$.

57. (a) $f(x) = x^2 + 1$

$f'(x) = 2x$

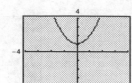

(b) $xf(x) = x^3 + x$

$\dfrac{d}{dx}[xf(x)] = 3x^2 + 1$

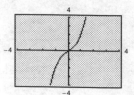

(c) $\dfrac{1}{f(x)} = \dfrac{1}{x^2 + 1}$

$\dfrac{d}{dx}\left[\dfrac{1}{f(x)}\right] = \dfrac{(x^2 + 1)(0) - (1)(2x)}{(x^2 + 1)^2} = -\dfrac{2x}{(x^2 + 1)^2}$

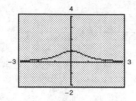

Since (a) and (b) are increasing functions, only (c) could represent a demand function.

59. $C = 100\left(\dfrac{200}{x^2} + \dfrac{x}{x + 30}\right), \quad x \geq 1$

$C' = 100\left[-2(200x^{-3}) + \dfrac{(x + 30) - x}{(x + 30)^2}\right] = 100\left[-\dfrac{400}{x^3} + \dfrac{30}{(x + 30)^2}\right]$

$C'(10) = 100\left(-\dfrac{400}{10^3} + \dfrac{30}{40^2}\right) = -38.125$

$C'(15) \approx -10.37$

$C'(20) \approx -3.8$

Increasing the order size reduces the cost per item.

61. $\dfrac{dC}{dt} = \dfrac{(1 - 0.18t + 0.01t^2)(-5.3 + 0.6t) - (27.9 - 5.3t + 0.3t^2)(-0.18 + 0.02t)}{(1 - 0.18t + 0.01t^2)}$

$= \dfrac{-10(t^2 - 42t + 278)}{(t^2 - 18t + 100)^2}$

For $t = 0$, $C' = -0.28$.

For $t = 5$, $C' = -0.76$.

For $t = 10$, $C' = 1.05$.

For $t = 15$, $C' = 0.42$.

In 1980, the cost per mile of owning a car was decreasing at a rate of 0.28 cents/mile/year. Similarly, in 1995, the cost was increasing at a rate of 0.42 cents/mile/year.

Section 2.5   The Chain Rule

$\underline{y = f(g(x))}$	$\underline{u = g(x)}$	$\underline{y = f(u)}$	$\underline{y = f(g(x))}$	$\underline{u = g(x)}$	$\underline{y = f(u)}$
1. $y = (6x - 5)^4$	$u = 6x - 5$	$y = u^4$	**3.** $y = (4 - x^2)^{-1}$	$u = 4 - x^2$	$y = u^{-1}$
5. $y = \sqrt{5x - 2}$	$u = 5x - 2$	$y = \sqrt{u}$	**7.** $y = \dfrac{1}{3x + 1}$	$u = 3x + 1$	$y = u^{-1}$

9. $f(x) = \dfrac{2}{1 - x^3} - 2(1 - x^3)^{-1}$

is most efficiently done by the General Power Rule (c).

11. $f(x) = \sqrt[3]{8^2}$ is most efficiently done by the Constant Rule (b).

13. $y' = 3(2x - 7)^2(2) = 6(2x - 7)^2$

15. $g'(x) = 3(4 - 2x)^2(-2) = -6(4 - 2x)^2$

17. $h'(x) = 2(6x - x^3)(6 - 3x^2) = 6x(6 - x^2)(2 - x^2)$

19. $f'(x) = \dfrac{2}{3}(x^2 - 9)^{-1/3}(2x) = \dfrac{4x}{3(x^2 - 9)^{1/3}}$

21. $f(t) = \sqrt{t + 1} = (t + 1)^{1/2}$

$f'(t) = \dfrac{1}{2}(t + 1)^{-1/2}(1) = \dfrac{1}{2\sqrt{t + 1}}$

23. $s(t) = \sqrt{2t^2 + 5t + 2} = (2t^2 + 5t + 2)^{1/2}$

$s'(t) = \dfrac{1}{2}(2t^2 + 5t + 2)^{-1/2}(4t + 5) = \dfrac{4t + 5}{2\sqrt{2t^2 + 5t + 2}}$

25. $y = \sqrt[3]{9x^2 + 4} = (9x^2 + 4)^{1/3}$

$y' = \dfrac{1}{3}(9x^2 + 4)^{-2/3}(18x) = \dfrac{6x}{(9x^2 + 4)^{2/3}}$

27. $f(x) = -3(2 - 9x)^{1/4}$

$f'(x) = -3\left(\dfrac{1}{4}\right)(2 - 9x)^{-3/4}(-9) = \dfrac{27}{4(2 - 9x)^{3/4}}$

29. $h'(x) = -\dfrac{4}{3}(4 - x^3)^{-7/3}(-3x^2) = \dfrac{4x^2}{(4 - x^3)^{7/3}}$

31. $f'(x) = 2(3)(x^2 - 1)^2(2x) = 12x(x^2 - 1)^2$

$f'(2) = 24(3^2) = 216$

$f(2) = 54$

$y - 54 = 216(x - 2)$

$y = 216x - 378$

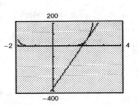

33. $f(x) = \sqrt{4x^2 - 7} = (4x^2 - 7)^{1/2}$

Point: $(2, f(2)) = (2, 3)$

$f'(x) = \dfrac{1}{2}(4x^2 - 7)^{-1/2}(8x) = \dfrac{4x}{\sqrt{4x^2 - 7}}$

When $x = 2$, the slope is $f'(2) = \frac{8}{3}$, and the equation of the tangent line is

$$y - 3 = \dfrac{8}{3}(x - 2)$$

$$y = \dfrac{8}{3}x - \dfrac{7}{3}$$

$$3y - 8x + 7 = 0.$$

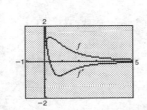

35. $f'(x) = \dfrac{1}{2}(x^2 - 2x + 1)^{-1/2}(2x - 2)$

$= \dfrac{x - 1}{\sqrt{x^2 - 2x + 1}}$

$= \dfrac{x - 1}{|x - 1|}$

$f'(2) = 1$

$f(2) = 1$

$y - 1 = 1(x - 2)$

$y = x - 1$

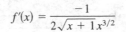

37. $f(x) = \dfrac{\sqrt{x} + 1}{x^2 + 1}$

$f'(x) = \dfrac{1 - 3x^2 - 4x^{3/2}}{2\sqrt{x}(x^2 + 1)^2}$

f has a horizontal tangent when $f' = 0$.

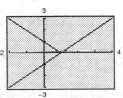

39. $f(x) = \sqrt{\dfrac{x + 1}{x}}$

$f'(x) = \dfrac{-1}{2\sqrt{x + 1}\, x^{3/2}}$

f' is never 0.

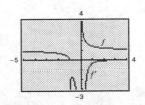

41. $y = (x - 2)^{-1}$

$$y' = (-1)(x - 2)^{-2}(1) = -\frac{1}{(x - 2)^2}$$

43. $y = -4(t + 2)^{-2}$

$$y' = 8(t + 2)^{-3} = \frac{8}{(t + 2)^3}$$

45. $f(x) = (x^2 - 3x)^{-2}$

$$f'(x) = -2(x^2 - 3x)^{-3}(2x - 3) = \frac{6 - 4x}{(x^2 - 3x)^3}$$

47. $g(t) = (t^2 - 2)^{-1}$

$$g'(t) = -(t^2 - 2)^{-2}(2t) = \frac{-2t}{(t^2 - 2)^2}$$

49. $y = x(2x + 3)^{1/2}$

$$y' = x\left[\frac{1}{2}(2x + 3)^{-1/2}(2)\right] + (2x + 3)^{1/2}$$

$$= (2x + 3)^{-1/2}[x + (2x + 3)]$$

$$= \frac{3(x + 1)}{\sqrt{2x + 3}}$$

51. $y = t^2(t - 2)^{1/2}$

$$y' = t^2\left[\frac{1}{2}(t - 2)^{-1/2}(1)\right] + 2t(t - 2)^{1/2}$$

$$= \frac{1}{2}(t - 2)^{-1/2}[t^2 + 4t(t - 2)]$$

$$= \frac{t^2 + 4t(t - 2)}{2\sqrt{t - 2}}$$

$$= \frac{t(5t - 8)}{2\sqrt{t - 2}}$$

53. $f'(x) = x(3)(3x - 9)^2(3) + (3x - 9)^3(1)$

$$= (3x - 9)^2[9x + (3x - 9)]$$

$$= 9(x - 3)^2(12x - 9)$$

$$= 27(x - 3)^2(4x - 3)$$

55. $f(x) = \sqrt{\frac{3 - 2x}{4x}} = \sqrt{\frac{3}{4}x^{-1} - \frac{1}{2}} = \left(\frac{3}{4}x^{-1} - \frac{1}{2}\right)^{1/2}$

$$f'(x) = \frac{1}{2}\left(\frac{3}{4}x^{-1} - \frac{1}{2}\right)^{-1/2}\left(-\frac{3}{4}x^{-2}\right)$$

$$= \frac{-3}{8x^2\sqrt{(3/4x) - (1/2)}}$$

$$= \frac{-3}{4\sqrt{3 - 2x}\,x^{3/2}}$$

57. $f(x) = \frac{\sqrt[3]{x^3 + 1}}{x} = \frac{(x^3 + 1)^{1/3}}{x}$

$$f'(x) = \frac{x(1/3)(x^3 + 1)^{-2/3}(3x^2) - (x^3 + 1)^{1/3}(1)}{x^2}$$

$$= \frac{(x^3 + 1)^{-2/3}[x^3 - (x^3 + 1)]}{x^2}$$

$$= \frac{-1}{x^2(x^3 + 1)^{2/3}}$$

59. $y = \left(\frac{6 - 5x}{x^2 - 1}\right)^2$

$$y' = 2\left(\frac{6 - 5x}{x^2 - 1}\right)\left[\frac{(x^2 - 1)(-5) - (6 - 5x)(2x)}{(x^2 - 1)^2}\right]$$

$$= \frac{2(6 - 5x)(5x^2 - 12x + 5)}{(x^2 - 1)^3}$$

61. $f(t) = 36(3 - t)^{-2}$

$$f'(t) = -72(3 - t)^{-3}(-1) = \frac{72}{(3 - t)^3}$$

$$f'(0) = \frac{72}{27} = \frac{8}{3}$$

$$y - 4 = \frac{8}{3}(t - 0)$$

$$y = \frac{8}{3}t + 4$$

63. $f(t) = (t^2 - 9)(t + 2)^{1/2}$

$$f'(t) = (t^2 - 9)\frac{1}{2}(t + 2)^{-1/2} + (t + 2)^{1/2}(2t)$$

$$f'(-1) = (-8)\frac{1}{2} + (-2) = -6$$

$$y - (-8) = -6[x - (-1)]$$

$$y = -6x - 14$$

65. $f(x) = \dfrac{x + 1}{2x - 3}$

$f'(x) = \dfrac{(2x - 3) - (x + 1)(2)}{(2x - 3)^2} = \dfrac{-5}{(2x - 3)^2}$

$f'(2) = -5$

$y - 3 = -5(x - 2)$

$y = -5x + 13$

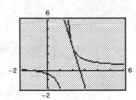

67. $A = 1000\left(1 + \dfrac{r}{12}\right)^{60}$

$A' = 1000(60)\left(1 + \dfrac{r}{12}\right)^{59}\left(\dfrac{1}{12}\right) = 5000\left(1 + \dfrac{r}{12}\right)^{59}$

(a) $A'(0.08) = 50\left(1 + \dfrac{0.08}{12}\right)^{59} \approx \$74.00$ per percentage point

(b) $A'(0.10) = 50\left(1 + \dfrac{0.10}{12}\right)^{59} \approx \$81.59$ per percentage point

(c) $A'(0.12) = 50\left(1 + \dfrac{0.12}{12}\right)^{59} \approx \$89.94$ per percentage point

69. $N = 400[1 - 3(t^2 + 2)^{-2}] = 400 - 1200(t^2 + 2)^{-2}$

$\dfrac{dN}{dt} = 2400(t^2 + 2)^{-3}(2t) = \dfrac{4800t}{(t^2 + 2)^3}$

The rate of growth of N is decreasing.

t	0	1	2	3	4
$\dfrac{dN}{dt}$	0	177.78	44.44	10.82	3.29

71. (a) $V = \dfrac{k}{\sqrt[3]{t + 1}}$

When $t = 0$, $V = 10,000 \Rightarrow k = 10,000$. Therefore, $V = \dfrac{10,000}{\sqrt[3]{t + 1}}$.

(b) When $t = 1$,

$\dfrac{dV}{dt} = -\dfrac{10,000}{3(2)^{4/3}} \approx -\$1322.83$ per year.

(c) When $t = 3$, $\dfrac{dV}{dt} = -\dfrac{10,000}{3(4)^{4/3}} \approx -\$524.97$ per year.

Section 2.6   Higher-Order Derivatives

1. $f'(x) = -4$

$f''(x) = 0$

3. $f'(x) = 2x + 7$

$f''(x) = 2$

5. $g'(t) = t^2 - 8t + 2$

$g''(t) = 2t - 8$

7. $f(t) = \dfrac{3}{4}t^{-2}$

$f'(t) = -\dfrac{3}{2}t^{-3}$

$f''(t) = \dfrac{9}{2}t^{-4} = \dfrac{9}{2t^4}$

9. $f(x) = 3(2 - x^2)^3$

$f'(x) = 9(2 - x^2)^2(-2x) = -18x(2 - x^2)^2$

$f''(x) = (-18x)2(2 - x^2)(-2x) + (2 - x^2)^2(-18)$

$\quad = 18(2 - x^2)[4x^2 - (2 - x^2)]$

$\quad = 18(2 - x^2)(5x^2 - 2)$

11. $f'(x) = \dfrac{(x-1)(1) - (x+1)(1)}{(x-1)^2}$ $x - 1 - x - 1 = -2$

$\qquad = -\dfrac{2}{(x-1)^2} = -2(x-1)^{-2}$

$\quad f''(x) = 4(x-1)^{-3}(1) = \dfrac{4}{(x-1)^3}$

13. $y = x^2(x^2 + 4x + 8) = x^4 + 4x^3 + 8x^2$

$\quad y' = 4x^3 + 12x^2 + 16x$

$\quad y'' = 12x^2 + 24x + 16$

15. $f'(x) = 5x^4 - 12x^3$

$\quad f''(x) = 20x^3 - 36x^2$

$\quad f'''(x) = 60x^2 - 72x$

17. $f(x) = 5x(x+4)^3$

$\qquad = 5x(x^3 + 12x^2 + 48x + 64)$

$\qquad = 5x^4 + 60x^3 + 240x^2 + 320x$

$\quad f'(x) = 20x^3 + 180x^2 + 480x + 320$

$\quad f''(x) = 60x^2 + 360x + 480$

$\quad f'''(x) = 120x + 360$

19. $f(x) = \dfrac{3}{(4x)^2} = \dfrac{3}{16}x^{-2}$

$\quad f'(x) = -\dfrac{3}{8}x^{-3}$

$\quad f''(x) = \dfrac{9}{8}x^{-4}$

$\quad f'''(x) = -\dfrac{9}{2}x^{-5} = -\dfrac{9}{2x^5}$

21. $g'(t) = 20t^3 + 20t$

$\quad g''(t) = 60t^2 + 20$

$\quad g''(2) = 60(4) + 20 = 260$

23. $f(x) = (4-x)^{1/2}$

$\quad f'(x) = -\dfrac{1}{2}(4-x)^{-1/2}$

$\quad f''(x) = -\dfrac{1}{4}(4-x)^{-3/2}$

$\quad f'''(x) = -\dfrac{3}{8}(4-x)^{-5/2} = \dfrac{-3}{8(4-x)^{5/2}}$

$\quad f'''(-5) = \dfrac{-3}{8(9)^{5/2}} = \dfrac{-1}{648}$

25. $f''(x) = 4x$

27. $f''(x) = \dfrac{2x-2}{x} = 2 - \dfrac{2}{x}$

$\quad f'''(x) = 0 - \left(-\dfrac{2}{x^2}\right) = \dfrac{2}{x^2}$

29. $f^{(5)}(x) = 2(x+1)$

$\quad f^{(6)}(x) = 2$

31. $f'(x) = 3x^2 - 18x + 27$

$\quad f''(x) = 6x - 18 = 0$

$\quad f''(x) = 0$ when $x = 3$.

33. $f(x) = (x+3)(x-4)(x+5)$

$\qquad = x^3 + 4x^2 - 17x - 60$

$\quad f'(x) = 3x^2 + 8x - 17$

$\quad f''(x) = 6x + 8$

$\quad f''(x) = 0$ when $6x + 8 = 0$

$\qquad\qquad\qquad x = -\frac{4}{3}.$

35. $f(x) = 3x^4 - 18x^2$

$f'(x) = 12x^3 - 36x$

$f''(x) = 36x^2 - 36 = 36(x^2 - 1) = 0$

$f''(x) = 0$ when $x = \pm 1$.

37. $f'(x) = \dfrac{(x^2 + 3)(1) - (x)(2x)}{(x^2 + 3)^2} = \dfrac{3 - x^2}{(x^2 + 3)^2} = (3 - x^2)(x^2 + 3)^{-2}$

$f''(x) = (3 - x^2)[-2(x^2 + 3)^{-3}(2x)] + (x^2 + 3)^{-2}(-2x)$

$\quad = -2x(x^2 + 3)^{-3}[2(3 - x^2) + (x^2 + 3)]$

$\quad = \dfrac{2x(9 - x^2)}{(x^2 + 3)^3}$

$\quad = \dfrac{2x(x^2 - 9)}{(x^2 + 3)^3}$

$f''(x) = 0$ when $2x(x^2 - 9) = 0$

$\qquad\qquad\qquad x = 0, \pm 3.$

39. (a) $s(t) = -16t^2 + 144t$

    (b) $v(t) = s'(t) = -32t + 144$

$\qquad a(t) = v'(t) = -32$

    (c) $v(t) = 0 = -32t + 144$ when $t = \frac{144}{32} = 4.5$ sec.

$\qquad s(4.5) = 324$ feet

    (d) $s(t) = 0$ when $16t^2 = 144t$, or $t = 0, 9$ sec.

$\qquad v(9) = -144$ ft/sec, which is the same speed as the initial velocity.

41. $\dfrac{d^2s}{dt^2} = \dfrac{(t + 10)(90) - (90t)(1)}{(t + 10)^2} = \dfrac{900}{(t + 10)^2}$

t	0	10	20	30	40	50	60
$\dfrac{ds}{dt}$	0	45	60	67.5	72	75	77.14
$\dfrac{d^2s}{dt^2}$	0	2.25	1	0.56	0.36	0.25	0.18

43. $f(x) = x^2 - 6x + 6$

$f'(x) = 2x - 6$

$f''(x) = 2$

The degrees of the successive derivatives decrease by 1.

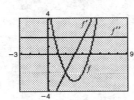

45. The degree of f is 3, and the degrees of the successive derivatives decrease by 1.

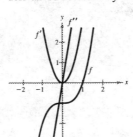

47. False. The Product Rule is

$\quad [f(x)g(x)]' = f'(x)g(x) + f(x)g'(x).$

49. True.

$\quad h'(c) = f'(c)g(c) + f(c)g'(c) = 0$

51. True

53. False. Let $f(x) = x^2$ and $g(x) = x^2 + 1$.

Section 2.7   Implicit Differentiation

1.
$$5xy = 1$$
$$5xy' + 5y = 0$$
$$5xy' = -5y$$
$$y' = -\frac{y}{x}$$

3.
$$y^2 = 1 - x^2$$
$$2yy' = -2x$$
$$y' = -\frac{x}{y}$$

5.
$$\frac{2 - x}{y - 3} = 5$$
$$2 - x = 5y - 15$$
$$-1 = 5y'$$
$$y' = -\frac{1}{5}$$

7.
$$x^2 + y^2 = 49$$
$$2x + 2yy' = 0$$
$$y' = -\frac{x}{y}$$
At $(0, 7)$, $y' = -\frac{0}{7} = 0$.

9.
$$y + xy = 4$$
$$\frac{dy}{dx} + x\frac{dy}{dx} + y = 0$$
$$\frac{dy}{dx}(1 + x) = -y$$
$$\frac{dy}{dx} = -\frac{y}{x + 1}$$
At $(-5, -1)$, $\frac{dy}{dx} = -\frac{1}{4}$.

11.
$$x^3 - xy + y^2 = 4$$
$$3x^2 - x\frac{dy}{dx} - y + 2y\frac{dy}{dx} = 0 \quad \text{(Product Rule)}$$
$$\frac{dy}{dx}(2y - x) = y - 3x^2$$
$$\frac{dy}{dx} = \frac{y - 3x^2}{2y - x}$$
At $(0, -2)$, $\frac{dy}{dx} = \frac{1}{2}$.

13.
$$x^3y^3 - y = x$$
$$3x^3y^2\frac{dy}{dx} + 3x^2y^3 - \frac{dy}{dx} = 1$$
$$\frac{dy}{dx}(3x^3y^2 - 1) = 1 - 3x^2y^3$$
$$\frac{dy}{dx} = \frac{1 - 3x^2y^3}{3x^3y^2 - 1}$$
At $(0, 0)$, $\frac{dy}{dx} = -1$.

15.
$$x^{1/2} + y^{1/2} = 9$$
$$\frac{1}{2}x^{-1/2} + \frac{1}{2}y^{-1/2}\frac{dy}{dx} = 0$$
$$x^{-1/2} + y^{-1/2}\frac{dy}{dx} = 0$$
$$\frac{dy}{dx} = \frac{-x^{-1/2}}{y^{-1/2}} = -\sqrt{\frac{y}{x}}$$
At $(16, 25)$, $\frac{dy}{dx} = -\frac{5}{4}$.

17.
$$x^{2/3} + y^{2/3} = 5$$
$$\frac{2}{3}x^{-1/3} + \frac{2}{3}y^{-1/3}\frac{dy}{dx} = 0$$
$$\frac{dy}{dx} = \frac{-x^{-1/3}}{y^{-1/3}} = -\frac{y^{1/3}}{x^{1/3}} = -\sqrt[3]{\frac{y}{x}}$$
At $(8, 1)$, $\frac{dy}{dx} = -\frac{1}{2}$.

19. $3x^2 - 2y + 5 = 0$
$$6x - 2\frac{dy}{dx} = 0$$
$$\frac{dy}{dx} = 3x$$
At $(1, 4)$, $\frac{dy}{dx} = 3$.

21. $4x^2 + 9y^2 = 36$

$8x + 18yy' = 0$

$$y' = -\frac{4x}{9y}$$

At $\left(\sqrt{5}, \frac{4}{3}\right)$, $y' = -\frac{4\sqrt{5}}{9(4/3)} = -\frac{\sqrt{5}}{3}$.

23. Implicitly: $2x + 2y\dfrac{dy}{dx} = 0$

$$\frac{dy}{dx} = -\frac{x}{y}$$

Explicitly: $y = \pm\sqrt{25 - x^2}$

$$\frac{dy}{dx} = \pm\left(\frac{1}{2}\right)(25 - x^2)^{-1/2}(-2x)$$

$$= \pm\frac{-x}{\sqrt{25 - x^2}} = -\frac{x}{\pm\sqrt{25 - x^2}} = -\frac{x}{y}$$

At $(-4, 3)$, $\dfrac{dy}{dx} = \dfrac{4}{3}$.

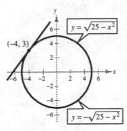

25. Implicitly: $18x + 32y\dfrac{dy}{dx} = 0$

$$\frac{dy}{dx} = -\frac{9x}{16y}$$

Explicitly: $y = \pm\left(\dfrac{1}{4}\right)\sqrt{144 - 9x^2}$

$$\frac{dy}{dx} = \pm\left(\frac{1}{8}\right)(144 - 9x^2)^{-1/2}(-18x)$$

$$= \pm\frac{-9x}{4\sqrt{144 - 9x^2}} = -\frac{9x}{16[\pm(1/4)\sqrt{144 - 9x^2}]} = -\frac{9x}{16y}$$

At $\left(2, \dfrac{3\sqrt{3}}{2}\right)$, $\dfrac{dy}{dx} = -\dfrac{\sqrt{3}}{4}$.

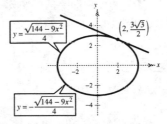

27. $x^2 + y^2 = 169$

$2x + 2y\dfrac{dy}{dx} = 0$

$$\frac{dy}{dx} = -\frac{x}{y}$$

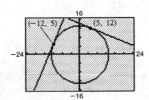

At $(5, 12)$:

$$m = -\frac{5}{12}$$

$$y - 12 = -\frac{5}{12}(x - 5)$$

$$5x + 12y - 169 = 0$$

At $(-12, 5)$:

$$m = \frac{12}{5}$$

$$y - 5 = \frac{12}{5}(x + 12)$$

$$12x - 5y + 169 = 0$$

29. $y^2 = 5x^3$

$2yy' = 15x^2$

$y' = \dfrac{15x^2}{2y}$

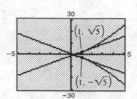

At $\left(1, \sqrt{5}\right)$:

$$y' = \frac{15}{2\sqrt{5}}$$

$$y - \sqrt{5} = \frac{15}{2\sqrt{5}}(x - 1)$$

$$2\sqrt{5}\,y - 10 = 15x - 15$$

$$2\sqrt{5}\,y - 15x + 5 = 0$$

At $\left(1, -\sqrt{5}\right)$:

$$y' = \frac{-15}{2\sqrt{5}}$$

$$y + \sqrt{5} = \frac{-15}{2\sqrt{5}}(x - 1)$$

$$2\sqrt{5}\,y + 10 = -15x + 15$$

$$2\sqrt{5}\,y + 15x - 5 = 0$$

31. $x^3 + y^3 = 8$

$3x^2 + 3y^2 y' = 0$

$3y^2 y' = -3x^2$

$y' = -\dfrac{x^2}{y^2}$

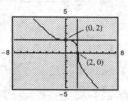

At $(0, 2)$, $y' = 0$ and $y = 2$ (tangent line).

At $(2, 0)$, y' is undefined and $x = 2$ (tangent line).

33. $p = 0.006x^4 + 0.02x^2 + 10, \quad x \geq 0$

$\dfrac{dp}{dp} = 1 = 0.024x^3 \dfrac{dx}{dp} + 0.04x \dfrac{dx}{dp}$

$1 = (0.024x^3 + 0.04x)\dfrac{dx}{dp}$

$\dfrac{dx}{dp} = \dfrac{1}{0.024x^3 + 0.04x}$

35. $p = \dfrac{200 - x}{2x} = \dfrac{100}{x} - \dfrac{1}{2} = 100x^{-1} - \dfrac{1}{2}$

$1 = -100x^{-2} \dfrac{dx}{dp}$

$\dfrac{dx}{dp} = \dfrac{-x^2}{100}$

37. (a)

$$100x^{0.75}y^{0.25} = 135{,}540$$

$$100x^{0.75}\left(0.25y^{-0.75}\frac{dy}{dx}\right) + y^{0.25}(75x^{-0.25}) = 0$$

$$\frac{25x^{0.75}}{y^{0.75}}\frac{dy}{dx} = -\frac{75y^{0.25}}{x^{0.25}}$$

$$\frac{dy}{dx} = -\frac{3y}{x}$$

When $x = 1500$ and $y = 1000$, $\dfrac{dy}{dx} = -2$.

(b)

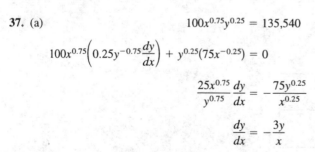

If more labor is used, then less capital is available.
If more capital is used, then less labor is available.

Section 2.8   Related Rates

1. $y = x^2 - \sqrt{x}$, $\dfrac{dy}{dt} = 2x\dfrac{dx}{dt} - \dfrac{1}{2\sqrt{x}}\dfrac{dx}{dt} = \left(2x - \dfrac{1}{2\sqrt{x}}\right)\dfrac{dx}{dt}$

(a) When $x = 4$ and $\dfrac{dx}{dt} = 8$, $\dfrac{dy}{dt} = \left(2(4) - \dfrac{1}{2\sqrt{4}}\right)8 = 62$.

(b) When $x = 16$ and $\dfrac{dy}{dt} = 12$, we have

$$12 = \left(2(16) - \dfrac{1}{2\sqrt{16}}\right)\dfrac{dx}{dt} = \left(32 - \dfrac{1}{8}\right)\dfrac{dx}{dt} = \dfrac{255}{8}\dfrac{dx}{dt}, \quad \dfrac{dx}{dt} = \dfrac{8 \cdot 12}{255} = \dfrac{32}{85}.$$

3. $xy = 4$, $x\dfrac{dy}{dt} + y\dfrac{dx}{dt} = 0$, $\dfrac{dy}{dt} = \left(-\dfrac{y}{x}\right)\dfrac{dx}{dt}$, $\dfrac{dx}{dt} = \left(-\dfrac{x}{y}\right)\dfrac{dy}{dt}$

(a) When $x = 8$, $y = \dfrac{1}{2}$, and $\dfrac{dx}{dt} = 10$, $\dfrac{dy}{dt} = -\dfrac{1/2}{8}(10) = -\dfrac{5}{8}$.

(b) When $x = 1$, $y = 4$, and $\dfrac{dy}{dt} = -6$, $\dfrac{dx}{dt} = -\dfrac{1}{4}(-6) = \dfrac{3}{2}$.

5. $A = \pi r^2$, $\dfrac{dr}{dt} = 2$, $\dfrac{dA}{dt} = 2\pi r\dfrac{dr}{dt}$

(a) When $r = 6$, $\dfrac{dA}{dt} = 2\pi(6)(2) = 24\pi$ in²/min.

(b) When $r = 24$, $\dfrac{dA}{dt} = 2\pi(24)(2) = 96\pi$ in²/min.

7. $A = \pi r^2$, $\dfrac{dA}{dt} = 2\pi r\dfrac{dr}{dt}$

If $\dfrac{dr}{dt}$ is constant, then $\dfrac{dA}{dt}$ is not constant; $\dfrac{dA}{dt}$ is proportional to r.

9. $V = \dfrac{4}{3}\pi r^3$, $\dfrac{dV}{dt} = 20$, $\dfrac{dV}{dt} = 4\pi r^2\dfrac{dr}{dt}$, $\dfrac{dr}{dt} = \left(\dfrac{1}{4\pi r^2}\right)\dfrac{dV}{dt}$

(a) When $r = 1$, $\dfrac{dr}{dt} = \dfrac{1}{4\pi(1)^2}(20) = \dfrac{5}{\pi}$ ft/min.

(b) When $r = 2$, $\dfrac{dr}{dt} = \dfrac{1}{4\pi(2)^2}(20) = \dfrac{5}{4\pi}$ ft/min.

11. $S = 2250 + 50x + 0.35x^2$

$\dfrac{dS}{dt} = 50\dfrac{dx}{dt} + 0.70x\dfrac{dx}{dt}$

$\dfrac{dS}{dt} = 50(125) + 0.70(1500)(125) = \$137,500$ per week

13. $V = x^3$, $\dfrac{dx}{dt} = 3$, $\dfrac{dV}{dt} = 3x^2\dfrac{dx}{dt}$

(a) When $x = 1$, $\dfrac{dV}{dt} = 3(1)^2(3) = 9$ cm³/sec.

(b) When $x = 10$, $\dfrac{dV}{dt} = 3(10)^2(3) = 900$ cm³/sec.

15. $y = x^2$, $\dfrac{dx}{dt} = 2$, $\dfrac{dy}{dt} = 2x\dfrac{dx}{dt}$

(a) When $x = -3$, $\dfrac{dy}{dt} = 2(-3)(2) = -12$ cm/min.

(b) When $x = 0$, $\dfrac{dy}{dt} = 2(0)(2) = 0$ cm/min.

(c) When $x = 1$, $\dfrac{dy}{dt} = 2(1)(2) = 4$ cm/min.

(d) When $x = 3$, $\dfrac{dy}{dt} = 2(3)(2) = 12$ cm/min.

17. $x^2 + y^2 = 25^2$, $2x\dfrac{dx}{dt} + 2y\dfrac{dy}{dt} = 0$, $\dfrac{dy}{dt} = \dfrac{-x}{y}\dfrac{dx}{dt} = \dfrac{-2x}{y}$ since $\dfrac{dx}{dt} = 2$.

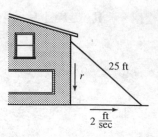

(a) When $x = 7$, $y = \sqrt{576} = 24$, $\dfrac{dy}{dt} = \dfrac{-2(7)}{24} = \dfrac{-7}{12}$ ft/sec.

(b) When $x = 15$, $y = \sqrt{400} = 20$, $\dfrac{dy}{dt} = \dfrac{-2(15)}{20} = \dfrac{-3}{2}$ ft/sec.

(c) When $x = 24$, $y = 7$, $\dfrac{dy}{dt} = \dfrac{-2(24)}{7} = \dfrac{-48}{7}$ ft/sec.

19. (a) $L^2 = x^2 + y^2$, $\dfrac{dx}{dt} = -450$, $\dfrac{dy}{dt} = -600$, and $\dfrac{dL}{dt} = \dfrac{x(dx/dt) + y(dy/dt)}{L}$

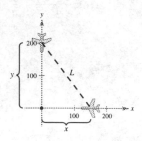

When $x = 150$ and $y = 200$, $\dfrac{dL}{dt} = \dfrac{150(-450) + 200(-600)}{250} = -750$ mph.

(b) $t = \dfrac{250}{750} = \dfrac{1}{3}$ hr = 20 min

21. $s^2 = 90^2 + x^2$, $x = 26$, $\dfrac{dx}{dt} = -30$

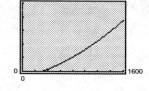

$2s\dfrac{ds}{dt} = 2x\dfrac{dx}{dt} \Rightarrow \dfrac{ds}{dt} = \dfrac{x}{s}\dfrac{dx}{dt}$

When $x = 26$,

$\quad s = \sqrt{90^2 + 26^2} \approx 93.68$

$\quad \dfrac{ds}{dt} = -8.85$ ft/sec $\dfrac{26}{93.68}(-30) \approx -8.33$ ft/sec.

23. $P = R - C$

$\quad = xp - C$

$\quad = x[50 - 0.01x] - (4000 + 40x - 0.02x^2)$

$\quad = 50x - 0.01x^2 - 4000 - 40x + 0.02x^2$

$\quad = 0.01x^2 + 10x - 4000$

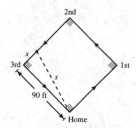

$\dfrac{dP}{dt} = 0.02x\dfrac{dx}{dt} + 10\dfrac{dx}{dt}$

When $x = 800$ and $\dfrac{dx}{dt} = 25$, $\dfrac{dP}{dt} = 0.02(800)(25) + (10)(25) = \$650$/week.

25. (a) $\dfrac{dC}{dp} = \dfrac{(100 - p)528 - 528p(-1)}{(100 - p)^2} = \dfrac{52{,}800}{(100 - p)^2}$

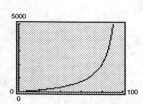

When $p = 30$, $\dfrac{dC}{dp} = \dfrac{52{,}800}{(100 - 30)^2} = \dfrac{528}{49} \approx 10.78\,\dfrac{\text{dollars}}{\text{percent}}$.

(b) From part (a), when $p = 60$, $\dfrac{dC}{dp} = \dfrac{52{,}800}{(100 - 60)^2} = 33\,\dfrac{\text{dollars}}{\text{percent}}$.

As $p \to 100^+$, C increases without bound.

Review Exercises for Chapter 2

1. Slope $\approx \dfrac{-4}{2} = -2$

3. Slope ≈ 0

5. When $t = 2$, slope $\approx \$70$ million per year (revenues are increasing).

When $t = 10$, slope $\approx \$300$ million per year (revenues are increasing).

7. $f(x) = -3x - 5, \quad (-2, 1)$

$$m = \lim_{\Delta x \to 0} \frac{f(-2 + \Delta x) - f(-2)}{\Delta x}$$

$$= \lim_{\Delta x \to 0} \frac{[-3(-2 + \Delta x) - 5] - 1}{\Delta x}$$

$$= \lim_{\Delta x \to 0} \frac{-3\Delta x}{\Delta x}$$

$$= -3$$

9. $f(x) = \sqrt{x + 9}, \quad (-5, 2)$

$$m = \lim_{\Delta x \to 0} \frac{\sqrt{-5 + \Delta x + 9} - 2}{\Delta x}$$

$$= \lim_{\Delta x \to 0} \frac{\sqrt{4 + \Delta x} - 2}{\Delta x} \cdot \frac{\sqrt{4 + \Delta x} + 2}{\sqrt{4 + \Delta x} + 2}$$

$$= \lim_{\Delta x \to 0} \frac{\Delta x}{\Delta x(\sqrt{4 + \Delta x} + 2)}$$

$$= \lim_{\Delta x \to 0} \frac{1}{\sqrt{4 + \Delta x} + 2}$$

$$= \frac{1}{4}$$

11. $f'(x) = \lim_{\Delta x \to 0} \dfrac{f(x + \Delta x) - f(x)}{\Delta x}$

$$= \lim_{\Delta x \to 0} \frac{7(x + \Delta x) + 3 - (7x + 3)}{\Delta x}$$

$$= \lim_{\Delta x \to 0} \frac{7\Delta x}{\Delta x}$$

$$= 7$$

13. $f'(x) = \lim_{\Delta x \to 0} \dfrac{f(x + \Delta x) - f(x)}{\Delta x}$

$$= \lim_{\Delta x \to 0} \frac{\dfrac{1}{x + \Delta x - 5} - \dfrac{1}{x - 5}}{\Delta x}$$

$$= \lim_{\Delta x \to 0} \frac{(x - 5) - (x + \Delta x - 5)}{(x + \Delta x - 5)(x - 5)\Delta x}$$

$$= \lim_{\Delta x \to 0} \frac{-1}{(x + \Delta x - 5)(x - 5)}$$

$$= \frac{-1}{(x - 5)^2}$$

15. $f(x) = 8 - 5x$

$f'(x) = -5$

$f'(3) = -5$

17. $f(x) = \sqrt{x} + 2 = x^{1/2} + 2$

$$f'(x) = \frac{1}{2}x^{-1/2} = \frac{1}{2\sqrt{x}}$$

$$f'(9) = \frac{1}{2\sqrt{9}} = \frac{1}{6}$$

19. $f(x) = \dfrac{x^2 + 3}{x} = x + 3x^{-1}$

$f'(x) = 1 - 3x^{-2}$

$f'(1) = 1 - 3 = -2$

Tangent line: $y - 4 = -2(x - 1)$

$$y = -2x + 6$$

21. $y = \dfrac{x + 1}{x - 1}$ is not differentiable at $x = 1$.

23. $y = \begin{cases} -x - 2, & x \le 0 \\ x^3 + 2, & x > 0 \end{cases}$ is not differentiable at $x = 0$.

25. $f'(x) = 0$

27. $y' = 5x^4$

29. $y' = \dfrac{1}{2\sqrt{x}}$

31. $f'(x) = 12x^3$

33. $g(t) = \dfrac{2}{3}t^{-2}$

$$g'(t) = -\dfrac{4}{3}t^{-3} = -\dfrac{4}{3t^3}$$

$$g'(1) = -\dfrac{4}{3}$$

$$y - \dfrac{2}{3} = -\dfrac{4}{3}(t - 1)$$

$$y = -\dfrac{4}{3}t + 2$$

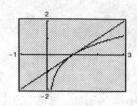

35. $y' = 44x^3 - 10x$

$y'(-1) = -34$

$y - 7 = -34(x + 1)$

$y = -34x - 27$

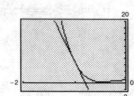

37. $f(x) = x^{1/2} - x^{-1/2}$

$$f'(x) = \dfrac{1}{2}x^{-1/2} + \dfrac{1}{2}x^{-3/2}$$

$$= \dfrac{1}{2\sqrt{x}} + \dfrac{1}{2x^{3/2}}$$

$$f'(1) = \dfrac{1}{2} + \dfrac{1}{2} = 1$$

$$y - 0 = 1(x - 1)$$

$$y = x - 1$$

39. $f(x) = x^2 + 3x - 4, \quad [0, 1]$

Average rate of change $= \dfrac{f(1) - f(0)}{1 - 0} = \dfrac{0 - (-4)}{1} = 4$

$f'(x) = 2x + 3$

$f'(0) = 3$

$f'(1) = 5$

41. (a) $\dfrac{R(6) - R(2)}{6 - 2} = \dfrac{3487.88 - 2941.36}{4} = \$136.63$ million per year

(b) $R'(t) = -7.9t^4 + 50.48t^3 + 54.36t^2 - 676.92t + 922.55$

$R'(2) = \$63.59$ million per year

$R'(6) = -\$516.73$ million per year

(c) Revenues were increasing in 1992 and decreasing in 1996.

43. (a) $P'(t) = 0.0125t^4 - 0.108t^3 + 0.27t^2 - 0.176t - 0.067$

(b) 1992: $P(2) \approx -\$0.003$ per pound

   1995: $P(5) \approx \$0.116$ per pound

(c)

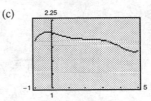

The price is increasing for $-1 < t < 0$ and decreasing for $0 < t < 5$.

45. (a) $s(t) = -16t^2 + 276$

(b) Average velocity $= \dfrac{s(2) - s(0)}{2 - 0} = \dfrac{-64}{2} = -32$ ft/sec

(c) $v(t) = -32t$

   $v(2) = -64$ ft/sec

   $v(3) = -96$ ft/sec

(d) $s(t) = -16t^2 + 276 = 0 \Rightarrow t^2 = \dfrac{276}{16} = 17.25$

   $t \approx 14.15$ sec

(e) $v(4.15) \approx -32(4.15) = -132.8$ velocity

   Speed $= 132.8$ ft/sec at impact

47. $R = 27.50x$

   $C = 15x + 2500$

   $P = R - C$

   $= 27.50x - (15x + 2500)$

   $= 12.50x - 2500$

49. $\dfrac{dC}{dx} = 320$

51. $R = \dfrac{35x}{\sqrt{x - 2}} = 35x(x - 2)^{-1/2}$

   $\dfrac{dR}{dx} = 35x\left[-\dfrac{1}{2}(x - 2)^{-3/2}\right] + 35(x - 2)^{-1/2}$

   $= \dfrac{35}{2}(x - 2)^{-3/2}[-x + 2(x - 2)]$

   $= \dfrac{35(x - 4)}{2(x - 2)^{3/2}}$

53. $\dfrac{dP}{dx} = -0.0006x^2 + 12x - 1$

55. $f(x) = x^3(5 - 3x^2) = 5x^3 - 3x^5$

   $f'(x) = 15x^2 - 15x^4 = 15x^2(1 - x^2)$

57. $y = (4x - 3)(x^3 - 2x^2) = 4x^4 - 11x^3 + 6x^2$

   $y' = 16x^3 - 33x^2 + 12x$

59. $f(x) = \dfrac{6x - 5}{x^2 + 1}$

   $f'(x) = \dfrac{(x^2 + 1)(6) - (6x - 5)(2x)}{(x^2 + 1)^2}$

   $= \dfrac{6 + 10x - 6x^2}{(x^2 + 1)^2}$

   $= \dfrac{2(3 + 5x - 3x^2)}{(x^2 + 1)^2}$

61. $f(x) = (5x^2 + 2)^3$

   $f'(x) = 3(5x^2 + 2)^2(10x)$

   $= 30x(5x^2 + 2)^2$

63. $h(x) = \dfrac{2}{\sqrt{x+1}} = 2(x+1)^{-1/2}$

$h'(x) = 2\left(-\dfrac{1}{2}\right)(x+1)^{-3/2}(1) = -\dfrac{1}{(x+1)^{3/2}}$

65. $g(x) = x\sqrt{x^2+1} = x(x^2+1)^{1/2}$

$g'(x) = x\left[\dfrac{1}{2}(x^2+1)^{-1/2}(2x)\right] + (1)(x^2+1)^{1/2}$

$= (x^2+1)^{-1/2}[x^2 + (x^2+1)]$

$= \dfrac{2x^2+1}{\sqrt{x^2+1}}$

67. $f'(x) = -2(2)(1-4x^2)(-8x) = 32x(1-4x^2)$

69. $h(x) = [x^2(2x+3)]^3 = x^6(2x+3)^3$

$h'(x) = x^6[3(2x+3)^2(2)] + 6x^5(2x+3)^3$

$= 6x^5(2x+3)^2[x + (2x+3)]$

$= 18x^5(2x+3)^2(x+1)$

71. $f(x) = x^2(x-1)^5$

$f'(x) = x^2 5(x-1)^4 + 2x(x-1)^5$

$= x(x-1)^4[5x + 2(x-1)]$

$= x(x-1)^4(7x-2)$

73. $h(t) = \dfrac{\sqrt{3t+1}}{(1-3t)^2}$

$h'(t) = \dfrac{(1-3t)^2(1/2)(3t+1)^{-1/2}(3) - (3t+1)^{1/2}(2)(1-3t)(-3)}{(1-3t)^4}$

$= \dfrac{(3t+1)^{-1/2}[(1-3t)(3/2) + (3t+1)6]}{(1-3t)^3}$

$= \dfrac{3(9t+5)}{2\sqrt{3t+1}\,(1-3t)^3}$

75. $T = \dfrac{1300}{t^2+2t+25} = 1300(t^2+2t+25)^{-1}$

$T'(t) = -1300(t^2+2t+25)^{-2}(2t+2) = \dfrac{-2600(t+1)}{(t^2+2t+25)^2}$

(a) $T'(1) = \dfrac{-325}{49} \approx -6.63$

$T'(3) = \dfrac{-13}{2} \approx -6.5$

$T'(5) = \dfrac{-13}{3} \approx -4.33$

$T'(10) = \dfrac{-1144}{841} \approx -1.36$

(b)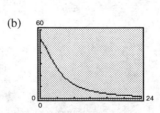

The rate of decrease is approaching zero.

77. $f(x) = 3x^2 + 7x + 1$

$f'(x) = 6x + 7$

$f''(x) = 6$

79. $f'''(x) = -6x^{-4}$

$f^{(4)}(x) = 24x^{-5}$

$f^{(5)}(x) = -120x^{-6} = \dfrac{-120}{x^6}$

81. $f'(x) = 7x^{5/2}$

$$f'''(x) = \frac{35}{2}x^{3/2}$$

83. $f''(x) = 6x^{1/3}$

$$f'''(x) = 2x^{-2/3} = \frac{2}{x^{2/3}}$$

85. (a) $s(t) = -16t^2 + 5t + 30$

(b) $s(t) = 0 = -16t^2 + 5t + 30$

Using the Quadratic Formula or a graphing utility,
$t \approx 1.534$ seconds.

(c) $v(t) = s'(t) = -32t + 5$

$v(1.534) \approx -44.09$ ft/sec

(d) $a(t) = v'(t) = -32$ ft/sec^2

87.
$$x^2 + 3xy + y^3 = 10$$

$$2x + 3x\frac{dy}{dx} + 3y + 3y^2\frac{dy}{dx} = 0$$

$$\frac{dy}{dx}(3x + 3y^2) = -2x - 3y$$

$$\frac{dy}{dx} = \frac{-2x - 3y}{3x + 3y^2} = -\frac{2x + 3y}{3(x + y^2)}$$

89.
$$y^2 - x^2 = 49$$

$$2y\frac{dy}{dx} - 2x = 0$$

$$\frac{dy}{dx} = \frac{2x}{2y} = \frac{x}{y}$$

91.
$$y^2 = x - y$$

$$2yy' = 1 - y'$$

$$2yy' + y' = 1$$

$$(2y + 1)y' = 1$$

$$y' = \frac{1}{2y + 1}$$

At $(2, 1)$, $y' = \frac{1}{3}$.

$$y - 1 = \frac{1}{3}(x - 2)$$

$$y = \frac{1}{3}x + \frac{1}{3}$$

93.
$$2x^{1/3} + 3y^{1/2} = 10$$

$$\tfrac{2}{3}x^{-2/3} + \tfrac{3}{2}y^{-1/2}y' = 0$$

At $(8, 4)$,

$$\tfrac{2}{3}(\tfrac{1}{4}) + \tfrac{3}{2}(\tfrac{1}{2})y' = 0$$

$$\tfrac{3}{4}y' = -\tfrac{1}{6}$$

$$y' = -\tfrac{2}{9}.$$

$$y - 4 = -\tfrac{2}{9}(x - 8)$$

$$y = -\tfrac{2}{9}x + \tfrac{52}{9}$$

95. $b = 8h,\quad 0 \le h \le 5$

$$V = \frac{1}{2}bh(20) = 10bh = 10(8h)h = 80h^2$$

$$\frac{dV}{dt} = 160h\frac{dh}{dt}$$

$$\frac{dh}{dt} = \frac{1}{160h}\frac{dV}{dt} = \frac{1}{16h}\left[\text{since } \frac{dV}{dt} - 10\right]$$

When $h = 4$, $\dfrac{dh}{dt} = \dfrac{1}{16(4)} = \dfrac{1}{64}$ ft/min.

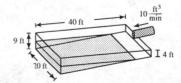

Practice Test for Chapter 2

1. Use the definition of the derivative to find the derivative of $f(x) = 2x^2 + 3x - 5$.

2. Use the definition of the derivative to find the derivative of $f(x) = \dfrac{1}{x - 4}$.

3. Use the definition of the derivative to find the equation of the tangent line to the graph of $f(x) = \sqrt{x - 2}$ at the point $(6, 2)$.

4. Find $f'(x)$ for $f(x) = 5x^3 - 6x^2 + 15x - 9$.

5. Find $f'(x)$ for $f(x) = \dfrac{6x^2 - 4x + 1}{x^2}$.

6. Find $f'(x)$ for $f(x) = \sqrt[3]{x^2} + \sqrt[5]{x^3}$.

7. Find the average rate of change of $f(x) = x^3 - 11$ over the interval $[0, 2]$. Compare this to the instantaneous rate of change at the endpoints of the interval.

8. Given the cost function $C = 6200 + 4.31x - 0.0001x^2$, find the marginal cost of producing x units.

9. Find $f'(x)$ for $f(x) = (x^3 - 4x)(x^2 + 7x - 9)$.

10. Find $f'(x)$ for $f(x) = \dfrac{x + 7}{x^2 - 8}$.

11. Find $f'(x)$ for $f(x) = x^3\left(\dfrac{x - 3}{x + 5}\right)$.

12. Find $f'(x)$ for $f(x) = \dfrac{\sqrt{x}}{x^2 + 4x - 1}$.

13. Find $f'(x)$ for $f(x) = (6x - 5)^{12}$.

14. Find $f'(x)$ for $f(x) = 8\sqrt{4 - 3x}$.

15. Find $f'(x)$ for $f(x) = -\dfrac{3}{(x^2 + 1)^3}$.

16. Find $f'(x)$ for $f(x) = \sqrt{\dfrac{10x}{x + 2}}$.

17. Find $f'''(x)$ for $f(x) = x^4 - 9x^3 + 17x^2 - 4x + 121$.

18. Find $f^{(4)}(x)$ for $f(x) = \sqrt{3 - x}$.

19. Use implicit differentiation to find $\dfrac{dy}{dx}$ for $x^5 + y^5 = 100$.

20. Use implicit differentiation to find $\dfrac{dy}{dx}$ for $x^2y^3 + 2x - 3y + 11 = 0$.

21. Use implicit differentiation to find $\dfrac{dy}{dx}$ for $\sqrt{xy + 4} = 5y - 4x$.

22. Use implicit differentiation to find $\dfrac{dy}{dx}$ for $y^3 = \dfrac{x^3 + 4}{x^3 - 4}$.

23. Let $y = 3x^2$. Find $\dfrac{dx}{dt}$ when $x = 2$ and $\dfrac{dy}{dt} = 5$.

24. The area A of a circle is increasing at a rate of 10 in.^{2}/min. Find the rate of change of the radius r when $r = 4$ inches.

25. The volume of a cone is $V = \left(\dfrac{1}{3}\right)\pi r^2 h$. Find the rate of change of the height when $\dfrac{dV}{dt} = 200$, $h = \dfrac{r}{2}$, and $h = 20$ inches.

Graphing Calculator Required

26. Graph $f(x) = \dfrac{x^2}{x - 2}$ and its derivative on the same set of coordinate axes. From the graph of $f(x)$, determine any points at which the graph has horizontal tangent lines. What is the value of $f'(x)$ at these points?

27. Use a graphing utility to graph $\sqrt[3]{x} + \sqrt[3]{y} = 3$. Then find and sketch the tangent line at the point $(8, 1)$.

C H A P T E R   3
Applications of the Derivative

CHAPTER 3
Applications of the Derivative

Section 3.1   Increasing and Decreasing Functions

Solutions to Odd-Numbered Exercises

1. $f'(x) = \dfrac{(x^2 + 4)(2x) - (x^2)(2x)}{(x^2 + 4)^2} = \dfrac{8x}{(x^2 + 4)^2}$

At $\left(-1, \frac{1}{5}\right)$, f is decreasing since $f'(-1) = -\frac{8}{25}$.

At $(0, 0)$, f has a critical number since $f'(0) = 0$.

At $\left(1, \frac{1}{5}\right)$, f is increasing since $f'(1) = \frac{8}{25}$.

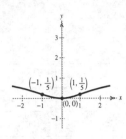

3. $f'(x) = \dfrac{2}{3}(x + 2)^{-1/3} = \dfrac{2}{3\sqrt[3]{x + 2}}$

At $(-3, 1)$, f is decreasing since $f'(-3) = -\frac{2}{3}$.

At $(-2, 0)$, f has a critical number since $f'(-2)$ is undefined.

At $(-1, 1)$, f is increasing since $f'(-1) = \frac{2}{3}$.

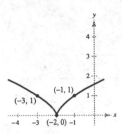

5. $f'(x) = -2(x + 1)$

f has a critical number at $x = -1$. Moreover, f is increasing on $(-\infty, -1)$ and decreasing on $(-1, \infty)$.

7. $f'(x) = 4x^3 - 4x = 4x(x^2 - 1)$

f has critical numbers at $x = 0, \pm 1$. Moreover, f is increasing on $(-1, 0)$, $(1, \infty)$ and decreasing on $(-\infty, -1)$, $(0, 1)$.

9. $f'(x) = 2$

Since the derivative is positive for all x, the function is increasing for all x. Thus, there are no critical numbers. Increasing on $(-\infty, \infty)$.

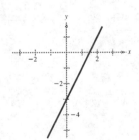

11. $g'(x) = -2(x - 1) = 0$

Critical number: $x = 1$

Interval	$-\infty < x < 1$	$1 < x < \infty$
Sign of g'	$g' > 0$	$g' < 0$
Conclusion	Increasing	Decreasing

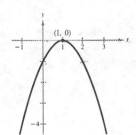

13. $y' = 2x - 5 = 0$

Critical number: $x = \frac{5}{2}$

Interval	$-\infty < x < \frac{5}{2}$	$\frac{5}{2} < x < \infty$
Sign of y'	$y' < 0$	$y' > 0$
Conclusion	Decreasing	Increasing

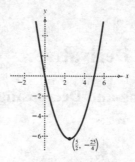

15. $y' = 3x^2 - 12x = 3x(x - 4) = 0$

Critical numbers: $x = 0$ and $x = 4$

Interval	$-\infty < x < 0$	$0 < x < 4$	$4 < x < \infty$
Sign of y'	$y' > 0$	$y' < 0$	$y' > 0$
Conclusion	Increasing	Decreasing	Increasing

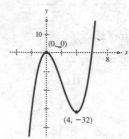

17. $f'(x) = -3(x + 1)^2 = 0$

Critical number: $x = -1$

Interval	$-\infty < x < -1$	$-1 < x < \infty$
Sign of f'	$f' < 0$	$f' < 0$
Conclusion	Decreasing	Decreasing

f is decreasing on $(-\infty, \infty)$.

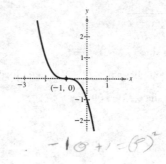

$-1 \otimes + 1 = (?)^2$

19. $f'(x) = -4x + 4 = 0$ $-4(x - 1) = 0$

Critical number: $x = 1$ $y = 1$

Interval	$-\infty < x < 1$	$1 < x < \infty$
Sign of f'	$f' > 0$	$f' < 0$
Conclusion	Increasing	Decreasing

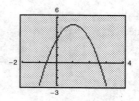

21. $y' = 9x^2 + 24x + 15 = 3(x + 1)(3x + 5) = 0$

Critical numbers: $x = -1, x = -\frac{5}{3}$

Interval	$-\infty < x < -\frac{5}{3}$	$-\frac{5}{3} < x < -1$	$-1 < x < \infty$
Sign of y'	$y' > 0$	$y' < 0$	$y' > 0$
Conclusion	Increasing	Decreasing	Increasing

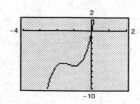

23. $h'(x) = \frac{2}{3}x^{-1/3} = \frac{2}{3\sqrt[3]{x}}$

Critical number: $x = 0$ (h' is undefined here.)

Interval	$-\infty < x < 0$	$0 < x < \infty$
Sign of h'	$h' < 0$	$h' > 0$
Conclusion	Decreasing	Increasing

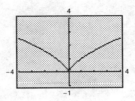

25. $f'(x) = 4x^3 - 6x^2 = 2x^2(2x - 3) = 0$

Critical numbers: $x = 0$ and $x = \frac{3}{2}$

Interval	$-\infty < x < 0$	$0 < x < \frac{3}{2}$	$\frac{3}{2} < x < \infty$
Sign of f'	$f' < 0$	$f' < 0$	$f' > 0$
Conclusion	Decreasing	Decreasing	Increasing

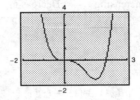

27. $f'(x) = \frac{(x^2 + 4)(1) - (x)(2x)}{(x^2 + 4)^2} = \frac{4 - x^2}{(x^2 + 4)^2}$ $(x-2)(x+2)$

Critical numbers: $x = \pm 2$

Interval	$-\infty < x < -2$	$-2 < x < 2$	$2 < x < \infty$
Sign of f'	$f' < 0$	$f' > 0$	$f' < 0$
Conclusion	Decreasing	Increasing	Decreasing

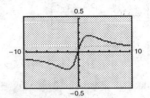

29. $f'(x) = 1 - \frac{1}{x^2} = \frac{x^2 - 1}{x^2}$

Critical numbers: $x = \pm 1$

Discontinuity: $x = 0$

Interval	$-\infty < x < -1$	$-1 < x < 0$	$0 < x < 1$	$1 < x < \infty$
Sign of f'	$f' > 0$	$f' < 0$	$f' < 0$	$f' > 0$
Conclusion	Increasing	Decreasing	Decreasing	Increasing

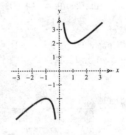

31. $f'(x) = \frac{(16 - x^2)2 - 2x(-2x)}{(16 - x^2)^2} = \frac{2x^2 + 32}{(16 - x^2)^2}$

No critical numbers

Discontinuities: $x = \pm 4$

Interval	$-\infty < x < -4$	$-4 < x < 4$	$4 < x < \infty$
Sign of f'	$f' > 0$	$f' > 0$	$f' > 0$
Conclusion	Increasing	Increasing	Increasing

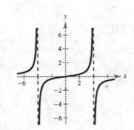

33. $f'(x) = \begin{cases} -2x, & x < 0 \\ -2, & x > 0 \end{cases}$

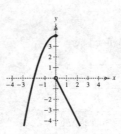

$f'(0)$ is undefined.

Discontinuity: $x = 0$

Critical number: $x = 0$

Interval	$-\infty < x < 0$	$0 < x < \infty$
Sign of f'	$f' > 0$	$f' < 0$
Conclusion	Increasing	Decreasing

35. $C = 10\left(\dfrac{1}{x} + \dfrac{x}{x + 3}\right), \quad 1 \le x$

(a) $\dfrac{dC}{dx} = 10\left[-x^{-2} + \dfrac{(x + 3)(1) - (x)(1)}{(x + 3)^2}\right]$

$= 10\left[-\dfrac{1}{x^2} + \dfrac{3}{(x + 3)^2}\right]$

$= 10\left[\dfrac{-(x + 3)^2 + 3x^2}{x^2(x + 3)^2}\right]$

$= 10\left[\dfrac{2x^2 - 6x - 9}{x^2(x + 3)^2}\right]$

By the Quadratic Formula, $2x^2 - 6x - 9 = 0$ when

$x = \dfrac{6 \pm \sqrt{108}}{4} = \dfrac{3 \pm 3\sqrt{3}}{2}.$

The only critical number in the domain is $x \approx 4.10$.
Thus, C is decreasing on the interval $[1, 4.10)$ and
increasing on $(4.10, \infty)$.

(b)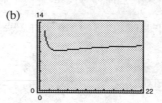

(c) $C = 9$ when $x = 2$ and $x = 15$. Use $x = 4$ to
minimize C.

37. Since $s'(t) = 96 - 32t = 0$, the critical number is $t = 3$. Therefore, the ball is moving up on the interval $(0, 3)$ and moving
down on $(3, 6)$.

39. (a)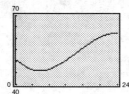

Decreasing: $(0, 6)$

Increasing: $(6, 24)$

(b) $y' = -0.25t^{1.5} + 1.82t - 2.94t^{1/2}$

$y' < 0$ on $(0, 5.85)$.

$y' > 0$ on $(5.85, 23.6)$.

Section 3.2   Extrema and the First-Derivative Test

1. $f'(x) = 4 - 4x = 4(1 - x)$

Critical number: $x = 1$

Interval	$(-\infty, 1)$	$(1, \infty)$
Sign of f'	+	−
f	Increasing	Decreasing

Relative maximum: $(1, 5)$

3. $f'(x) = 2x - 6 = 2(x - 3)$

Critical number: $x = 3$

Interval	$(-\infty, 3)$	$(3, \infty)$
Sign of f'	−	+
f	Decreasing	Increasing

Relative minimum: $(3, -9)$

5. $g'(x) = 18x^2 - 30x + 12$
$$= 6(3x^2 - 5x + 2)$$
$$= 6(x - 1)(3x - 2)$$

Critical numbers: $x = 1, \frac{2}{3}$

Interval	$\left(-\infty, \frac{2}{3}\right)$	$\left(\frac{2}{3}, 1\right)$	$(1, \infty)$
Sign of g'	+	−	+
g	Increasing	Decreasing	Increasing

Relative maximum: $\left(\frac{2}{3}, \frac{28}{9}\right)$

Relative minimum: $(1, 3)$

7. $h'(x) = -3(x + 4)^2$

Critical number: $x = -4$

Interval	$(-\infty, -4)$	$(-4, \infty)$
Sign of h'	−	−
h	Decreasing	Decreasing

No relative extrema

9. $f'(x) = 3x^2 - 12x = 3x(x - 4)$

Critical numbers: $x = 0, x = 4$

Interval	$(-\infty, 0)$	$(0, 4)$	$(4, \infty)$
Sign of f'	+	−	+
f	Increasing	Decreasing	Increasing

Relative maximum: $(0, 15)$

Relative minimum: $(4, -17)$

11. $f'(x) = 2x^2(2x - 3)$

Critical numbers: $x = 0, x = \frac{3}{2}$

Interval	$(-\infty, 0)$	$\left(0, \frac{3}{2}\right)$	$\left(\frac{3}{2}, \infty\right)$
Sign of f'	−	−	+
f	Decreasing	Decreasing	Increasing

Relative minimum: $\left(\frac{3}{2}, -\frac{27}{16}\right)$

13. $f'(x) = \frac{1}{5}x^{-4/5} = \frac{1}{5x^{4/5}}$

Critical number: $x = 0$

Interval	$(-\infty, 0)$	$(0, \infty)$
Sign of f'	+	+
f	Increasing	Increasing

No relative extrema

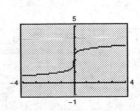

15. $g'(t) = \frac{2}{3}t^{-1/3} = \frac{2}{3\sqrt[3]{t}}$

Critical number: $t = 0$

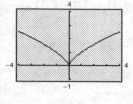

Interval	$(-\infty, 0)$	$(0, \infty)$
Sign of g'	$-$	$+$
g	Decreasing	Increasing

Relative minimum: $(0, 0)$

17. $f'(x) = \frac{(x + 1) - x}{(x + 1)^2} = \frac{1}{(x + 1)^2}$

No critical numbers

Discontinuity: $x = -1$

Interval	$(-\infty, -1)$	$(-1, \infty)$
Sign of f'	$+$	$+$
f	Increasing	Increasing

No relative extrema

19. $f'(x) = -2$

No critical numbers

x-value	Endpoint $x = -1$	Endpoint $x = 2$
$f(x)$	8	2
Conclusion	Maximum	Minimum

21. $f(x) = 5 - 2x^2$, $[0, 3]$

$f'(x) = -4x$

Critical number: $x = 0$ (also endpoint)

x-value	Endpoint $x = 0$	Endpoint $x = 3$
$f(x)$	5	-13
Conclusion	Maximum	Minimum

23. $f'(x) = 3x^2 - 6x = 3x(x - 2)$

Critical numbers: $x = 0$ and $x = 2$

x-value	Endpoint $x = -1$	Critical $x = 0$	Critical $x = 2$	Endpoint $x = 3$
$f(x)$	-4	0	-4	0
Conclusion	Minimum	Maximum	Minimum	Maximum

25. $h(s) = \frac{1}{3 - s} = (3 - s)^{-1}$, $[0, 2]$

$h'(s) = -(3 - s)^{-2}(-1) = \frac{1}{(3 - s)^2}$

No critical numbers

s-value	Endpoint $s = 0$	Endpoint $s = 2$
$h(s)$	$\frac{1}{3}$	1
Conclusion	Minimum	Maximum

27. $f'(x) = 3x^2 - 12 = 3(x^2 - 4) = 0$

Critical numbers: $x = \pm 2$

x-value	Endpoint $x = 0$	Critical $x = 2$	Endpoint $x = 4$
$f(x)$	0	-16	16
Conclusion		Minimum	Maximum

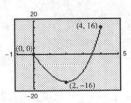

29. $h'(t) = \frac{2}{3}(t - 1)^{-1/3} = \frac{2}{3(t - 1)^{1/3}}$

t-value	Endpoint $t = -7$	Critical $t = 1$	Endpoint $t = 2$
$h(t)$	4	0	1
Conclusion	Maximum	Minimum	

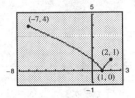

31. Maximum: $(5, 7)$

Minimum: $(2.69, -5.55)$

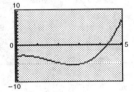

33. Maximum: $(1, 4.7)$

Minimum: $(0.4398, -1.0613)$

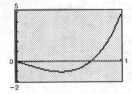

35. $f'(x) = \frac{(x^2 + 1)(4) - 4x(2x)}{(x^2 + 1)^2} = \frac{4(1 - x^2)}{(x^2 + 1)^2}$

Critical number: $x = 1$

x-value	Endpoint $x = 0$	Critical $x = 1$	Interval $(1, \infty)$
$f(x)$	0	2	$0 < f(x) < 2$
Conclusion	Minimum	Maximum	f is decreasing.

37. $f'(x) = \frac{(x^2 + 4)2 - 2x(2x)}{(x^2 + 4)^2}$

$= \frac{8 - 2x^2}{(x^2 + 4)^2}$

$= \frac{2(2 - x)(2 + x)}{(x^2 + 4)^2}$

On $[0, \infty)$, $x = 2$ is a critical number.

x-value	Endpoint $x = 0$	Critical $x = 2$
$f(x)$	0	$\frac{1}{2}$
Conclusion	Minimum	Maximum

39. $f'(x) = 15x^4 - 30x^2$

$f''(x) = 60x^3 - 60x$

$f'''(x) = 180x^2 - 60 = 60(3x^2 - 1)$

Critical numbers for f'' in $[0, 1]$: $x = \frac{1}{\sqrt{3}}$

x-value	Endpoint $x = 0$	Critical $x = \frac{1}{\sqrt{3}}$	Endpoint $x = 1$
$|f''(x)|$	0	$\frac{40}{\sqrt{3}}$	0
Conclusion		Maximum	

41. $f'(x) = 60x^3 - 6\left(\dfrac{2x-1}{2}\right)^5$

$f''(x) = 180x^2 - 30\left(\dfrac{2x-1}{2}\right)^4$

$f'''(x) = 360x - 120\left(\dfrac{2x-1}{2}\right)^3$

$f^{(4)}(x) = 360 - 360\left(\dfrac{2x-1}{2}\right)^2$

$f^{(5)}(x) = -720\left(\dfrac{2x-1}{2}\right)$

Critical number of $f^{(4)}$: $x = \dfrac{1}{2}$

x-value	Endpoint $x = 0$	Critical $x = \frac{1}{2}$	Endpoint $x = 1$
$|f^{(4)}(x)|$	270	360	270
Conclusion		Maximum	

45. Demand: $(6000, 0.80)$, $(5600, 1.00)$

$m = \dfrac{1 - 0.8}{5600 - 6000} = \dfrac{0.2}{-400} = -0.0005$

$p - 1 = -0.0005(x - 5600)$

$p = -0.0005x + 3.80$

Cost $= C = 5000 + 0.40x$

Profit $= P = R - C$

$\quad\quad\quad = xp - C$

$\quad\quad\quad = x(-0.0005x + 3.80) - (5000 + 0.40x)$

$\quad\quad\quad = -0.0005x^2 + 3.40x - 5000$

$P' = -0.001x + 3.40 = 0 \Longrightarrow x = 3400$

$p(3400) = \$2.10$ per can

47. $r'(t) = 0.00018t^2 + 0.0008t - 0.21 = 0$

$\quad\quad t \approx 32.0 \quad (1982)$

$r(32.0) \approx 94$

In 1982, there were about 94 males for every 100 females.

43. $C = 3x + 20{,}000x^{-1}, \quad 0 < x \le 200$

$C' = 3 - 20{,}000x^{-2}$

$\quad = \dfrac{3x^2 - 20{,}000}{x^2}$

Critical numbers: $x = \sqrt{\dfrac{20{,}000}{3}} \approx 81.65 \approx 82$ units

$C(82) \approx 489.90$, which is the minimum by the First-Derivative Test.

X	Y1	Y2
3100	735	2.25
3200	760	2.2
3300	775	2.15
3400	780	2.1
3500	775	2.05
3600	760	2
3700	735	1.95

X=3100

Section 3.3   Concavity and the Second-Derivative Test

1. $y' = 2x - 1$

$y'' = 2$

Concave upward on $(-\infty, \infty)$

3. $f(x) = \dfrac{x^2 - 1}{2x + 1}$

$f'(x) = \dfrac{(2x + 1)(2x) - (x^2 - 1)(2)}{(2x + 1)^2}$

$= \dfrac{2x^2 + 2x + 2}{(2x + 1)^2}$

$= (2x^2 + 2x + 2)(2x + 1)^{-2}$

$f''(x) = (2x^2 + 2x + 2)[-2(2x + 1)^{-3}(2)] + (2x + 1)^{-2}(4x + 2)$

$= -8(x^2 + x + 1)(2x + 1)^{-3} + 2(2x + 1)^{-2}(2x + 1)$

$= 2(2x + 1)^{-3}[-4(x^2 + x + 1) + (2x + 1)(2x + 1)]$

$= 2(2x + 1)^{-3}[-4x^2 - 4x - 4 + 4x^2 + 4x + 1]$

$= \dfrac{-6}{(2x + 1)^3}$

$f''(x) \neq 0$ for any value of x.

$x = -\frac{1}{2}$ is a discontinuity.

Concave upward on $\left(-\infty, -\frac{1}{2}\right)$

Concave downward on $\left(-\frac{1}{2}, \infty\right)$

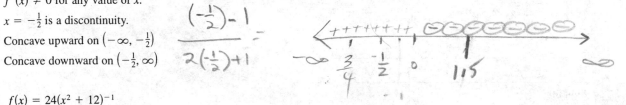

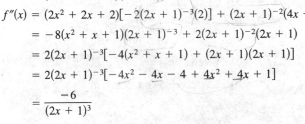

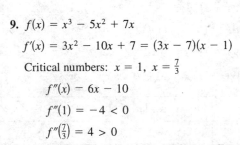

5. $f(x) = 24(x^2 + 12)^{-1}$

$f'(x) = -24(2x)(x^2 + 12)^{-2}$

$f''(x) = -48[x(-2)(2x)(x^2 + 12)^{-3} + (x^2 + 12)^{-2}]$

$= \dfrac{-48(-4x^2 + x^2 + 12)}{(x^2 + 12)^3}$

$= \dfrac{144(x^2 - 4)}{(x^2 + 12)^3}$

$f''(x) = 0$ when $x = \pm 2$.

Concave upward on $(-\infty, -2)$ and $(2, \infty)$

Concave downward on $(-2, 2)$

7. $f'(x) = 6 - 2x = 0$

Critical number: $x = 3$

$f''(x) = -2$

$f''(3) = -2 < 0$

Thus, $(3, 9)$ is a relative maximum.

9. $f(x) = x^3 - 5x^2 + 7x$

$f'(x) = 3x^2 - 10x + 7 = (3x - 7)(x - 1)$

Critical numbers: $x = 1$, $x = \frac{7}{3}$

$f''(x) = 6x - 10$

$f''(1) = -4 < 0$

$f''\left(\frac{7}{3}\right) = 4 > 0$

Thus, $(1, 3)$ is a relative maximum and $\left(\frac{7}{3}, 1.\overline{814}\right)$ is a relative minimum.

11. $f'(x) = \frac{2}{3}x^{-1/3} = \dfrac{2}{3\sqrt[3]{x}}$

Critical number: $x = 0$

The Second-Derivative Test does not apply, so we use the First-Derivative Test to conclude that $(0, -3)$ is a relative minimum.

13. $f'(x) = 1 - \dfrac{4}{x^2} = \dfrac{x^2 - 4}{x^2}$

Critical numbers: $x = \pm 2$

$f''(x) = \dfrac{8}{x^3}$

$f''(2) = 1 > 0$

$f''(-2) = -1 < 0$

Thus, $(2, 4)$ is a relative minimum and $(-2, -4)$ is a relative maximum.

15.

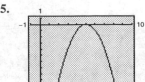

Relative maximum: $(5, 0)$

17.

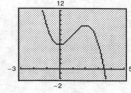

Relative minimum: $(0, 5)$

Relative maximum: $(2, 9)$

19. $f' > 0$ (increasing)

$f'' > 0$ (concave upward)

21. $f' < 0$ (decreasing)

$f'' < 0$ (concave downward)

23. $f(x) = x^3 - 9x^2 + 24x - 18$

$f'(x) = 3x^2 - 18x + 24$

$f''(x) = 6x - 18 = 0 \Rightarrow x = 3$

Interval	$(-\infty, 3)$	$(3, \infty)$
Sign of f''	$-$	$+$
Conclusion	Concave downward	Concave upward

Inflection point: $(3, 0)$

25. $f(x) = (x - 1)^3(x - 5)$

$f'(x) = (x - 1)^3(1) + (x - 5)(3)(x - 1)^2(1)$

$\quad = (x - 1)^2[(x - 1) + 3(x - 5)]$

$\quad = (x - 1)^2(4x - 16)$

$\quad = 4(x - 1)^2(x - 4)$

$f''(x) = 4(x - 1)^2(1) + (x - 4)8(x - 1)(1)$

$\quad = 4(x - 1)[(x - 1) + 2(x - 4)]$

$\quad = 4(x - 1)(3x - 9)$

$\quad = 12(x - 1)(x - 3) = 0 \Rightarrow x = 1 \text{ or } x = 3$

Interval	$(-\infty, 1)$	$(1, 3)$	$(3, \infty)$
Sign of f''	$+$	$-$	$+$
Conclusion	Concave upward	Concave downward	Concave upward

Inflection points: $(1, 0), (3, -16)$

27. $g(x) = 2x^4 - 8x^3 + 12x^2 + 12x$

$g'(x) = 8x^3 - 24x^2 + 24x + 12$

$g''(x) = 24x^2 - 48x + 24 = 24(x - 1)^2$

$g'' < 0$ on $(-\infty, 1)$ and $(1, \infty)$.

No inflection points

29. $h(x) = (x - 2)^3(x - 1)$

$h'(x) = (4x - 5)(x - 2)^2$

$h''(x) = 6(2x - 3)(x - 2)$

Inflection points: $\left(\frac{3}{2}, -\frac{1}{16}\right), (2, 0)$

31. $f'(x) = 3x^2 - 12 = 3(x^2 - 4)$

Critical numbers: $x = \pm 2$

$$f''(x) = 6x$$

$$f''(2) = 12 > 0$$

$$f''(-2) = -12 < 0$$

Relative maximum: $(-2, 16)$

Relative minimum: $(2, -16)$

$f''(x) = 0$ when $x = 0$.

$f''(x) < 0$ on $(-\infty, 0)$.

$f''(x) > 0$ on $(0, \infty)$.

Inflection point: $(0, 0)$

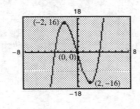

33. $f'(x) = 3x^2 - 12x + 12 = 3(x - 2)^2$

Critical number: $x = 2$

$$f''(x) = 6(x - 2)$$

$$f''(x) = 0 \text{ when } x = 2.$$

Since $f'(x) > 0$ when $x \neq 2$ and the concavity changes at $x = 2$, $(2, 8)$ is an inflection point.

No relative extrema

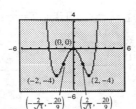

35. $f'(x) = x^3 - 4x = x(x + 2)(x - 2)$

Critical numbers: $x = \pm 2$, $x = 0$

$$f''(x) = 3x^2 - 4$$

$$f''(-2) = 8 > 0$$

$$f''(0) = -4 < 0$$

$$f''(2) = 8 > 0$$

Relative maximum: $(0, 0)$

Relative minima: $(\pm 2, -4)$

$f''(x) = 3x^2 - 4 = 0$ when $x = \pm\dfrac{2\sqrt{3}}{3}$.

$f''(x) > 0$ on $\left(-\infty, -\dfrac{2\sqrt{3}}{3}\right)$.

$f''(x) < 0$ on $\left(-\dfrac{2\sqrt{3}}{3}, \dfrac{2\sqrt{3}}{3}\right)$.

$f''(x) > 0$ on $\left(\dfrac{2\sqrt{3}}{3}, \infty\right)$.

Inflection points: $\left(-\dfrac{2\sqrt{3}}{3}, -\dfrac{20}{9}\right), \left(\dfrac{2\sqrt{3}}{3}, -\dfrac{20}{9}\right)$

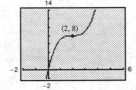

37. $g(x) = (x - 2)(x + 1)^2 = x^3 - 3x - 2$

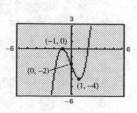

$g'(x) = 3x^2 - 3 = 3(x - 1)(x + 1)$

Critical numbers: $x = \pm 1$

$\qquad f''(x) = 6x$

$\quad f''(-1) = -6 < 0$

$\qquad f''(1) = 6 > 0$

Relative maximum: $(-1, 0)$

Relative minimum: $(1, -4)$

$\quad f''(x) = 6x = 0$ when $x = 0$.

$\quad f''(x) < 0$ on $(-\infty, 0)$.

$\quad f''(x) > 0$ on $(0, \infty)$.

Inflection point: $(0, 2)$

39. The domain of g is $[-3, \infty)$.

$$g'(x) = x\left[\frac{1}{2}(x + 3)^{-1/2}\right] + \sqrt{x + 3} = \frac{3x + 6}{2\sqrt{x + 3}}$$

Critical numbers: $x = -3$, $x = -2$

By the First-Derivative Test, $(-2, -2)$ is a relative minimum.

$$g''(x) = \frac{\left(2\sqrt{x + 3}\right)(3) - (3x + 6)\left(1/\sqrt{x + 3}\right)}{4(x + 3)} = \frac{3(x + 4)}{4(x + 3)^{3/2}}$$

$x = -4$ is not in the domain of g. On $[-3, \infty)$, $g''(x) > 0$ and is concave upward.

41. $f'(x) = \dfrac{-8x}{(1 + x^2)^2}$

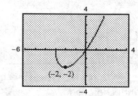

Critical number: $x = 0$

$\quad f''(x) = \dfrac{-8(1 - 3x^2)}{(1 + x^2)^3}$

$\quad f''(0) = -8 < 0$

Thus, $(0, 4)$ is a relative maximum.

$\quad f''(x) = 0$ when $1 - 3x^2 = 0$, $x = \pm\dfrac{\sqrt{3}}{3}$.

$\quad f''(x) > 0$ on $\left(-\infty, -\dfrac{\sqrt{3}}{3}\right)$.

$\quad f''(x) < 0$ on $\left(-\dfrac{\sqrt{3}}{3}, \dfrac{\sqrt{3}}{3}\right)$.

$\quad f''(x) > 0$ on $\left(\dfrac{\sqrt{3}}{3}, \infty\right)$.

Inflection points: $\left(\dfrac{\sqrt{3}}{3}, 3\right), \left(-\dfrac{\sqrt{3}}{3}, 3\right)$

43. The function has x-intercepts at $(2, 0)$ and $(4, 0)$. On $(-\infty, 3)$, f is decreasing, and on $(3, \infty)$, f is increasing. A relative minimum occurs when $x = 3$. The graph of f is concave upward.

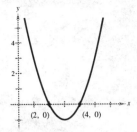

45. (a) $f'(x) > 0$ on $(-\infty, 0)$ where f is increasing.

(b) $f'(x) < 0$ on $(0, \infty)$ where f is decreasing.

(c) f' is not increasing. f is not concave upward.

(d) f' is decreasing on $(-\infty, \infty)$ where f is concave downward.

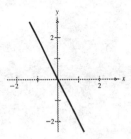

47. $R' = \dfrac{1}{50,000}(1200x - 3x^2)$

$R'' = \dfrac{1}{50,000}(1200 - 6x) = 0$ when $x = 200$.

$R'' > 0$ on $(0, 200)$.

$R'' < 0$ on $(200, 400)$.

Since $(200, 320)$ is a point of inflection, it is the point of diminishing returns.

49. $C = 0.5x^2 + 15x + 5000$

$\overline{C} = 0.5x + 15 + \dfrac{5000}{x}$

$\overline{C}' = 0.5 - \dfrac{5000}{x^2}$

Critical numbers: $x = \pm 100$

$x = 100$ units

51. $N(t) = -0.12t^3 + 0.54t^2 + 8.22t, \quad 0 \leq t \leq 4$

$N'(t) = -0.36t^2 + 1.08t + 8.22$

$N''(t) = -0.72t + 1.08 = 0$

$t = \dfrac{1.08}{0.72} = 1.5$ hours

Thus, the time is 8:30 P.M.

53. $x = 10,000\left[\dfrac{t^2}{9 + t^2}\right]$

$x = 10,000\left[\dfrac{(9 + t^2)(2t) - (t^2)(2t)}{(9 + t^2)^2}\right]$

$= 10,000\left[\dfrac{18t}{(9 + t^2)^2}\right]$

$= 180,000\left[\dfrac{t}{(9 + t^2)^2}\right]$

$x = 180,000\left[\dfrac{(9 + t^2)^2(1) - t(2)(9 + t^2)(2t)}{(9 + t^2)^4}\right]$

$= 180,000\left[\dfrac{9 - 3t^2}{(9 + t^2)^3}\right] = 0$

$9 = 3t^2 \implies t = \sqrt{3} \approx 1.732$ years

55. Point of inflection: $(4, 5000)$

The sales of the new product are increasing at the greatest rate when $t = 4$.

57. $f(x) = \frac{1}{2}x^3 - x^2 + 3x - 5, \quad [0, 3]$

$f'(x) = \frac{3}{2}x^2 - 2x + 3$

$f''(x) = 3x - 2$

Minimum: $(0, -5)$

Maximum: $(3, 8.5)$

Point of inflection: $\left(\frac{2}{3}, -3.2963\right)$

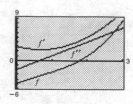

59.

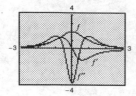

Relative maximum: $(0, 2)$

Inflection points: $(0.58, 1.5), (-0.58, 1.5)$

61.

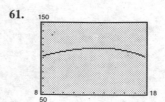

Section 3.4   Optimization Problems

1. Let x be the first number and y be the second number. Then $x + y = 110$ and $y = 110 - x$. Thus, the product of x and y is given by the following.

$$P = xy = x(110 - x)$$

$$P' = 110 - 2x$$

$P' = 0$ when $x = 55$. Since $P''(55) = -2 < 0$, the product is a maximum when $x = 55$ and $y = 110 - 55 = 55$.

3. Let x be the first number and y be the second number. Then $x + 2y = 36$ and $x = 36 - 2y$. The product of x and y is given by the following.

$$P = xy = (36 - 2y)y$$

$$P' = 36 - 4y$$

$P' = 0$ when $y = 9$. Since $P''(9) = -4 < 0$, the product is a maximum when $y = 9$ and $x = 36 - 2(9) = 18$.

5. Let x be the first number and y be the second number. Then $xy = 192$ and $y = 192/x$. The sum of x and y is given by the following.

$$S = x + y = x + \frac{192}{x}$$

$$S' = 1 - \frac{192}{x^2}$$

$S' = 0$ when $x = \sqrt{192}$. Since $S''(x) > 0$ when $x > 0$, S is minimum when $x = \sqrt{192} = 8\sqrt{3}$ and $y = 192/\sqrt{192} = \sqrt{192} = 8\sqrt{3}$.

7. $S = x + \dfrac{1}{x}, \quad x > 0$

$$S' = 1 - \frac{1}{x^2} = \frac{x^2 - 1}{x^2}$$

Critical number: $x = 1$

$$S'' = \frac{2}{x^3}$$

Since $S''(1) = 2 > 0$, $(1, 2)$ is a relative minimum and the sum is a minimum when $x = 1$.

9. (a) $A = 4(11)(3) + 2(3)(3) = 150$ square inches

     $V = 11(3)(3) = 99$ cubic inches

(b) $A = 6(5)(5) = 150$ square inches

     $V = (5)(5)(5) = 125$ cubic inches

(c) $A = 4(6)(3.25) + 2(6)(6) = 150$ square inches

     $V = 6(6)(3.25) = 117$ cubic inches

11. Let x be the length and y be the width of the rectangle. Then $2x + 2y = 100$ and $y = 50 - x$. The area is given by the following.

$$A = xy = x(50 - x)$$

$$A' = 50 - 2x$$

$A' = 0$ when $x = 25$. Since $A''(25) = -2 < 0$, A is maximum when $x = 25$ feet and $y = 50 - 25 = 25$ feet.

13. Let x and y be the lengths shown in the figure. Then $4x + 3y = 200$ and $y = (200 - 4x)/3$. The area of the corrals is given by the following.

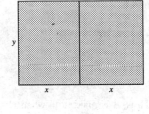

$$A = 2xy = 2x\left(\frac{200 - 4x}{3}\right) = \frac{8}{3}(50x - x^2)$$

$$A' = \frac{8}{3}(50 - 2x)$$

$A' = 0$ when $x = 25$. Since $A''(25) = -\frac{16}{3} < 0$, A is maximum when $x = 25$ feet and $y = \frac{100}{3}$ feet.

15. Let x be the length shown in the figure. Then the volume of the box is given by the following.

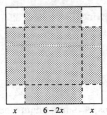

$$V = x(6 - 2x)^2, \quad 0 < x < 3$$

$$V' = 12(x - 1)(x - 3)$$

$V' = 0$ when $x = 3$ and $x = 1$. Since $V = 0$ when $x = 3$ and $V = 16$ when $x = 1$, we conclude that the volume is maximum when $x = 1$. The corresponding volume is $V = 16$ cubic inches.

17. Let x and y be the lengths shown in the figure. Then $x^2 y = 250/3$ and $y = 250/3x^2$. The surface area of the enclosure is given by the following.

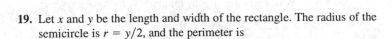

$$A = 3xy + x^2 = 3x\left(\frac{250}{3x^2}\right) + x^2 = \frac{250}{x} + x^2$$

$$A' = -\frac{250}{x^2} + 2x = \frac{2x^3 - 250}{x^2}$$

$A' = 0$ when $x = 5$. The surface area is minimum when $x = 5$ meters and

$$y = \frac{250}{3(5)^2} = \frac{10}{3} \text{ meters.}$$

19. Let x and y be the length and width of the rectangle. The radius of the semicircle is $r = y/2$, and the perimeter is

$$200 = 2x + 2\pi r = 2x + 2\pi\left(\frac{y}{2}\right) = 2x + \pi y$$

which implies that $y = (200 - 2x)/\pi$. The area of the rectangle is given by the following.

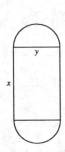

$$A = xy = x\left[\frac{200 - 2x}{\pi}\right] = \frac{2}{\pi}(100x - x^2)$$

$$A' = \frac{2}{\pi}(100 - 2x)$$

$A' = 0$ when $x = 50$. Thus, A is maximum when $x = 50$ meters and

$$y = \frac{200 - 2(50)}{\pi} = \frac{100}{\pi} \text{ meters.}$$

21. The area of the rectangle is

$$A = xy = x\left(\frac{6 - x}{2}\right) = \frac{1}{2}(6x - x^2)$$

$$A' = \frac{1}{2}(6 - 2x).$$

$A' = 0$ when $x = 3$. Thus, A is maximum when $x = 3$ and

$$y = \frac{6 - 3}{2} = \frac{3}{2}.$$

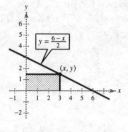

23. The area is given by the following.

$$A = 2xy = 2x\sqrt{25 - x^2}$$

$$A' = 2\left(\frac{25 - 2x^2}{\sqrt{25 - x^2}}\right)$$

$A' = 0$ when $x = 5/\sqrt{2}$. Thus, A is maximum when the length is

$$2x = \frac{10}{\sqrt{2}} \approx 7.07$$

and the width is

$$y = \sqrt{25 - \left(\frac{5}{\sqrt{2}}\right)^2} = \frac{5}{\sqrt{2}} \approx 3.54.$$

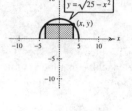

25. The volume of the cylinder is

$$V = \pi r^2 h = 12(1.80469) \approx 21.66$$

which implies that $h = 21.66/\pi r^2$. The surface area of the cylinder is

$$S = 2\pi r^2 + 2\pi rh$$

$$= 2\pi r^2 + 2\pi r\left(\frac{21.66}{\pi r^2}\right)$$

$$= 2\left(\pi r^2 + \frac{21.66}{r}\right)$$

$$S' = 2\left(2\pi r - \frac{21.66}{r^2}\right).$$

$S' = 0$ when $2\pi r^3 - 21.66 = 0$, which implies that

$$r = \sqrt[3]{\frac{21.66}{2\pi}} \approx 1.51 \text{ inches}$$

$$h = \frac{21.66}{\pi(1.51)^2} \approx 3.02 \text{ inches.}$$

(Note that in the solution, $h = 2r$.)

27. The distance between a point (x, y) on the graph and the point $(0, 4)$ is

$$d = \sqrt{(x - 0)^2 + (y - 4)^2}$$

$$= \sqrt{x^2 + (x^2 + 1 - 4)^2}$$

$$= \sqrt{x^2 + (x^2 - 3)^2}.$$

We can minimize d by minimizing its square $L = d^2$.

$$L = x^2 + (x^2 - 3)^2 = x^4 - 5x^2 + 9$$

$$L' = 4x^3 - 10x = 2x(2x^2 - 5) \Longrightarrow x = 0, \pm\sqrt{\tfrac{5}{2}}$$

Hence, the points are $\left(\pm\sqrt{\tfrac{5}{2}}, \tfrac{7}{2}\right)$.

29. The length and girth is $4x + y = 108$. Thus, $y = 108 - 4x$. The volume is

$$V = x^2 y = x^2(108 - 4x) = 108x^2 - 4x^3$$

$$V' = 216x - 12x^2 = 12x(18 - x).$$

$V' = 0$ when $x = 0$ and $x = 18$. Thus, the maximum volume occurs when $x = 18$ inches and $y = 36$ inches. The dimensions are 18 inches by 18 inches by 36 inches.

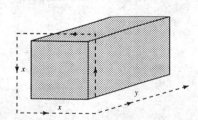

31. Let x be the length of a side of the square and r be the radius of the circle. Then the combined perimeter is $4x + 2\pi r = 16$, which implies that

$$x = \frac{16 - 2\pi r}{4} = 4 - \frac{\pi r}{2}.$$

The combined area of the circle and square is

$$A = x^2 + \pi r^2 = \left(4 - \frac{\pi r}{2}\right)^2 + \pi r^2$$

$$A' = 2\left(4 - \frac{\pi r}{2}\right)\left(-\frac{\pi}{2}\right) + 2\pi r = \frac{1}{2}(\pi^2 r + 4\pi r - 8\pi).$$

$A' = 0$ when

$$r = \frac{8\pi}{\pi^2 + 4\pi} = \frac{8}{\pi + 4}$$

and the corresponding x-value is

$$x = 4 - \frac{\pi[8/(\pi + 4)]}{2} = \frac{16}{\pi + 4}.$$

This is a minimum by the Second-Derivative Test $[A'' = \frac{1}{2}(\pi^2 + 4\pi) > 0]$.

33. Using the formula Distance = (Rate)(Time), we have $T = D/R$.

$$T = T_{\text{rowed}} + T_{\text{walked}} = \frac{D_{\text{rowed}}}{R_{\text{rowed}}} + \frac{D_{\text{walked}}}{R_{\text{walked}}} = \frac{\sqrt{x^2 + 4}}{2} + \frac{\sqrt{1 + (3 - x)^2}}{4}$$

$$T' = \frac{x}{2\sqrt{x^2 + 4}} - \frac{3 - x}{4\sqrt{1 + (3 - x)^2}}$$

By setting $T' = 0$, we have the following.

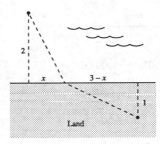

$$\frac{x^2}{4(x^2 + 4)} = \frac{(3 - x)^2}{16[1 + (3 - x)^2]}$$

$$\frac{x^2}{x^2 + 4} = \frac{9 - 6x + x^2}{4(10 - 6x + x^2)}$$

$$4(x^4 - 6x^3 + 10x^2) = (x^2 + 4)(9 - 6x + x^2)$$

$$x^4 - 6x^3 + 9x^2 + 8x - 12 = 0$$

Using a graphing utility, the solution on $[0, 3]$ is $x = 1$ mile.

Possible rational roots: $\pm 1, \pm 2, \pm 3, \pm 4, \pm 6, \pm 12$

By testing, we find that $x = 1$ mile.

35. Since $h^2 + w^2 = 24^2$, we have $h^2 = 24^2 - w^2$.

$$S = kh^2 w = k(24^2 - w^2)w = k(576w - w^3)$$

$$S' = k(576 - 3w^2)$$

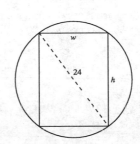

$S' = 0$ when $w = \sqrt{192} = 8\sqrt{3}$. Thus, S is maximum when

$$h^2 = 24^2 - 192 = 384 \Rightarrow h = \sqrt{384} = 8\sqrt{6}.$$

The dimensions are $w = 8\sqrt{3} \approx 13.856$ inches and $h = 8\sqrt{6} \approx 19.596$ inches.

37. (a) Area $= x^2 + \pi r^2$

$$4 = 4x + 2\pi r \Rightarrow r = \frac{2 - 2x}{\pi}$$

$$A(x) = x^2 + \pi \left(\frac{2 - 2x}{\pi}\right)^2 = x^2 + \frac{(2 - 2x)^2}{\pi}$$

(b) Domain: $0 \le x \le 1$

(c)

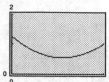

(d) A is minimum if $x = 0.5601$ and $r = 0.28$.

(e) A is maximum if $x = 0$ and $r = 2/\pi$ (all the wire for the circle).

Section 3.5   Business and Economics Applications

1. $R = 800x - 0.2x^2$

$R' = 800 - 0.4x$

$R' = 0$ when $x = \frac{800}{0.4} = 2000$. R is maximum when $x = 2000$.

3. $R = 400x - x^2$

$R' = 400 - 2x = 0$ when $x = 200$.

Thus, R is maximum when $x = 200$ units.

5. $\overline{C} = 1.25x + 255 + \dfrac{8000}{x}$

$\overline{C}' = 1.25 - \dfrac{8000}{x^2}$

$\overline{C}' = 0$ when

$$1.25x^2 = 8000$$

$$x^2 = 6400$$

$$x = 80 \text{ units.}$$

7. $\overline{C} = 2x + 255 + \dfrac{5000}{x}$

$\overline{C}' = 2 - \dfrac{5000}{x^2}$

$\overline{C}' = 0$ when $2 = 5000/x^2$, which implies that $x^2 = 2500$ and $x = 50$ units.

9. $P = xp - C$

$= (90x - x^2) - (100 + 30x)$

$= -x^2 + 60x - 100$

$P' = -2x + 60$

$P' = 0$ when $x = 30$. Thus, the maximum profit occurs at $x = 30$ units and $p = 90 - 30 = \$60$.

11. $P = xp - C$

$= x(70 - 0.001x) - (8000 + 50x + 0.03x^2)$

$= -0.031x^2 + 20x - 8000$

$P' = -0.062x + 20$

$P' = 0$ when $x \approx 322.58$ and $p \approx \$69.68$.

13. $\overline{C} = 2x + 5 + \dfrac{18}{x}$

$\overline{C}' = 2 - \dfrac{18}{x^2}$

$\overline{C}' = 0$ when $x = 3$. Thus, the average cost is minimum when $x = 3$ units and $\overline{C}(3) = \$17$ per unit.

$$C' = \text{Marginal cost} = 4x + 5$$

$$C'(3) = 17 = \overline{C}(3)$$

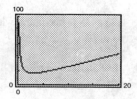

15. (a) $P = xp - C$

$$= x\left(100 - \frac{1}{2}x^2\right) - (40x + 37.5)$$

$$= -\frac{1}{2}x^3 + 60x - 37.5$$

$$P' = -\frac{3}{2}x^2 + 60$$

$P' = 0$ when $x = \sqrt{40} = 2\sqrt{10} \approx 6.32$ units. The price is $p = 100 - \frac{1}{2}(40) = \$80$.

(b) $\overline{C} = 40 + \dfrac{37.5}{x}$

When $x = 2\sqrt{10}$, the average price is

$$\overline{C}(2\sqrt{10}) = 40 + \frac{37.5}{2\sqrt{10}} \approx \$45.93.$$

17. $P' = -6s^2 + 70s - 100$

$$= -2(3s^2 - 35s + 50)$$

$$= -2(3s - 5)(s - 10)$$

Critical numbers: $s = \frac{5}{3}$ and $s = 10$

$$P'' = -12s + 70$$

$$P''\left(\frac{5}{3}\right) = 50 > 0 \Longrightarrow \text{Minimum}$$

$$P''(10) = -50 < 0 \Longrightarrow \text{Maximum}$$

$P'' = -12s + 70 = 0$ when $s = \frac{35}{6}$. The maximum profit occurs when $s = 10$ (or \$10,000) and the point of diminishing returns occurs at $s = \frac{35}{6}$ (or \$5833.33).

19. Let x = number of units purchased, p = price per unit, and P = profit.

$$p = 90 - (0.10)(x - 100)$$

$$= 100 - 0.10x, \quad x \geq 100$$

$$P = xp - C$$

$$= x(100 - 0.10x) - 60x$$

$$= 40x - 0.10x^2$$

$$P' = 40 - 0.20x$$

$P' = 0$ when $x = 200$ radios.

21. $(40, 300), (45, 275)$

$$\text{Slope} = \frac{275 - 300}{45 - 40} = -5$$

$$x - 300 = -5(p - 40)$$

$$x = -5p + 500$$

$$R = xp = (-5p + 500)p = -5p^2 + 500p$$

$$R' = -10p + 500 = 0 \Longrightarrow p = \$50$$

23. Let T be the total cost.

$$T = 8(5280)\sqrt{x^2 + \frac{1}{4}} + 6(5280)(6 - x)$$

$$= 2(5280)\left[4\sqrt{x^2 + \frac{1}{4}} + 18 - 3x\right]$$

$$= 2(5280)\left[2\sqrt{4x^2 + 1} + 18 - 3x\right]$$

$$\frac{dT}{dx} = 2(5280)\left[\frac{2(8x)}{2\sqrt{4x^2 + 1}} - 3\right] = 2(5280)\left(\frac{8x - 3\sqrt{4x^2 + 1}}{\sqrt{4x^2 + 1}}\right)$$

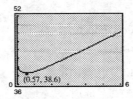

$dT/dx = 0$ when $8x = 3\sqrt{4x^2 + 1}$ which implies that

$$64x^2 = 9(4x^2 + 1)$$

$$28x^2 = 9$$

$$x^2 = \frac{9}{28}$$

$$x = \frac{3}{2\sqrt{7}} \approx 0.57 \text{ mile.}$$

25. $C = \left(\dfrac{v^2}{600} + 10\right)\left(\dfrac{110}{v}\right) = \dfrac{11}{60}v + \dfrac{1100}{v}$

$C' = \dfrac{11}{60} - \dfrac{1100}{v^2} = \dfrac{11v^2 - 66{,}000}{60v^2}$

$C' = 0$ when $v = \sqrt{\frac{66{,}000}{11}} \approx 77.46$ mph.

27. Since $dp/dx = -3$, the price elasticity of demand is

$$\eta = \dfrac{p/x}{dp/dx} = \dfrac{(400 - 3x)/x}{-3} = 1 - \dfrac{400}{3x}.$$

When $x = 20$, we have

$$\eta = 1 - \dfrac{400}{3(20)} = -\dfrac{17}{3}.$$

Since $|\eta(20)| = \frac{17}{3} > 1$, the demand is elastic.

Elastic: $\left(0, \frac{200}{3}\right)$

Inelastic: $\left(\frac{200}{3}, \frac{400}{3}\right)$

29. $p = 300 - 0.2x^2 \cdot \dfrac{dp}{dx} = -0.4x$

$$\eta = \dfrac{p/x}{dp/dx} = \dfrac{(300 - 0.2x^2)/x}{-0.4x} = 0.5 - \dfrac{750}{x^2}$$

When $x = 30$,

$$\eta = 0.5 - \dfrac{750}{30^2} = -0.3\overline{3}.$$

Since $|\eta(30)| = \frac{1}{3} < 1$, the demand is inelastic.

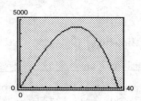

Elastic: $(0, 22.4)$

Inelastic: $(22.4, 38.7)$

31. Since $dp/dx = -200/x^3$, the price elasticity of demand is

$$\eta = \dfrac{p/x}{dp/dx} = \dfrac{[(100/x^2) + 2]/x}{-(200/x^3)} = -\dfrac{1}{2} - \dfrac{x^2}{100}.$$

When $x = 10$, we have

$$\eta = -\dfrac{1}{2} - \dfrac{(10)^2}{100} = -\dfrac{3}{2}.$$

Since $|\eta(10)| = \frac{3}{2} > 1$, the demand is elastic.

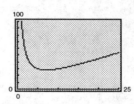

Elastic: $\left(5\sqrt{2}, \infty\right)$

Inelastic: $\left(0, 5\sqrt{2}\right)$

33. (a) If $p = 2$, $x = 12$, and p increases by 5%,

$$p = 2 + 2(0.05) = 2.1$$
$$x = 20 - 2(2.1)^2 = 11.18.$$

The percentage increase in x is

$$\dfrac{11.18 - 12}{12} = -\dfrac{41}{600} \approx -6.83\%.$$

(c) The exact elasticity of demand at $(2, 12)$ is

$$\eta = \left(\dfrac{p}{x}\right)\left(\dfrac{dx}{dp}\right)$$
$$= \left(\dfrac{p}{x}\right)(-4p)$$
$$= \left(\dfrac{2}{12}\right)[-4(2)]$$
$$= -\dfrac{4}{3}.$$

(b) At $(2, 12)$, the average elasticity of demand is

$$\dfrac{\%\ \text{change in } x}{\%\ \text{change in } p} = \dfrac{-41/600}{0.05} = \dfrac{-41}{30} \approx -1.37.$$

(d) The total revenue is

$$R = xp = (20 - 2p^2)p = 20p - 2p^3$$
$$\dfrac{dR}{dp} = 20 - 6p^2.$$

$dR/dp = 0$ when $6p^2 = 20$ which implies that

$$p = \sqrt{\dfrac{10}{3}} \approx \$1.83$$
$$x = 20 - 2\left(\sqrt{\dfrac{10}{3}}\right)^2 = \dfrac{40}{3}\ \text{units}.$$

35. (a) $\dfrac{dp}{dx} = -\dfrac{1}{2\sqrt{16-x}}$

$\eta = \dfrac{p/x}{dp/dx} = \dfrac{(16-x)^{1/2}/x}{-[1/(2\sqrt{16-x})]} = -\dfrac{2(16-x)}{x}$

When $x = 9$, $\eta = -\dfrac{14}{9}$.

(c) $\left| \eta\left(\dfrac{32}{3}\right) \right| = \left| -\dfrac{2[16-(32/3)]}{32/3} \right|$

$= \left| -\dfrac{2(16/3)}{32/3} \right| = |-1| = 1$

(b) $R = px = x(16-x)^{1/2}$

$R' = x\left(-\dfrac{1}{2\sqrt{16-x}} \right) + (16-x)^{1/2} = \dfrac{32-3x}{2\sqrt{16-x}}$

$R' = 0$ when $x = \dfrac{32}{3}$ and $p = \dfrac{4\sqrt{3}}{3}$.

37. $x = 800 - 40p$

$p = 20 - \dfrac{x}{40}$

$\dfrac{dp}{dx} = \dfrac{-1}{40}$

When $p = 5$, $x = 600$; $\eta = \dfrac{p/x}{dp/dx}$. Since $|\eta| < 1$, the demand is inelastic.

39. (a) $R'(t) = -439.14t^2 + 12{,}028.38t - 75{,}778.84$

$R'(t) = 0$ for $t \approx 9.82$ and $t \approx 17.57$. For the interval $8 \le t \le 16$, use $t = 9.82$ which corresponds to around 1990.

(c) 1990: $R(10) \approx \$30{,}477.83$ million

1996: $R(16) \approx \$60{,}796.31$ million

(b) Because $R(8) \approx 36{,}965$, $R(9.82) \approx 30{,}426$, and $R(16) \approx 60{,}796$, the revenue was greatest in 1996.

(d)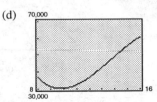

41. (a) Demand function   (b) Cost function

     (c) Revenue function   (d) Profit function

Section 3.6   Asymptotes

1. A horizontal asymptote occurs at $y = 1$ since

$\displaystyle\lim_{x\to\infty} \dfrac{x^2+1}{x^2} = 1$, $\displaystyle\lim_{x\to-\infty} \dfrac{x^2+1}{x^2} = 0$.

A vertical asymptote occurs at $x = 0$ since

$\displaystyle\lim_{x\to 0^-} \dfrac{x^2+1}{x^2} = \infty$, $\displaystyle\lim_{x\to 0^+} \dfrac{x^2+1}{x^2} = \infty$.

3. A horizontal asymptote occurs at $y = 1$ since

$\displaystyle\lim_{x\to\infty} \dfrac{x^2-2}{x^2-x-2} = 1$, $\displaystyle\lim_{x\to-\infty} \dfrac{x^2-2}{x^2-x-2} = 1$.

Vertical asymptotes occur at $x = -1$ and $x = 2$ since

$\displaystyle\lim_{x\to -1^-} \dfrac{x^2-2}{x^2-x-2} = -\infty$,

$\displaystyle\lim_{x\to -1^+} \dfrac{x^2-2}{x^2-x-2} = \infty$,

$\displaystyle\lim_{x\to 2^-} \dfrac{x^2-2}{x^2-x-2} = -\infty$,

$\displaystyle\lim_{x\to 2^+} \dfrac{x^2-2}{x^2-x-2} = \infty$.

5. There are no horizontal asymptotes. Vertical asymptotes occur at $x = -1$ and $x = 1$ since

$$\lim_{x\to-1^-} \frac{x^3}{x^2-1} = -\infty, \quad \lim_{x\to-1^+} \frac{x^3}{x^2-1} = \infty,$$

$$\lim_{x\to1^-} \frac{x^3}{x^2-1} = -\infty, \quad \lim_{x\to1^+} \frac{x^3}{x^2-1} = \infty.$$

7. A horizontal asymptote occurs at $y = \frac{1}{2}$ since

$$\lim_{x\to\infty} \frac{x^2-1}{2x^2-8} = \lim_{x\to-\infty} \frac{x^2-1}{2x^2-8} = \frac{1}{2}.$$

Vertical asymptotes occur at $x = \pm2$ since

$$\lim_{x\to2^-} \frac{x^2-1}{2x^2-8} = -20, \quad \lim_{x\to2^+} \frac{x^2-1}{2x^2-8} = \infty,$$

$$\lim_{x\to-2^-} \frac{x^2-1}{2x^2-8} = \infty, \quad \lim_{x\to-2^+} \frac{x^2-1}{2x^2-8} = -\infty.$$

9. The graph of f has a horizontal asymptote at $y = 3$. It matches graph (f).

11. The graph of f has a horizontal asymptote at $y = 0$. It matches graph (c).

13. The graph of f has a horizontal asymptote at $y = 5$. It matches graph (e).

15. $\lim_{x\to-2^-} \frac{1}{(x+2)^2} = \infty$

17. $\lim_{x\to3^+} \frac{x-4}{x-3} = -\infty$

19. $\lim_{x\to4^-} \frac{x^2}{x^2-16} = -\infty$

21. $\lim_{x\to0^-} \left(1 + \frac{1}{x}\right) = -\infty$

23. $\lim_{x\to\infty} \frac{2x-1}{3x+2} = \frac{2}{3}$

25. $\lim_{x\to\infty} \frac{3x}{4x^2-1} = 0$

27. $\lim_{x\to-\infty} \frac{5x^2}{x+3} = -\infty$

29. $\lim_{x\to\infty} \left(2x - \frac{1}{x^2}\right) = \lim_{x\to\infty} \frac{2x^3-1}{x^2} = \infty$

31. $\lim_{x\to-\infty} \left(\frac{2x}{x-1} + \frac{3x}{x+1}\right) = \lim_{x\to-\infty} \frac{5x^2-x}{x^2-1} = 5$

33.

x	10^0	10^1	10^2	10^3	10^4	10^5	10^6
$f(x)$	2.000	0.348	0.101	0.032	0.010	0.003	0.001

$$\lim_{x\to\infty} \frac{x+1}{x\sqrt{x}} = 0$$

35.

x	-10^6	-10^4	-10^2	10^0	10^2	10^4	10^6
$f(x)$	-2	-2	-1.9996	0.8944	1.9996	2	2

$$\lim_{x\to\infty} \frac{2x}{\sqrt{x^2+4}} = 2, \quad \lim_{x\to-\infty} \frac{2x}{\sqrt{x^2+4}} = -2$$

37. x-intercept: $(-2, 0)$

y-intercept: $(0, 2)$

Horizontal asymptote: $y = -1$

Vertical asymptote: $x = 1$

$$y' = \frac{3}{(1-x)^2}$$

No relative extrema

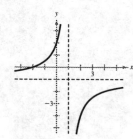

39. Intercept: $(0, 0)$

Horizontal asymptote: $y = 1$

$$f'(x) = \frac{18x}{(x^2 + 9)^2}$$

Relative minimum: $(0, 0)$

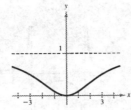

41. Intercept: $(0, 0)$

Horizontal asymptote: $y = 1$

Vertical asymptotes: $x = \pm 4$

$$g'(x) = \frac{-32x}{(x^2 - 16)^2}$$

Relative maximum: $(0, 0)$

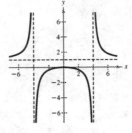

43. $x = \dfrac{4}{y^2}$

No intercepts

Horizontal asymptote: $y = 0$

Vertical asymptote: $x = 0$

No relative extrema

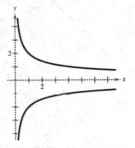

45. Intercept: $(0, 0)$

Horizontal asymptote: $y = -2$

$$y = \frac{2x}{1 - x}$$

Vertical asymptote: $x = 1$

$$y' = \frac{2}{(1 - x)^2}$$

No relative extrema

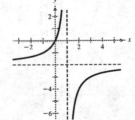

47. x-intercepts: $(\pm 1, 0)$

Horizontal asymptote: $y = 3$

$$y = \frac{3(x^2 - 1)}{x^2}$$

Vertical asymptote: $x = 0$

$$y' = \frac{6}{x^3}$$

No relative extrema

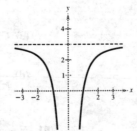

49. y-intercept: $\left(0, -\dfrac{1}{2}\right)$

Horizontal asymptote: $y = 0$

$$f(x) = \frac{1}{(x + 1)(x - 2)}$$

Vertical asymptotes: $x = -1, x = 2$

$$f'(x) = -\frac{2x - 1}{(x^2 - x - 2)^2}$$

Relative maximum: $\left(\dfrac{1}{2}, -\dfrac{4}{9}\right)$

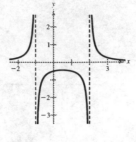

51. $g(x) = \dfrac{x^2 - x - 2}{x - 2} = x + 1$ for $x \neq 2$

x-intercept: $(-1, 0)$

y-intercept: $(0, 1)$

No asymptotes

No relative extrema

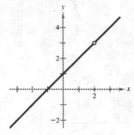

53. $y = \dfrac{2(x^2 - 3)}{x^2 - 2x + 1}$

x-intercepts: $(\pm\sqrt{3}, 0)$

y-intercept: $(0, -6)$

Vertical asymptote: $x = 1$

Horizontal asymptote: $y = 2$

$$y' = \frac{4(3 - x)}{(x - 1)^3}$$

Relative maximum: $(3, 3)$

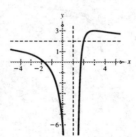

55. (a) $\overline{C} = 1.35 + \dfrac{4570}{x}$

When $x = 100, \overline{C} = \$47.05$. When $x = 1000$, $\overline{C} = \$5.92$.

(b) $\displaystyle\lim_{x \to \infty} \left(1.35 + \frac{4570}{x}\right) = 1.35 + 0 = \$1.35$

57. (a) $C(25) = \dfrac{528(25)}{100 - 25} = \$176$ million

(b) $C(50) = \dfrac{528(50)}{100 - 50} = \$528$ million

(c) $C(75) = \dfrac{528(75)}{100 - 75} = \$1584$ million

(d) $\displaystyle\lim_{p \to 100^-} \frac{528p}{100 - p} = \infty$

59. $\displaystyle\lim_{n \to \infty} \frac{\theta an - \theta a + b}{\theta n - \theta + 1} = \frac{\theta a}{\theta} = a$

61. $N = \dfrac{10(3 + 4t)}{1 + 0.1t} = \dfrac{40t + 30}{0.1t + 1} = \dfrac{400t + 300}{t + 10}$

(a) $N(5) = 153.3\overline{3} \approx 153$ elk

$N(10) = 215$ elk

$N(25) \approx 294.29 \approx 294$ elk

(b) $\displaystyle\lim_{t \to \infty} N(t) = 400$ elk

63.

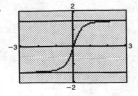

Horizontal asymptotes: $y = \pm\frac{3}{2}$

Section 3.7   Curve Sketching: A Summary

1. $y = -x^2 - 2x + 3 = -(x + 3)(x - 1)$

$y' = -2x - 2 = -2(x + 1)$

$y'' = -2$

Intercepts: $(0, 3), (1, 0), (-3, 0)$

Relative maximum: $(-1, 4)$

Concave downward

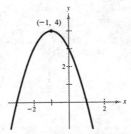

3. $y = x^3 - 4x^2 + 6$

$y' = 3x^2 - 8x = x(3x - 8)$

$y'' = 6x - 8 = 2(3x - 4)$

Relative maximum: $(0, 6)$

Relative minimum: $\left(\frac{8}{3}, -3.\overline{481}\right)$

Point of inflection: $\left(\frac{4}{3}, 1.\overline{259}\right)$

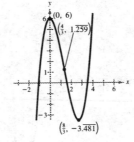

5. $y = 2 - x - x^3$

$y' = -1 - 3x^2$

$y'' = -6x$

No relative extrema

Point of inflection: $(0, 2)$

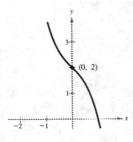

7. $y = 3x^3 - 9x + 1$

$y' = 9x^2 - 9$

$\quad = 9(x - 1)(x + 1)$

$y'' = 18x$

Relative maximum: $(-1, 7)$

Relative minimum: $(1, -5)$

Point of inflection: $(0, 1)$

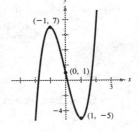

9. $y = 3x^4 + 4x^3 = x^3(3x + 4)$

$y' = 12x^3 + 12x^2 = 12x^2(x + 1)$

$y'' = 36x^2 + 24x = 12x(3x + 2)$

Intercepts: $(0, 0), \left(-\frac{4}{3}, 0\right)$

Relative minimum: $(-1, -1)$

Points of inflection: $(0, 0), \left(-\frac{2}{3}, -\frac{16}{27}\right)$

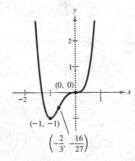

11. $y = (x + 1)(x - 2)(x - 5)$

$\quad = x^3 - 6x^2 + 3x + 10$

$\quad y' = 3x^2 - 12x + 3 = 3(x^2 - 4x + 1)$

$\quad y'' = 6x - 12 = 6(x - 2)$

Intercepts: $(0, 10), (-1, 0), (2, 0), (5, 0)$

Relative maximum: $\left(2 - \sqrt{3}, 10.392\right)$

Relative minimum: $\left(2 + \sqrt{3}, -10.392\right)$

Point of inflection: $(2, 0)$

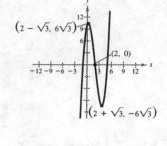

13. $y = x^4 - 8x^3 + 18x^2 - 16x + 5$

$\quad y' = 4x^3 - 24x^2 + 36x - 16$

$\quad\quad = 4(x^3 - 6x^2 + 9x - 4)$

$\quad\quad = 4(x - 4)(x - 1)^2$

$\quad y'' = 12x^2 - 48x + 36$

$\quad\quad = 12(x^2 - 4x + 3)$

$\quad\quad = 12(x - 1)(x - 3)$

Intercepts: $(0, 5), (1, 0), (5, 0)$

Relative minimum: $(4, -27)$

Points of inflection: $(1, 0), (3, -16)$

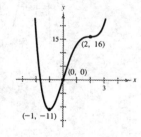

15. $y = x^4 - 4x^3 + 16x$

$\quad y' = 4x^3 - 12x^2 + 16 = 4(x + 1)(x - 2)^2$

$\quad y'' = 12x^2 - 24x = 12x(x - 2)$

Relative minimum: $(-1, -11)$

Points of inflection: $(0, 0), (2, 16)$

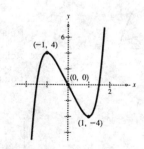

17. $y = x^5 - 5x$

$\quad y' = 5x^4 - 5 = 5(x + 1)(x - 1)(x^2 + 1)$

$\quad y'' = 20x^3$

Intercepts: $(0, 0), \left(\pm \sqrt[4]{5}, 0\right)$

Relative maximum: $(-1, 4)$

Relative minimum: $(1, -4)$

Point of inflection: $(0, 0)$

19. $y = |2x - 3|$

$\quad y' = (2)\dfrac{2x - 3}{|2x - 3|}$

$\quad y'' = 0$

Relative minimum: $\left(\dfrac{3}{2}, 0\right)$

No points of inflection

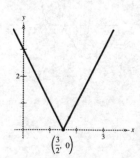

21. $y = \dfrac{x^2 + 2}{x^2 + 1}$

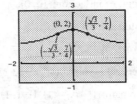

$y' = \dfrac{-2x}{(x^2 + 1)^2}$

$y'' = \dfrac{2(3x^2 - 1)}{x^2 + 1}$

Relative maximum: $(0, 2)$

Points of inflection: $x = \left(\dfrac{\sqrt{3}}{3}, \dfrac{7}{4}\right), \left(-\dfrac{\sqrt{3}}{3}, \dfrac{7}{4}\right)$

Horizontal asymptote: $y = 1$

23. $y = 3x^{2/3} - 2x$

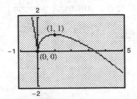

$y' = 2x^{-1/3} - 2 = 2(x^{-1/3} - 1)$

$y'' = -\dfrac{2}{3x^{4/3}}$

Intercepts: $(0, 0), \left(\dfrac{27}{8}, 0\right)$

Relative maximum: $(1, 1)$

Relative minimum: $(0, 0)$

25. $y = 1 - x^{2/3}$

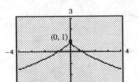

$y' = -\dfrac{2}{3}x^{-1/3} = -\dfrac{2}{3\sqrt[3]{x}}$

$y'' = \dfrac{2}{9}x^{-4/3} = \dfrac{2}{9\sqrt[3]{x^4}}$

Intercepts: $(0, 1), (\pm 1, 0)$

Relative maximum: $(0, 1)$

27. $y = x^{1/3} + 1$

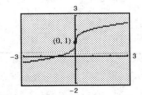

$y' = \dfrac{1}{3}x^{-2/3} = \dfrac{1}{3\sqrt[3]{x^2}}$

$y'' = -\dfrac{2}{9}x^{-5/3} = -\dfrac{2}{9\sqrt[3]{x^5}}$

Intercepts: $(0, 1), (-1, 0)$

Point of inflection: $(0, 1)$

29. $y = x^{5/3} - 5x^{2/3} = x^{2/3}(x - 5)$

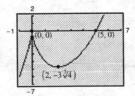

$y' = \dfrac{5}{3}x^{2/3} - \dfrac{10}{3}x^{-1/3} = \dfrac{5}{3}x^{-1/3}(x - 2)$

$y'' = \dfrac{10}{9}x^{-1/3} + \dfrac{10}{9}x^{-4/3} = \dfrac{10}{3}x^{-4/3}(x + 1)$

Intercepts: $(0, 0), (5, 0)$

Relative maximum: $(0, 0)$

Relative minimum: $\left(2, -3\sqrt[3]{4}\right)$

Point of inflection: $(-1, -4)$

31. $y = \dfrac{1}{x - 2} - 3$

$y' = -\dfrac{1}{(x - 2)^2}$

$y'' = \dfrac{2}{(x - 2)^3}$

Intercepts: $\left(\dfrac{7}{3}, 0\right), \left(0, -\dfrac{7}{2}\right)$

No relative extrema

Horizontal asymptote: $y = -3$

Vertical asymptote: $x = 2$

Domain: $(-\infty, 2), (2, \infty)$

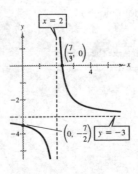

33. $y = \dfrac{2x}{x^2 - 1}$

$y' = \dfrac{-2(x^2 + 1)}{(x^2 - 1)^2}$

$y'' = \dfrac{4x(x^2 + 3)}{(x^2 - 1)^3}$

Point of inflection: $(0, 0)$

Intercept: $(0, 0)$

Horizontal asymptote: $y = 0$

Vertical asymptotes: $x = \pm 1$

Domain: $(-\infty, -1), (-1, 1), (1, \infty)$

Symmetry with respect to the origin

No relative extrema

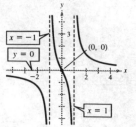

35. $y = x\sqrt{4 - x}$

$y' = \dfrac{8 - 3x}{2\sqrt{4 - x}}$

$y'' = \dfrac{3x - 16}{4(4 - x)^{3/2}}$

Intercepts: $(0, 0), (4, 0)$

Relative maximum: $\left(\dfrac{8}{3}, \dfrac{16}{3\sqrt{3}}\right)$

Domain: $(-\infty, 4]$

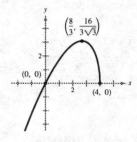

37. $y = \dfrac{x - 3}{x} = 1 - \dfrac{3}{x}$

$y' = \dfrac{3}{x^2}$

$y'' = \dfrac{-6}{x^3}$

Intercept: $(3, 0)$

Horizontal asymptote: $y = 1$

Vertical asymptote: $x = 0$

Domain: $(-\infty, 0), (0, \infty)$

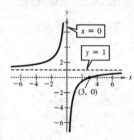

39. $y = \dfrac{x^3}{x^3 - 1}$

$y' = \dfrac{-3x^2}{(x^3 - 1)^2}$

$y'' = \dfrac{6x(2x^3 + 1)}{(x^3 - 1)^3}$

Points of inflection: $(0, 0), \left(-\dfrac{1}{\sqrt[3]{2}}, \dfrac{1}{3}\right)$

Intercept: $(0, 0)$

Horizontal asymptote: $y = 1$

Vertical asymptote: $x = 1$

Domain: $(-\infty, 1), (1, \infty)$

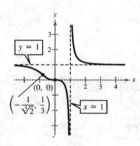

41. Since the graph rises as $x \to -\infty$ and falls as $x \to \infty$, a must be negative.

$$f(x) = -x^3 + x^2 + x + 1$$

(Solution not unique.)

43. Since the graph falls as $x \to -\infty$ and rises as $x \to \infty$, a must be positive.

$$f(x) = x^3 + 1$$

(Solution not unique.)

45. Since $f'(x) = 2$, the graph of f is a line with a slope of 2.

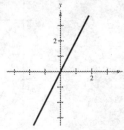

47. Since $f''(x) = 2$, the graph of f' is a line with a slope of 2, and the graph of f is a parabola opening upward.

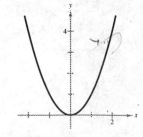

49.

x	$-\infty < x < -1$	$-1 < x < 0$	$0 < x < \infty$
$f'(x)$	$+$	$-$	$+$
$f(x)$	Increasing	Decreasing	Increasing

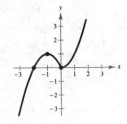

Relative maximum: $(-1, f(-1))$

Relative minimum: $(0, f(0)) = (0, 0)$

Intercepts: $(-2, 0), (0, 0)$

51. (a) $C = 1.20\left(\dfrac{100}{500/s}\right) + 9\left(\dfrac{100}{s}\right)$

$= 0.24s + \dfrac{900}{s}, \quad 40 \leq s \leq 65$

(b) $C' = 0.24 - \dfrac{900}{s^2}$

$= 0$ when $s \approx 61.2$ mph.

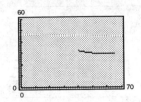

53. (a)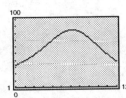

(b) $T(1) \approx 32.9°$

$T(12) \approx 37.8°$

$T(7.2) \approx 84.2°$

The maximum is $84.2°$ at $t = 7.2$.
The minimum is $32.9°$ at $t = 1$.

(c) The model says that the average daily high temperature is lowest in January and highest in July.

Section 3.8   Differentials and Marginal Analysis

1. $dy = 6x\, dx$

3. $dy = 3(4x - 1)^2(4)\, dx$

$= 12(4x - 1)^2\, dx$

5. $dy = \dfrac{1}{2}(x^2 + 1)^{-1/2}(2x)\, dx = \dfrac{x}{\sqrt{x^2 + 1}}\, dx$

7. $f(x) = 5x^2 - 1, \quad x = 1, \quad \Delta x = 0.01$

$\Delta y = f(x + \Delta x) - f(x)$

$= [5(1.01)^2 - 1] - [5(1)^2 - 1]$

$= 0.1005$

9. $f(x) = \dfrac{4}{x^{1/3}}, \quad x = 1, \quad \Delta x = 0.01$

$\Delta y = f(x + \Delta x) - f(x)$

$= \dfrac{4}{(1.01)^{1/3}} - \dfrac{4}{1}$

≈ -0.013245

11. $dy = 3x^2\, dx = 0.3$

$\Delta y = (1.1)^3 - 1^3 = 0.331$

13. $dy = 4x^3\, dx = -0.04$

$\Delta y = [(-0.99)^4 + 1] - [(-1)^4 + 1] \approx -0.0394$

15. $dy = 2x\,dx$

$dx = \Delta x$	dy	Δy	$\Delta y - dy$	$\dfrac{dy}{\Delta y}$
1.0000	4.0000	5.0000	1.0000	0.8000
0.5000	2.0000	2.2500	0.2500	0.8889
0.1000	0.4000	0.4100	0.0100	0.9756
0.0100	0.0400	0.0401	0.0001	0.9975
0.0010	0.0040	0.0040	0.0000	0.9998

17. $dy = 5x^4\,dx$

$dx = \Delta x$	dy	Δy	$\Delta y - dy$	$\dfrac{dy}{\Delta y}$
1.0000	80.0000	211.0000	131.0000	0.3791
0.5000	40.0000	65.6562	25.6562	0.6092
0.1000	8.0000	8.8410	0.8410	0.9049
0.0100	0.8000	0.8080	0.0080	0.9901
0.0010	0.0800	0.0801	0.0001	0.9990

19. $f(x) = \dfrac{x}{x^2 + 1}$

$f'(x) = \dfrac{(x^2 + 1)(1) - x(2x)}{(x^2 + 1)^2} = \dfrac{1 - x^2}{(x^2 + 1)^2}$

$f'(0) = 1$

$y - 0 = 1(x - 0)$

$\qquad y = x \qquad\qquad$ Tangent line

$f(x + \Delta x) = f(0 + 0.01) = f(0.01) \approx 0.009999$

$f(x - \Delta x) = f(0 - 0.01) = f(-0.01) \approx -0.009999$

$y(x \pm \Delta x) = y(0 \pm 0.01) = y(\pm 0.01) = \pm 0.01$

21. $f(x) = 2x^3 - x^2 + 1$

$f'(x) = 6x^2 - 2x$

$f'(-2) = 24 + 4 = 28$

$y + 19 = 28(x + 2)$

$\qquad y = 28x + 37 \qquad$ Tangent line

$f(x + \Delta x) = f(-2 + 0.01) = f(-1.99) \approx -18.72$

$f(x - \Delta x) = f(-2 - 0.01) = f(-2.01) \approx -19.28$

$y(x + \Delta x) = y(-1.99) = -18.72$

$y(x - \Delta x) = y(-2.01) = -19.28$

23. $p = 75 - 0.25x, \quad dp = -0.25$

(a) $x = 7, \ \Delta x = 1$

$\Delta p = [75 - 0.25(8)] - [75 - 0.25(7)]$

$\qquad = -0.25 = dp$

(b) $x = 70, \quad \Delta x = 1$

$\Delta p = [75 - 0.25(71)] - [75 - 0.25(70)]$

$\qquad = -0.25 = dp$

25. $x = 12, \quad dx = \Delta x = 1$

$\Delta C \approx dC = (0.10x + 4)\,dx$

$\qquad = [0.10(12) + 4](1)$

$\qquad = \$5.20$

27. $x = 50, \quad dx = \Delta x = 1$

$\Delta P \approx dP = (-1.5x^2 + 2500)\,dx$

$\qquad = [-1.5(50)^2 + 2500](1)$

$\qquad = -1250$

29. $(150, 50), (120, 60)$

$$m = \frac{60 - 50}{120 - 150} = -\frac{1}{3}$$

$$p - 50 = -\frac{1}{3}(x - 150)$$

$$p = -\frac{1}{3}x + 100$$

$$R = xp = -\frac{1}{3}x^2 + 100x$$

When $x = 141$ and $dx = \Delta x = 1$, we have

$$\Delta R \approx dR = \left(-\frac{2}{3}x + 100\right)dx = \left[-\frac{2}{3}(141) + 100\right](1) = \$6.00.$$

$\boxed{R = -\frac{1}{3}x^2 + 100x}$

$\boxed{y = 6x + 6627}$

$\left(142, 7478\frac{2}{3}\right)$

dR ΔR

$(141, 7473)$

31. $(30{,}000, 25), (40{,}000, 20)$

$$m = \frac{20 - 25}{40{,}000 - 30{,}000} = \frac{-5}{10{,}000} = \frac{-1}{2000}$$

$$p - 25 = \frac{-1}{2000}(x - 30{,}000)$$

$$p = \frac{-1}{2000}x + 40$$

$$C = 275{,}000 + 17x$$

$$P = R - C = xp - C$$

$$= \left(\frac{-1}{2000}x^2 + 40x\right) - (275{,}000 + 17x)$$

$$= \frac{-1}{2000}x^2 + 23x - 275{,}000$$

$\boxed{y = -5x + 117{,}000}$

$(28{,}000, -23{,}000)$

$(28{,}001, -23{,}005)$

$\Delta R \{$ $dR \{$

$\boxed{P = -\frac{1}{2000}x^2 + 23x - 275{,}000}$

When $x = 28{,}000$ and $dx = \Delta x = 1$, we have $\Delta P \approx dP = \frac{-1}{1000}x + 23 = \frac{-1}{1000}(28{,}000) + 23 = -\$5.00.$

33. (a) $dA = 2x\,dx = 2x\Delta x$

$$\Delta A = (x + \Delta x)^2 - x^2$$

$$= 2x\Delta x + (\Delta x)^2$$

(b) See graph.

(c) $\Delta A - dA = (\Delta x)^2$

(See graph.)

$\Delta A - dA$

$\Delta x \{$

dA

$x \{$

x Δx

35. $A = \pi r^2,\quad dr = \Delta r = \frac{1}{8},\quad r = 10$

$$dA = 2\pi r \cdot dr = 2\pi(10)\left(\pm\frac{1}{8}\right) = \pm\frac{5}{2}\pi\ \text{in.}^2$$

Relative error $= \dfrac{dA}{A} = \dfrac{\pm\frac{5}{2}\pi}{100\pi} = \pm\dfrac{1}{40}$

37. Let $\Delta r = dr = 0.02$ inch.

$$V = \frac{4}{3}\pi r^3$$

$$dV = 4\pi r^2\,dr = 4\pi(6)^2(\pm 0.02) = \pm 2.88\pi\ \text{in.}^3$$

When $r = 6$, the relative error is

$$\frac{dV}{V} = \frac{4\pi r^2\,dr}{(4/3)\pi r^3} = \frac{3\,dr}{r} = \frac{3(\pm 0.02)}{6} = \pm 0.01.$$

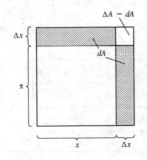

$5(x + \Delta x)^2 - (5x^2)$

$x^2 + 2\Delta x + \Delta x^2 - 5x^2$

$2\Delta x + \Delta x^2$

$\Delta x(2 + \Delta x)$

$\Delta x = 2$

Review Exercises for Chapter 3

1. $f(x) = -x^2 + 2x + 4$

$f'(x) = -2x + 2 = 0$ when $x = 1$.

Critical number: $x = 1$

3. $f(x) = x^{3/2} - 3x^{1/2}$ where $x \geq 0$.

$f'(x) = \dfrac{3}{2}x^{1/2} - \dfrac{3}{2}x^{-1/2} = \dfrac{3}{2}x^{-1/2}(x - 1) = \dfrac{3(x - 1)}{2\sqrt{x}}$

Critical numbers: $x = 0$ and $x = 1$

5. $f(x) = x^2 + x - 2$

$f'(x) = 2x + 1$

Critical number: $x = -\frac{1}{2}$

Increasing on $\left(-\frac{1}{2}, \infty\right)$

Decreasing on $\left(-\infty, -\frac{1}{2}\right)$

7. $h(x) = \dfrac{x^2 - 3x - 4}{(x - 3)}$

$h'(x) = \dfrac{x^2 - 6x + 13}{(x - 3)^2} > 0$ for all $x \neq 3$.

Increasing on $(-\infty, 3)$, $(3, \infty)$

9. $T = 0.036t^4 - 0.909t^3 + 5.874t^2 - 2.599t + 37.789$

$T' = 0.144t^3 - 2.727t^2 + 11.748t - 2.599$

$T' = 0$ for $t \approx 0.23$, $t \approx 6.15$.

(a) T increasing on $(0.23, 6.15)$

(b) T decreasing on $(0, 0.23)$, $(6.15, 12)$

(c) The maximum daily temperature is rising from early January to June.

(d)

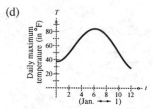

11. $f(x) = 4x^3 - 6x^2 - 2$

$f'(x) = 12x^2 - 12x = 12x(x - 1)$

Critical numbers: $x = 0$, $x = 1$

Relative maximum: $(0, -2)$

Relative minimum: $(1, -4)$

Interval	$(-\infty, 0)$	$(0, 1)$	$(1, \infty)$
Sign of f'	$+$	$-$	$+$
Conclusion	Increasing	Decreasing	Increasing

13. $g(x) = x^2 - 16x + 12$

$g'(x) = 2x - 16 = 2(x - 8)$

Critical number: $x = 8$

Increasing on $(8, \infty)$

Decreasing on $(-\infty, 8)$

Relative minimum: $(8, -32)$

Interval	$(-\infty, 8)$	$(8, \infty)$
Sign of g'	$-$	$+$
Conclusion	Decreasing	Increasing

15. $h(x) = 2x^2 - x^4$

$h'(x) = 4x - 4x^3$

$\qquad = 4x(1 - x^2) = 4x(1 - x)(1 + x)$

Critical numbers: $x = 0, 1, -1$

Relative maxima: $(-1, 1), (1, 1)$

Relative minimum: $(0, 0)$

Interval	$(-\infty, -1)$	$(-1, 0)$	$(0, 1)$	$(1, \infty)$
Sign of h'	$+$	$-$	$+$	$-$
Conclusion	Increasing	Decreasing	Increasing	Decreasing

17. $f(x) = \dfrac{6}{x^2 + 1}$

$f'(x) = \dfrac{-12x}{(x^2 + 1)^2}$

Interval	$(-\infty, 0)$	$(0, \infty)$
Sign of f'	$+$	$-$
Conclusion	Increasing	Decreasing

Critical number: $x = 0$

Relative maximum: $(0, 6)$

19. $h(x) = \dfrac{x^2}{x - 2}$

$h'(x) = \dfrac{x(x - 4)}{(x - 2)^2}$

Interval	$(-\infty, 0)$	$(0, 2)$	$(2, 4)$	$(4, 8)$
Sign of h'	$+$	$-$	$-$	$+$
Conclusion	Increasing	Decreasing	Decreasing	Increasing

Critical numbers: $x = 0, 4$

Discontinuity: $x = 2$

Relative maximum: $(0, 0)$

Relative minimum: $(4, 8)$

21. $f(x) = x^2 + 5x + 6, \quad [-3, 0]$

$f'(x) = 2x + 5$

Critical number: $x = -\dfrac{5}{2}$

x	$f(x)$	
-3	0	
0	6	Maximum
$-\frac{5}{2}$	$-\frac{1}{4}$	Minimum

23. $f(x) = x^3 - 12x + 1, \quad [-4, 4]$

$f'(x) = 3x^2 - 12 = 3(x - 2)(x + 2)$

Critical numbers: $x = \pm 2$

x	$f(x)$	
-4	-15	Minimum
4	17	Maximum
-2	17	Maximum
2	-15	Minimum

25. $f(x) = 3x^4 - 6x^2 + 2, \quad [0, 2]$

$f'(x) = 12x^3 - 12x = 12x(x - 1)(x + 1)$

Critical numbers: $x = 0, 1$

x	$f(x)$	
0	2	
2	26	Maximum
1	-1	Minimum

27. $f(x) = \dfrac{2x}{x^2 + 1}, \quad [-1, 2]$

$f'(x) = \dfrac{-2(x^2 - 1)}{(x^2 + 1)^2}$

Critical numbers: $x = 1, -1$

x	$f(x)$	
-1	-1	Minimum
2	$\frac{4}{5}$	
1	1	Maximum

29. $S = 2\pi r^2 + 50r^{-1}$

$S' = 4\pi r - 50r^{-2} = 4\pi r - \dfrac{50}{r^2}$

$S' = 0 \Longrightarrow 4\pi r = \dfrac{50}{r^2}$

$r^3 = \dfrac{50}{4\pi}$

$r \approx 1.58$ inches

[*Note:* You can check that $h = 2r$.]

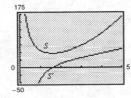

31. $f(x) = (x - 2)^3$

$f'(x) = 3(x - 2)^2$

$f''(x) = 6(x - 2)$

$f''(x) > 0$ for $x > 2$: concave upward on $(2, \infty)$

$f''(x) < 0$ for $x < 2$: concave downward on $(-\infty, 2)$

33. $g(x) = \frac{1}{4}(-x^4 + 8x^2 - 12)$

$g'(x) = -x^3 + 4x$

$g''(x) = -3x^2 + 4$

Because g changes concavity at $x \approx \pm 1.1547$, the points of inflection are $(-1.1547, -0.7778)$ and $(1.1547, -0.7778)$. The graph is concave downward on $(-\infty, -1.1547)$ and $(1.1547, \infty)$. The graph is concave upward on $(-1.1547, 1.1547)$.

35. $f(x) = \frac{1}{2}x^4 - 4x^3$

$f'(x) = 2x^3 - 12x^2$

$f''(x) = 6x^2 - 24x = 6x(x - 4)$

$f''(x) = 0$ when $x = 0, 4$. Because f changes concavity at $x = 0$ and $x = 4$, the points of inflection are $(0, 0)$ and $(4, -128)$.

37. $f(x) = x^3(x - 3)^2$

$f'(x) = x^2(5x - 9)(x - 3)$

$f''(x) = 2x(10x^2 - 36x + 27)$

Because f changes concavity at $x = 0$, $x = 1.0652$, and $x = 2.5348$, the points of inflection are $(0, 0)$, $(1.0652, 4.5244)$, and $(2.5348, 3.5246)$.

39. $f(x) = x^5 - 5x^3$

$f'(x) = 5x^4 - 15x^2 = 5x^2(x^2 - 3)$

$f''(x) = 20x^3 - 30x$

Critical numbers: $x = 0, \pm\sqrt{3}$

$f''(\sqrt{3}) > 0 \Rightarrow (\sqrt{3}, -6\sqrt{3})$ is a relative minimum.

$f''(-\sqrt{3}) < 0 \Rightarrow (-\sqrt{3}, 6\sqrt{3})$ is a relative maximum.

$f''(0) = 0 \Rightarrow$ test fails.

By the First-Derivative Test, $(0, 0)$ is not a relative extremum.

41. $f(x) = (x - 1)^3(x + 4)^2$

$f'(x) = 5(x + 4)(x + 2)(x - 1)^2$

$f''(x) = 10(x - 1)(2x^2 + 8x + 5)$

Critical numbers: $x = -4, -2, 1$

$f''(-4) < 0 \Rightarrow (-4, 0)$ is a relative maximum.

$f''(-2) > 0 \Rightarrow (-2, -108)$ is a relative minimum.

$f''(1) = 0 \Rightarrow$ test fails.

By the First-Derivative Test, $(1, 0)$ is not a relative extremum.

43. $R = \frac{1}{1500}(150x^2 - x^3), \quad 0 \le x \le 100$

$R' = \frac{1}{1500}(300x - 3x^2)$

$R'' = \frac{1}{1500}(300 - 6x)$

$R'' = 0$ when $x = 50$. The point of diminishing returns is $(50, 166\frac{2}{3})$.

45. $xy = 169 \quad$ (product is 169)

$\text{Sum} = S = x + y = x + \dfrac{169}{x}$

$S' = 1 - \dfrac{169}{x^2} = 0 \Rightarrow x = 13, \ y = 13$

47. $P = 0.0014x^2 - 0.1529x + 5.855, \quad 5 \le x \le 100$

$P' = 0.0028x - 0.1529$

$P' = 0$ when $x = 54.607$.

(a) $x = 54.607$ ($54,607) corresponds to the least percent.

(b) $x = 5$ ($5000) corresponds to the greatest percent.

(c)

49. $y' = -0.009x^2 + 0.274x + 0.458$

$y'' = -0.018x + 0.274 = 0 \Rightarrow x = \frac{137}{9} \approx 15.2$ years

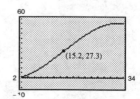

51. $s'(r) = -2cr = 0 \Rightarrow r = 0$

$s''(r) = -2c < 0$ for all r.

By the Second-Derivative Test, $r = 0$ yields a maximum.

53. $p = 15 - 0.1(n - 20) = 17 - 0.1n, \quad 20 \le n \le N$

Let x be the number of people over 20 in the group.

$$\text{Revenue} = R = (x + 20)[15 - 0.1(n - 20)]$$

$$= (x + 20)(15 - 0.1x), \quad x + 20 = n$$

$$R' = 13 - 0.2x = 0 \Rightarrow x = 65$$

Thus, $N = 65 + 20 = 85$ people would maximize revenue.

55. $\dfrac{dC}{dx} = -\dfrac{Qs}{x^2} + \dfrac{r}{2}$

$\dfrac{dC}{dx} = 0$ when $x^2 = \dfrac{2Qs}{r}$ or $x = \sqrt{\dfrac{2Qs}{r}}$.

Since $Q = 10{,}000$, $s = 4.5$ and $r = 5.76$, $x = 125$.

57. $\dfrac{dp}{dx} = -x$

$$\eta = \frac{p/x}{dp/dx} = \frac{\dfrac{100 - (1/2)x^2}{x}}{-x} = \frac{1}{2} - \frac{100}{x^2}$$

$$|\eta| = 1 = \left|\frac{1}{2} - \frac{100}{x^2}\right| \Rightarrow \frac{100}{x^2} = \frac{3}{2} \Rightarrow x^2 = \frac{200}{3} \Rightarrow x = \frac{10\sqrt{6}}{3} \approx 8.16$$

$0 < x < \dfrac{10\sqrt{6}}{3}$, elastic

$\dfrac{10\sqrt{6}}{3} < x < 10\sqrt{2}$, inelastic

$x = \dfrac{10\sqrt{6}}{3}$, unit elasticity

59. Since $x = 4$ is an infinite discontinuity, it is a vertical asymptote. Since

$$\lim_{x \to \infty} \frac{2x + 3}{x - 4} = \frac{2}{1} = 2$$

$y = 2$ is a horizontal asymptote.

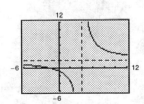

61. Since $x = 0$ is an infinite discontinuity, it is a vertical asymptote. Since

$$\lim_{x \to \infty} \left(\frac{1 - 3x}{x}\right) = -3$$

$y = -3$ is a horizontal asymptote.

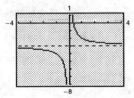

63. $f(x) = \dfrac{3}{x^2 - 5x + 4} - \dfrac{3}{(x - 4)(x - 1)}$

$x = 1$ and $x = 4$ are vertical asymptotes. $y = 0$ is the horizontal asymptote.

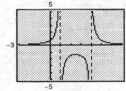

65. $\displaystyle\lim_{x \to 0^+} \left(x - \frac{1}{x^3}\right) = -\infty$

67. $\lim\limits_{x\to\infty} \dfrac{5x^2 + 3}{2x^2 - x + 1} = \lim\limits_{x\to\infty} \dfrac{5 + (3/x^2)}{2 - (1/x) + (1/x^2)} = \dfrac{5}{2}$

69. $\lim\limits_{x\to-\infty} \dfrac{3x^2}{x + 2} = -\infty$

71. $T = \dfrac{0.37s + 23.8}{s}, \quad 0 < s \le 120$

(a)

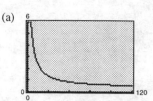

(b) $\lim\limits_{s\to\infty} T = \lim\limits_{s\to\infty} \dfrac{0.37s + 23.8}{s} = 0.37$

73. $f(x) = 4x - x^2 = x(4 - x)$

$f'(x) = 4 - 2x = -2(x - 2)$

$f''(x) = -2$

Intercepts: $(0, 0), (4, 0)$

Domain: $(-\infty, \infty)$

Range: $(-\infty, 4]$

Relative maximum: $(2, 4)$

Concave downward on $(-\infty, \infty)$

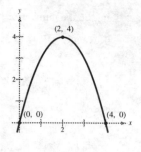

75. $f(x) = x\sqrt{16 - x^2}$

$f'(x) = \dfrac{16 - 2x^2}{\sqrt{16 - x^2}}$

$f''(x) = \dfrac{2x(x^2 - 24)}{(16 - x^2)^{3/2}}$

Domain: $[-4, 4]$

Range: $[-8, 8]$

Intercepts: $(0, 0), (4, 0), (-4, 0)$

Relative maximum: $\left(2\sqrt{2}, 8\right)$

Relative minimum: $\left(-2\sqrt{2}, -8\right)$

Point of inflection: $(0, 0)$

Concave upward on $(-4, 0)$

Concave downward on $(0, 4)$

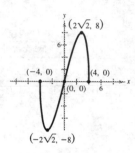

77. $f(x) = \dfrac{x + 1}{x - 1}$

$f'(x) = \dfrac{-2}{(x - 1)^2}$

$f''(x) = \dfrac{4}{(x - 1)^3}$

Domain: all real numbers except 1

Range: all real numbers except 1

Intercepts: $(-1, 0), (0, -1)$

Horizontal asymptote: $y = 1$

Vertical asymptote: $x = 1$

Concave upward on $(1, \infty)$

Concave downward on $(-\infty, 1)$

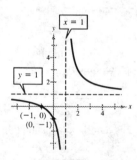

79. $f(x) = \dfrac{2x}{1 + x^2}$

$f'(x) = \dfrac{(1 + x^2)2 - 2x(2x)}{(1 + x^2)^2} = \dfrac{2 - 2x^2}{(1 + x^2)^2}$

$f''(x) = \dfrac{(1 + x^2)^2(-4x) - (2 - 2x^2)2(1 + x^2)2x}{(1 + x^2)^4}$

$= \dfrac{-4x(1 + x^2) - 4x(2 - 2x^2)}{(1 + x^2)^3}$

$= \dfrac{-12x + 4x^3}{(1 + x^2)^3}$

$= \dfrac{4x(x^2 - 3)}{(1 + x^2)^3}$

Domain: all real numbers

Intercept: $(0, 0)$

Asymptote: $y = 0$

Relative maxima: $(1, 1), (-1, -1)$

Points of inflection: $(0, 0),$

$$\left(\sqrt{3}, \frac{\sqrt{3}}{2}\right), \left(-\sqrt{3}, -\frac{\sqrt{3}}{2}\right)$$

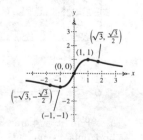

81. $y = 6x^2 - 5$

$dy = 12x \, dx$

83. $y = \dfrac{-5}{x^{1/3}} = -5x^{-1/3}$

$dy = \dfrac{5}{3}x^{-4/3} \, dx = \dfrac{5}{3x^{4/3}} \, dx$

85. $C = 40x^2 + 1225, \quad x = 10$

$dC = 80x \cdot dx = 80(10)(1) = 800$

87. $P = 0.003x^2 + 0.019x - 1200, \quad x = 750$

$dP = (0.006x + 0.019) \, dx = 4.519$

89. (a) $S = -0.0286t^2 + 0.6159t + 1.2962$

(c)

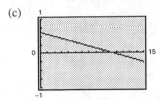

$S'(t) = -0.0572t + 0.6159$

(e) No, the model shows a maximum in 1990, when actually sales were greater in 1988 and 1994.

(b)

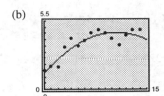

(d) The derivative shows that sales were decreasing from mid-1990 to 1994.

(f) $S'' < 0$ always. Therefore, the rate of change was decreasing.

91. The radius is 9 inches with a possible error of 0.025 inch.

$S = 4\pi r^2$

$dS = 8\pi r \, dr$

$V = \dfrac{4}{3}\pi r^3$

$dV = 4\pi r^2 \, dr$

When $r = 9$ and $dr = \pm 0.025$, we have the following.

$dS = 8\pi(9)(\pm 0.025) = \pm 1.8\pi \text{ in.}^2$

$dV = 4\pi(9)^2(\pm 0.025) = \pm 8.1\pi \text{ in.}^3$

93.

Quantity	Price	Total revenue	Marginal revenue
1	14.00	14.00	10.00
2	12.00	24.00	6.00
3	10.00	30.00	4.00
4	8.50	34.00	1.00
5	7.00	35.00	-2.00
6	5.50	33.00	

(a) $R = -1.43x^2 + 13.77x + 1.8$

(b) $\dfrac{dR}{dx} = -2.86x + 13.77$

x	1	2	3	4	5	6
$R'(x)$	10.91	8.05	5.19	2.33	-0.53	-3.39

(c) The revenue is maximized for 5 units of output. The quadratic model is maximized at $x = 4.8$.

Practice Test for Chapter 3

1. Find the critical numbers and the intervals on which f is increasing or decreasing for $f(x) = x^3 - 6x^2 + 5$.

2. Find the critical numbers and the intervals on which f is increasing or decreasing for $f(x) = 2x\sqrt{1 - x}$.

3. Find the relative extrema of $f(x) = x^4 - 32x + 3$.

4. Find the relative extrema of $f(x) = (x + 3)^{4/3}$.

5. Find the extrema of $f(x) = x^2 - 4x - 5$ on $[0, 5]$.

6. Find the points of inflection of $f(x) = 3x^4 - 24x + 2$.

7. Find the points of inflection of $f(x) = \dfrac{x^2}{1 + x^2}$.

8. Find two positive numbers whose product is 200 such that the sum of the first plus three times the second is a minimum.

9. Three rectangular fields are to be enclosed by 3000 feet of fencing, as shown in the accompanying figure. What dimensions should be used so that the enclosed area will be a maximum?

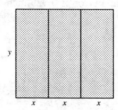

10. Find the number of units that produces a maximum revenue for $R = 400x^2 - 0.02x^3$.

11. Find the price per unit p that produces a maximum profit P given the cost function $C = 300x + 45,000$ and the demand function $p = 21,000 - 0.03x^2$.

12. Given the demand function $p = 600 - 0.02x^2$, find η (the price elasticity of demand) when $x = 100$.

13. Find $\displaystyle\lim_{x \to 3^-} \dfrac{x + 4}{x - 3}$.

14. Find $\displaystyle\lim_{x \to \infty} \dfrac{4x^3 - 9x^2 + 1}{1 - 2x^3}$.

15. Sketch the graph of $f(x) = \dfrac{x^2}{x^2 - 9}$.

16. Sketch the graph of $f(x) = \dfrac{x + 2}{x^2 + 5}$.

17. Sketch the graph of $f(x) = x^3 + 3x^2 + 3x - 1$.

18. Sketch the graph of $f(x) = |4 - 2x|$.

19. Sketch the graph of $f(x) = (2 - x)^{2/3}$.

20. Use differentials to approximate $\sqrt[3]{65}$.

Graphing Calculator Required

21. Graph $y = \dfrac{5x}{\sqrt{x^2 + 4}}$ on a graphing calculator and find any asymptotes that exist. Are there any relative extrema?

22. Graph $y = \dfrac{2x^4 - 5x + 1}{x^4 + 1}$ on a graphing calculator. Is it possible for a graph to cross its horizontal asymptote?

CHAPTER 4
Exponential and Logarithmic Functions

CHAPTER 4
Exponential and Logarithmic Functions

Section 4.1   Exponential Functions

Solutions to Odd-Numbered Exercises

1. (a) $5(5^3) = 5^4 = 625$

(b) $27^{2/3} = \left(\sqrt[3]{27}\right)^2 = 3^2 = 9$

(c) $64^{3/4} = (2^6)^{3/4} = 2^{9/2} = \sqrt{2^9} = 2^4\sqrt{2} = 16\sqrt{2}$

(d) $81^{1/2} = \sqrt{81} = 9$

(e) $25^{3/2} = \left(\sqrt{25}\right)^3 = 5^3 = 125$

(f) $32^{2/5} = \left(\sqrt[5]{32}\right)^2 = 2^2 = 4$

3. (a) $(5^2)(5^3) = 5^5 = 3125$

(b) $(5^2)(5^{-3}) = 5^{-1} = \dfrac{1}{5}$

(c) $(5^2)^2 = 5^4 = 625$

(d) $5^{-3} = \dfrac{1}{5^3} = \dfrac{1}{125}$

5. (a) $\dfrac{5^3}{25^2} = \dfrac{5^3}{(5^2)^2} = \dfrac{5^3}{5^4} = \dfrac{1}{5}$

(b) $(9^{2/3})(3)(3^{2/3}) = (3^2)^{2/3}(3)(3^{2/3})$

$$= (3^{4/3})(3^{5/3})$$

$$= 3^{9/3} = 3^3 = 27$$

(c) $[(25^{1/2})t^2] = [5 \cdot 5^2]^{1/3} = [5^3]^{1/3} = 5$

(d) $(8^2)(4^3) = (64)(64) = 4096$

7. (a) $e^3(e^4) = e^7$

(b) $(e^3)^4 = e^{12}$

(c) $(e^3)^{-2} = e^{-6} = \dfrac{1}{e^6}$

(d) $e^0 = 1$

9. $3^x = 81 = 3^4$

$x = 4$

11. $\left(\frac{1}{3}\right)^{x-1} = 27$

$3^{-(x-1)} = 3^3$

$-(x-1) = 3$

$-x + 1 = 3$

$x = -2$

13. $4^3 = (x+2)^3$

$4 = x + 2$

$x = 2$

15. $x = 8^{4/3} = \left(\sqrt[3]{8}\right)^4 = 2^4 = 16$

17. $e^{-3x} = e$

$-3x = 1$

$x = -\frac{1}{3}$

19. $e^{\sqrt{x}} = e^3$

$\sqrt{x} = 3$

$x = 9$

21. $x^{2/3} = \sqrt[3]{e^2} = e^{2/3}$

$x = e$

23. The graph of $f(x) = 3^x$ is an exponential curve with the following characteristics.

Passes through $(0, 1)$, $(1, 3)$, $\left(-1, \frac{1}{3}\right)$

Horizontal asymptote: x-axis

Therefore, it matches graph (e).

25. The graph of $f(x) = -3^x = (-1)(3^x)$ is an exponential curve with the following characteristics.

Passes through $(0, -1), (1, -3), \left(-1, -\frac{1}{3}\right)$

Horizontal asymptote: x-axis

Therefore, it matches graph (a).

27. The graph of $f(x) = 3^{-x} - 1 = \left(\frac{1}{3}\right)^x - 1$ is an exponential curve with the following characteristics.

Passes through $(0, 0), \left(1, -\frac{2}{3}\right), (-1, 2)$

Horizontal asymptote: $y = -1$

Therefore, it matches graph (d).

29.

x	-2	-1	0	1	2
$f(x)$	$\frac{1}{36}$	$\frac{1}{6}$	1	6	36

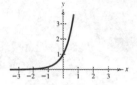

31.

x	-2	-1	0	1	2
$f(x)$	25	5	1	$\frac{1}{5}$	$\frac{1}{25}$

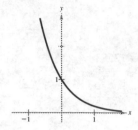

33.

x	-2	-1	0	1	2
y	$\frac{1}{81}$	$\frac{1}{3}$	1	$\frac{1}{3}$	$\frac{1}{81}$

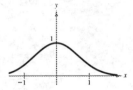

35.

x	-2	-1	0	1	2
y	$\frac{1}{9}$	$\frac{1}{3}$	1	$\frac{1}{3}$	$\frac{1}{9}$

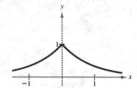

37. $s(t) = \dfrac{3^{-t}}{4} = \dfrac{1}{4(3^t)}$

t	-2	-1	0	1	2
$s(t)$	$\frac{9}{4}$	$\frac{3}{4}$	$\frac{1}{4}$	$\frac{1}{12}$	$\frac{1}{36}$

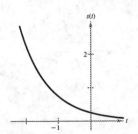

39.

x	-1	0	1	2	3
$h(x)$	0.050	0.135	0.368	1	2.718

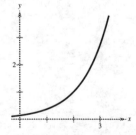

41.

t	-5	0	5	10	20
$N(t)$	1359.1	500	183.9	67.7	9.16

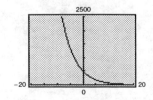

43.

x	-2	-1	0	1	2
$g(x)$	0.036	0.538	1	0.538	0.036

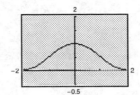

45. $P = 1000,\quad r = 0.03,\quad t = 10$

$$A = P\left(1 + \frac{r}{n}\right)^{nt} = 1000\left(1 + \frac{0.03}{n}\right)^{10n}$$

Continuous compounding: $A = Pe^{rt} = 1000e^{(0.03)(10)}$

n	1	2	4	12	365	Continuous
A	1343.92	1346.86	1348.35	1349.35	1349.84	1349.86

47. $P = 1000,\quad r = 0.03,\quad t = 40$

$$A = P\left(1 + \frac{r}{n}\right)^{nt} = 1000\left(1 + \frac{0.03}{n}\right)^{40n}$$

Continuous compounding: $A = Pe^{rt} = 1000e^{(0.03)(40)}$

n	1	2	4	12	365	Continuous
A	3264.04	3290.66	3305.28	3315.15	3319.95	3320.12

49. $A = Pe^{rt},\quad A = 100{,}000,\quad r = 0.04 \Rightarrow P = 100{,}000e^{-0.04t}$

t	1	10	20	30	40	50
P	96,078.94	67,032.00	44,932.90	30,119.42	20,189.65	13,533.53

51. $A = P\left(1 + \dfrac{r}{n}\right)^{nt},\quad A = 100{,}000,\quad r = 0.05,\quad n = 12 \Rightarrow P = \dfrac{100{,}000}{\left(1 + \dfrac{0.05}{12}\right)^{12}}$

t	1	10	20	30	40	50
P	95,152.82	60,716.10	36,864.45	22,382.66	13,589.88	8251.24

53. (a) $p(100) = 5000\left(1 - \dfrac{4}{4 + e^{-0.002(100)}}\right) \approx \$849.53$

   (b) $p(500) = 5000\left(1 - \dfrac{4}{4 - e^{-0.002(500)}}\right) \approx \$421.12$

   $\lim\limits_{x\to\infty} p = 0$

55. (a) $P\left(\dfrac{1}{2}\right) = 1 - e^{-1/6} \approx 0.1535 = 15.35\%$

   (b) $P(2) = 1 - e^{-2/3} \approx 0.4866 = 48.66\%$

   (c) $P(5) - 1 - e^{-5/3} \approx 0.8111 = 81.11\%$

57. (a)

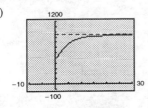

   (b) $\lim\limits_{t\to\infty} \dfrac{925}{1 + e^{-0.3t}} = \dfrac{925}{1 + 0} = 925$

   (c) $\lim\limits_{t\to\infty} \dfrac{1000}{1 + e^{-0.3t}} = 1000$

   Models of the form

   $$y = \frac{a}{1 + e^{-ct}},\quad c > 0,$$

   have a limit of a as $t \to \infty$.

59. $p = \dfrac{0.83}{1 + e^{-0.2n}}$

    (a) $n = 10 \Rightarrow P = 0.731$

    (b) $P = 0.75 \Rightarrow n \approx 11.19$ or 11 trials

    (c) $\displaystyle\lim_{n \to \infty} \dfrac{0.83}{1 + e^{-0.2n}} = \dfrac{0.83}{1 + 0} = 0.83$

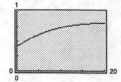

61.

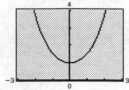

$\displaystyle\lim_{x \to \pm\infty} \dfrac{e^x + e^{-x}}{2} = \infty$

No horizontal asymptotes

Continuous on the entire real line

63.

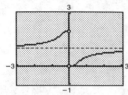

$\displaystyle\lim_{x \to \pm\infty} \dfrac{2}{1 + e^{1/x}} = \dfrac{2}{1 + 1} = 1$

Horizontal asymptote: $y = 1$

Discontinuous at $x = 0$

Section 4.2   Derivatives of Exponential Functions

1. $y' = 3e^{3x}$

    $y'(0) = 3$

3. $y' = -e^{-x}$

    $y'(0) = -1$

5. $y' = 4e^{4x}$

7. $y' = (-2x)e^{-x^2} = -2xe^{-x^2}$

9. $f'(x) = e^{-1/x^2} \cdot \dfrac{d}{dx}(-x^{-2})$

        $= 2x^{-3}e^{-1/x^2} = \dfrac{2}{x^3}e^{-1/x^2}$

11. $f'(x) = (x^2 + 1)4e^{4x} + (2x)e^{4x} = e^{4x}(4x^2 + 2x + 4)$

13. $f(x) = \dfrac{2}{e^x + e^{-x}} = 2(e^x + e^{-x})^{-1}$

     $f'(x) = -2(e^x + e^{-x})^{-2}(e^x - e^{-x}) = -\dfrac{2(e^x - e^{-x})}{(e^x + e^{-x})^2}$

15. $y' = xe^x + e^x + 4e^{-x}$

17. $y' = (-2 + 2x)e^{-2x + x^2}$

    $y'(0) = -2$

    $y - 1 = -2(x - 0)$

        $y = -2x + 1$

19. $y' = x^2(-e^{-x}) + 2xe^{-x} = xe^{-x}(-x + 2)$

    $y'(2) = 0$

    $y - \dfrac{4}{e^2} = 0(x - 2)$

       $y = \dfrac{4}{e^2} = 4e^{-2}$ (horizontal line)

21.
$$xe^x + 2ye^x = 0$$

$$xe^x + e^x + 2ye^x + 2\frac{dy}{dx}e^x = 0$$

$$x + 1 + 2y + 2\frac{dy}{dx} = 0$$

$$\frac{dy}{dx} = \frac{1}{2}(-x - 1 - 2y)$$

Alternatively,

$$x + 2y = 0$$

$$y = -\frac{x}{2}$$

$$\frac{dy}{dx} = -\frac{1}{2}.$$

These answers are the same since $2ye^x = -xe^x$ implies $2y = -x$.

23. $f'(x) = 6e^{3x} - 6e^{-2x}$

$f''(x) = 18e^{3x} + 12e^{-2x} = 6(3e^{3x} + 2e^{-2x})$

25. $f'(x) = -5e^{-x} + 10e^{-5x}$

$f''(x) = 5e^{-x} - 50e^{-5x}$

27. $f(x) - \dfrac{1}{2 - e^{-x}}$

$$f'(x) = \frac{-e^{-x}}{(2 - e^{-x})^2}$$

$$f''(x) = \frac{e^{-x}(2 + e^{-x})}{(2 - e^{-x})^3}$$

Horizontal asymptote to the right: $y = \dfrac{1}{2}$

Horizontal asymptote to the left: $y = 0$

Vertical asymptote when $2 = e^{-x} \Longrightarrow x \approx -0.693$

No relative extrema or inflection points

29. $f(x) = x^2 e^{-x}$

$f'(x) = -x^2 e^{-x} + 2xe^{-x} = xe^{-x}(2 - x)$

$f'(x) = 0$ when $x = 0$ and $x = 2$. Since $f''(0) > 0$ and $f''(2) < 0$, we have the following.

Relative minimum: $(0, 0)$

Relative maximum: $\left(2, \dfrac{4}{e^2}\right)$

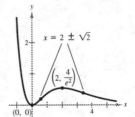

Since $f''(x) = 0$ when $x = 2 \pm \sqrt{2}$, the inflection points occur at $\left(2 - \sqrt{2}, 0.191\right)$ and $\left(2 + \sqrt{2}, 0.384\right)$.

x	-2	-1	0	1	2	3
$f(x)$	29.556	2.718	0	0.368	0.541	0.448

31. $f(x) = \dfrac{8}{1 + e^{-0.5x}}$

Horizontal asymptotes:

$\quad y = 8 \;(\text{as } x \to \infty)$

$\quad y = 0 \;(\text{as } x \to -\infty)$

No vertical asymptotes

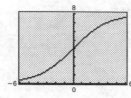

33. (a)

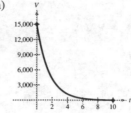

(b) $V'(1) = -9429e^{-0.6286} \approx -\$5028.84$ per year

(c) $V'(5) = -9429e^{-0.6286(5)} \approx -\$406.89$ per year

(d) In this model, the initial rate of depreciation is greater than in a linear model.

35. $p = \dfrac{300}{3 + 17e^{-1.57x}}$

(a)

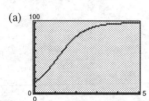

(b) When $x = 2, p = 80.3\%$.

(c) p' is increasing most rapidly when $x \approx 1.1$ (point of inflection).

37. $A' = 400^{0.08t}$

(a) $A'(1) \approx \$433.31$ per year

(b) $A'(10) \approx \$890.22$ per year

(c) $A'(50) \approx \$21,839.26$ per year

39. $V' = \dfrac{364.36648e^{-0.0458/t}}{t^2}$

When $t = 5$, $V' = 14.44$.

When $t = 10$, $V' = 3.63$.

When $t = 25$, $V' = 0.58$.

41. $f(x) = \dfrac{1}{\sigma\sqrt{\pi}}e^{-(x-\mu)^2/2\sigma^2} = \dfrac{1}{\sigma\sqrt{2\pi}}e^{-x^2/2\sigma^2}$

For larger σ, the graph becomes flatter.

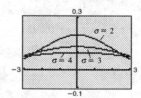

43. (a) $\dfrac{dh}{dt} = 50(-1.6)e^{-1.6t} - 20 = -80e^{-1.6t} - 20$

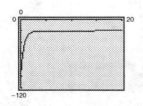

(b)

t	0	1	5	10	20
h'	-100	-36.15	-20.03	-20.00	-20.00

(c) The values in (b) are rates of descent in feet per second. As time increases, the rate is approximately constant at -20 ft/sec.

Section 4.3   Logarithmic Functions

1. $e^{0.6931\cdots} = 2$

3. $e^{-1.6094\cdots} = 0.2$

5. $\ln 1 = 0$

7. $\ln 0.0498 = -3$

9. The graph is a logarithmic curve that passes through the point $(1, 2)$ with a vertical asymptote at $x = 0$. Therefore, it matches graph (c).

11. The graph is a logarithmic curve that passes through the point $(-1, 0)$ with a vertical asymptote at $x = -2$. Therefore, it matches graph (b).

13.

x	1.5	2	3	4	5
y	-0.69	0	0.69	1.10	1.39

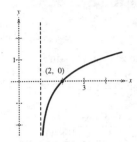

15.

x	0.25	0.5	1	3	5
y	-0.69	0	0.69	1.79	2.30

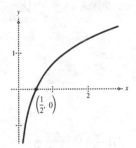

17.

x	0.5	1	2	3	4
y	-2.08	0	2.08	3.30	4.16

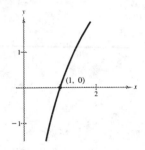

19. $g(x) = \ln \sqrt{x} = \frac{1}{2} \ln x$

$f(g(x)) = f\left(\frac{1}{2} \ln x\right) = e^{2(1/2 \ln x)} = e^{\ln x} = x$

$g(f(x)) = g(e^{2x}) = \frac{1}{2} \ln e^{2x} = \frac{1}{2}(2x) \ln e - x$

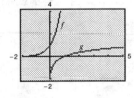

21. $f(g(x)) = f\left(\frac{1}{2} + \ln \sqrt{x}\right) = e^{2\left[(1/2) + \ln \sqrt{x}\right] - 1} = e^{2 \ln x^{1/2}} = e^{\ln x} = x$

$g(f(x)) = g(e^{2x-1}) = \frac{1}{2} + \ln \sqrt{e^{2x-1}} = \frac{1}{2} + \frac{1}{2} \ln e^{2x-1} = \frac{1}{2} + \frac{1}{2}(2x - 1) = x$

23. $\ln e^{x^2} = x^2$

25. $e^{\ln(5x+2)} = 5x + 2$

27. $e^{\ln \sqrt{x}} = \sqrt{x}$

29. (a) $\ln 6 = \ln(2 \cdot 3)$

$\qquad = \ln 2 + \ln 3 = 0.6931 + 1.0986 = 1.7917$

(b) $\ln \frac{3}{2} = \ln 3 - \ln 2 = 1.0986 - 0.6931 = 0.4055$

(c) $\ln 81 = \ln 3^4 = 4 \ln 3 = 4(1.0986) = 4.3944$

(d) $\ln \sqrt{3} = \left(\frac{1}{2}\right) \ln 3 = \left(\frac{1}{2}\right)(1.0986) = 0.5493$

31. $\ln \frac{2}{3} = \ln 2 - \ln 3$

33. $\ln xyz = \ln x + \ln y + \ln z$

35. $\ln\sqrt{x^2 + 1} = \ln(x^2 + 1)^{1/2} = \frac{1}{2}\ln(x^2 + 1)$

37. $\ln\dfrac{2x}{\sqrt{x^2 - 1}} = \ln 2x - \ln\sqrt{x^2 - 1}$

$\qquad\qquad = \ln 2 + \ln x - \dfrac{1}{2}\ln[(x + 1)(x - 1)]$

$\qquad\qquad = \ln 2 + \ln x - \dfrac{1}{2}[\ln(x + 1) + \ln(x - 1)]$

$\qquad\qquad = \ln 2 + \ln x - \dfrac{1}{2}\ln(x + 1) - \dfrac{1}{2}\ln(x - 1)$

39. $\ln\dfrac{3x(x + 1)}{(2x + 1)^2} = \ln[3x(x + 1)] - \ln(2x + 1)^2$

$\qquad\qquad = \ln 3 + \ln x + \ln(x + 1) - 2\ln(2x + 1)$

41. $\ln(x - 2) - \ln(x + 2) = \ln\dfrac{x - 2}{x + 2}$

43. $3\ln x + 2\ln y - 4\ln z = \ln x^3 + \ln y^2 - \ln z^4$

$\qquad\qquad = \ln\left(\dfrac{x^3 y^2}{z^4}\right)$

45. $3[\ln x + \ln(x + 3) - \ln(x + 4)] = 3\ln\dfrac{x(x + 3)}{x + 4}$

$\qquad\qquad = \ln\left[\dfrac{x(x + 3)}{x + 4}\right]^3$

47. $\dfrac{3}{2}[\ln x(x^2 + 1) - \ln(x + 1)] = \dfrac{3}{2}\ln\dfrac{x(x^2 + 1)}{x + 1}$

$\qquad\qquad = \ln\left[\dfrac{x(x^2 + 1)}{x + 1}\right]^{3/2}$

49. $2[\ln x - \ln(x + 1)] - 3[\ln x - \ln(x - 1)] = 2\ln x - 2\ln(x + 1) - 3\ln x + 3\ln(x - 1)$

$\qquad\qquad = \ln(x - 1)^3 - \ln(x + 1)^2 - \ln x$

$\qquad\qquad = \ln\dfrac{(x - 1)^3}{(x + 1)^2 x}$

51. $x = 4$

53. $x = e^0 = 1$

55. $x + 1 = \ln 4$

$\quad\;\; x = (\ln 4) - 1 \approx 0.3863$

57. $300e^{-0.2t} = 700$

$\qquad e^{-0.2t} = \dfrac{7}{3}$

$\qquad -0.2t = \ln 7 - \ln 3$

$\qquad\qquad t = \dfrac{\ln 7 - \ln 3}{-0.2} \approx -4.2365$

59. $2x\ln 5 = \ln 15$

$\qquad x = \dfrac{\ln 15}{2\ln 5} \approx 0.8413$

61. $500(1.07)^t = 1000$

$\qquad t\ln 1.07 = \ln 2$

$\qquad\qquad t = \dfrac{\ln 2}{\ln 1.07} \approx 10.2448$

63. $1000\left(1 + \dfrac{0.07}{12}\right)^{12t} = 3000$

$\qquad \left(1 + \dfrac{0.07}{12}\right)^{12t} = 3$

$\qquad 12t\ln\left(1 + \dfrac{0.07}{12}\right) = \ln 3$

$\qquad\qquad t = \dfrac{\ln 3}{12\ln[1 + (0.07/12)]} \approx 15.7402$

65. $P = 1000, \quad r = 0.05, \quad A = 2000$

(a) $2000 = 1000(1 + 0.05)^t$

$$t = \frac{\ln 2}{\ln 1.05} \approx 14.2 \text{ years}$$

(b) $2000 = 1000\left(1 + \dfrac{0.05}{12}\right)^{12t}$

$$t \approx \frac{\ln 2}{12 \ln 1.00417} \approx 13.88 \text{ years}$$

(c) $2000 = 1000\left(1 + \dfrac{0.05}{365}\right)^{365t}$

$$t \approx \frac{\ln 2}{365 \ln 1.000137} \approx 13.87 \text{ years}$$

(d) $2000 = 1000e^{0.05t}$

$$t = \frac{\ln 2}{0.05} \approx 13.86 \text{ years}$$

67. $3P = Pe^{rt}$

$3 = e^{rt}$

$\ln 3 = rt$

$t = \dfrac{\ln 3}{r}$

r	2%	4%	6%	8%	10%	12%	14%
t	54.93	27.47	18.31	13.73	10.99	9.16	7.85

69. $P = 243{,}000e^{0.01737t}$

$400{,}000 = 243{,}000e^{0.01737t}$

$\dfrac{400}{243} = e^{0.01737t}$

$\ln\left(\dfrac{400}{243}\right) = 0.01737t$

$t = \dfrac{\ln(400/243)}{0.01737} \approx 28.69 \text{ years}$

The city will have a population of 400,000 in the year 1999 (1970 + 29). Alternatively, you could use a graphing utility to find the point of intersection of $y_1 = 243{,}000e^{0.01737t}$ and $y_2 = 400{,}000$, which is $t \approx 28.69$.

71. $0.32 \times 10^{-12} = 10^{-12}\left(\dfrac{1}{2}\right)^{t/5700}$

$0.32 = \left(\dfrac{1}{2}\right)^{t/5700}$

$\ln 0.32 = \dfrac{t}{5700} \ln \dfrac{1}{2}$

$t = \dfrac{5700 \ln 0.32}{\ln(1/2)} \approx 9370 \text{ years}$

73. $0.22 \times 10^{-12} = 10^{-12}\left(\dfrac{1}{2}\right)^{t/5700}$

$0.22 = \left(\dfrac{1}{2}\right)^{t/5700}$

$\ln 0.22 = \dfrac{t}{5700} \ln \dfrac{1}{2}$

$t = \dfrac{5700 \ln 0.22}{\ln(1/2)} \approx 12{,}451 \text{ years}$

75. (a) $S(0) = 80 - 14 \ln 1 = 80$

(b) $S(4) = 80 - 14 \ln 5 \approx 57.5$

(c) $46 = 80 - \ln(t + 1) \ln(t + 1) = \frac{34}{14} \Longrightarrow \approx 10 \text{ months}$

77.

x	y	$\dfrac{\ln x}{\ln y}$	$\ln \dfrac{x}{y}$	$\ln x - \ln y$
1	2	0.0000	-0.6931	-0.6931
3	4	0.7925	-0.2877	-0.2877
10	5	1.4307	0.6931	0.6931
4	0.5	-2.0000	2.0794	2.0794

79.

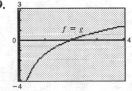

The graphs appear to be identical.

Section 4.4   Derivatives of Logarithmic Functions

1. $y = \ln x^3 = 3 \ln x$

$$y' = \frac{3}{x}$$

$$y'(1) = 3$$

3. $y = \ln x^2 = 2 \ln x$

$$y' = \frac{2}{x}$$

$$y'(1) = 2$$

5. $y = \ln x^2 = 2 \ln x$

$$y' = \frac{2}{x}$$

7. $f(x) = \ln 2x$

$$f'(x) = \frac{2}{2x} = \frac{1}{x}$$

9. $y = \ln \sqrt{x^4 - 4x} = \frac{1}{2} \ln(x^4 - 4x)$

$$y' = \frac{1}{2}\left(\frac{4x^3 - 4}{x^4 - 4x}\right) = \frac{2(x^3 - 1)}{x(x^3 - 4)}$$

11. $y = \frac{1}{2}(\ln x)^6$

$$y' = \frac{1}{2} \cdot 6(\ln x)^5 \cdot \frac{1}{x} = \frac{3}{x}(\ln x)^5$$

13. $y = x^2 \ln x$

$$y' = x^2\left(\frac{1}{x}\right) + 2x \ln x = x + 2x \ln x = x(1 + \ln x^2)$$

15. $y = \ln x^2 + \ln(x + 1)^3 = 2 \ln x + 3 \ln(x + 1)$

$$y' = 2\left(\frac{1}{x}\right) + 3\left(\frac{1}{x + 1}\right) = \frac{5x + 2}{x(x + 1)}$$

17. $y = \ln x - \ln(x + 1)$

$$y' = \frac{1}{x} - \frac{1}{x + 1} = \frac{1}{x(x + 1)}$$

19. $y = \ln\left(\frac{x - 1}{x + 1}\right)^{1/3} = \frac{1}{3}[\ln(x - 1) - \ln(x + 1)]$

$$y' = \frac{1}{3}\left[\frac{1}{x - 1} - \frac{1}{x + 1}\right] = \frac{1}{3}\left[\frac{2}{x^2 - 1}\right] = \frac{2}{3(x^2 - 1)}$$

21. $y = \ln \frac{\sqrt{4 + x^2}}{x} = \frac{1}{2} \ln(4 + x^2) - \ln x$

$$y' = \frac{1}{2}\left(\frac{2x}{4 + x^2}\right) - \frac{1}{x} = -\frac{4}{x(4 + x^2)}$$

23. $g(x) = e^{-x} \ln x$

$$g'(x) = e^{-x}\left(\frac{1}{x}\right) + (-e^{-x}) \ln x = e^{-x}\left(\frac{1}{x} - \ln x\right)$$

25. $f(x) = \ln e^{x^2} = x^2$

$$f'(x) = 2x$$

27. $2^x = e^{x(\ln 2)}$

29. $\log_4 x = \frac{1}{\ln 4} \ln x$

31. $\log_2 4 = 2$ because $2^2 = 4$.

33. $\log_3 \frac{1}{2} = \frac{1}{\ln 3} \ln \frac{1}{2} \approx -0.63093$ (calculator)

35. $\log_{10} 31 = \frac{1}{\ln 10} \cdot \ln 31 \approx 1.49136$ (calculator)

37. $y = 3^x$

$$y' = (\ln 3)3^x$$

39. $f(x) = \log_2 x$

$$f'(x) = \frac{1}{\ln 2} \cdot \frac{1}{x} = \frac{1}{x \ln 2}$$

41. $h(x) = 4^{2x-3}$

$$h'(x) = (\ln 4)4^{2x-3}(2) = 2 \ln 4 \cdot 4^{2x-3}$$

43. $y = \log_{10}(x^2 + 6x)$

$$y' = \frac{1}{\ln 10} \frac{1}{x^2 + 6x}(2x + 6) = \frac{2x + 6}{(x^2 + 6x) \ln 10}$$

45. $y = x2^x$

$y' = x(\ln 2)2^x + 2^x = 2^x(1 + x \ln 2)$

47. $y' = x\left(\dfrac{1}{x}\right) + (1) \ln x = 1 + \ln x$

$y'(1) = 1$

$y - 0 = 1(x - 1)$

$y = x - 1$ Tangent line

49. $y' = \dfrac{1}{\ln 3}\dfrac{1}{3x + 7}(3)$

$y'\left(\dfrac{2}{3}\right) = \dfrac{1}{\ln 3}\dfrac{1}{9}(3) = \dfrac{1}{3 \ln 3}$

$y - 2 = \dfrac{1}{3 \ln 3}\left(x - \dfrac{2}{3}\right)$

$y = \dfrac{1}{3 \ln 3}x + 2 - \dfrac{2}{9 \ln 3}$

51. $x^2 - 3 \ln y + y^2 = 10$

$2x - 3\left(\dfrac{1}{y}\right)\dfrac{dy}{dx} + 2y\dfrac{dy}{dx} = 0$

$2x = \dfrac{dy}{dx}\left(\dfrac{3}{y} - 2y\right)$

$2x = \dfrac{dy}{dx}\left(\dfrac{3 - 2y^2}{y}\right)$

$\dfrac{2xy}{3 - 2y^2} = \dfrac{dy}{dx}$

53. $4x^3 + \ln y^2 + 2y = 2x$

$12x^2 + 2\dfrac{y'}{y} + 2y' = 2$

$\left(\dfrac{2}{y} + 2\right)y' = 2 - 12x^2$

$y' = \dfrac{2 - 12x^2}{(2/y) + 2}$

$= \dfrac{1 - 6x^2}{(1/y) + 1} = \dfrac{(1 - 6x^2)}{1 + y}y$

55. $f(x) = x \ln \sqrt{x} + 2x = \dfrac{1}{2}x \ln x + 2x$

$f'(x) = \dfrac{1}{2}x\left(\dfrac{1}{x}\right) + \dfrac{1}{2} \ln x + 2 = \dfrac{1}{2} \ln x + \dfrac{5}{2}$

$f''(x) = \dfrac{1}{2x}$

57. $f(x) = 5^x$

$f'(x) = (\ln 5)5^x$

$f''(x) = (\ln 5)(\ln 5)5^x(\ln 5)^2 5^x$

59. $\beta = 10 \log_{10} I - 10 \log_{10}(10^{-16})$

$\dfrac{d\beta}{dI} = \dfrac{10}{(\ln 10)I}$

For $I = 10^{-4}$,

$\dfrac{d\beta}{dI} = \dfrac{10}{(\ln 10)10^{-4}}$

$= \dfrac{10^5}{\ln 10} \approx 43{,}429.4$ decibels per watt per cm^2.

61. $f(x) = 1 + 2x \ln x$

$f'(x) = 2x\left(\dfrac{1}{x}\right) + 2 \ln x = 2 + 2 \ln x$

At $(1, 1)$, the slope of the tangent line is $f'(1) = 2$.

Tangent line:

$y - 1 = 2(x - 1)$

$y = 2x - 1$

63. $f(x) = \ln \dfrac{5(x + 2)}{x} = \ln 5 + \ln(x + 2) - \ln x$

$f'(x) = \dfrac{1}{x + 2} - \dfrac{1}{x}$

At $(-2.5, 0)$, the slope of the tangent line is

$f'(-2.5) = \dfrac{1}{-2.5 + 2} - \dfrac{1}{-2.5} = -2 + \dfrac{2}{5} = -\dfrac{8}{5}.$

Tangent line: $y - 0 = -\dfrac{8}{5}\left(x + \dfrac{5}{2}\right)$

$y = -\dfrac{8}{5}x - 4$

65. $f(x) = \log_2 x$

$$f'(x) = \frac{1}{\ln 2} \cdot \frac{1}{x}$$

At the point $(1, 0)$, the slope of the tangent line is

$$f'(1) = \frac{1}{\ln 2}.$$

Tangent line:

$$y - 0 = \frac{1}{\ln 2}(x - 1)$$

$$y = \frac{1}{\ln 2}x - \frac{1}{\ln 2}$$

67. $y = x - \ln x$

$$y' = 1 - \frac{1}{x} = \frac{x - 1}{x}$$

$y' = 0$ when $x = 1$.

$$y'' = \frac{1}{x^2}$$

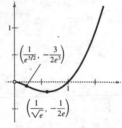

Since $y''(1) = 1 > 0$, there is a relative minimum at $(1, 1)$. Moreover, since $y'' > 0$ on $(0, \infty)$, it follows that the graph is concave upward on its domain and there are no inflection points.

69. The domain of the function

$$y = \frac{\ln x}{x}$$

is $(0, \infty)$.

$$y' = \frac{1 - \ln x}{x^2}$$

$y' = 0$ when $x = e$.

$$y'' = \frac{2 \ln x - 3}{x^3}$$

Since $y''(e) < 0$, it follows that $(e, 1/e)$ is a relative maximum. Since $y'' = 0$ when $2 \ln x - 3 = 0$ and $x = e^{3/2}$, there is an inflection point at $(e^{3/2}, 3/(2e^{3/2}))$.

71. $y = x^2 \ln x$

$$y' = x(1 + 2 \ln x)$$

$y' = 0$ when $x = e^{-1/2}$.

$$y'' = 3 + 2 \ln x$$

($x = 0$ is not in the domain.)

Since $y''(e^{-1/2}) > 0$, it follows that there is a relative minimum at $(1/\sqrt{e}, -1/(2e))$. Since $y'' = 0$ when $x = e^{-3/2}$, it follows that there is an inflection point at $(1/e^{3/2}, -3/(2e^3))$.

73. $x = \ln \frac{1000}{p} = \ln 1000 - \ln p$

$$\frac{dx}{dp} = 0 - \frac{1}{p} = -\frac{1}{p}$$

When $p = 10$,

$$\frac{dy}{dp} = -\frac{1}{10}.$$

75. $x = \frac{500}{\ln(p^2 + 1)} = 500[\ln(p^2 + 1)]^{-1}$

$$\frac{dx}{dp} = -500[\ln(p^2 + 1)]^{-2}\frac{1}{p^2 + 1}(2p)$$

$$= \frac{-1000p}{(p^2 + 1)[\ln(p^2 + 1)]^2}$$

If $p = 10$,

$$\frac{dx}{dp} = \frac{-10,000}{101[\ln(101)]^2} \approx -4.65.$$

77. $x = \ln \frac{1000}{p}$

$$e^x = \frac{1000}{p}$$

$$p = 1000e^{-x}$$

$$\frac{dp}{dx} = -1000e^{-x}$$

When $p = 10$, $x = \ln \frac{1000}{10} = \ln 100$ and $\frac{dp}{dx} = -1000e^{-\ln 100} = \frac{-1000}{100} = -10.$

Note that $\frac{dp}{dx}$ and $\frac{dx}{dp}$ are reciprocals of each other.

79. $C = 500 + 300x - 300 \ln x, \quad x \geq 1$

$\overline{C} = \text{average cost} = \dfrac{C}{x} = \dfrac{500}{x} + 300 - 300 \dfrac{\ln x}{x}$

Using a graphing utility, we determine that the minimum average cost is $\overline{C} \approx 279.15$ when $x \approx 14.39$.

$\overline{C}' = \dfrac{-500}{x^2} - 300\left[\dfrac{x(1/x) - \ln x}{x^2}\right] = \dfrac{-500}{x^2} - \dfrac{300}{x^2}(1 - \ln x)$

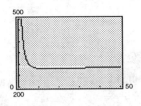

Setting $\overline{C}' = 0$,

$500 = -300(1 - \ln x)$

$\dfrac{5}{3} = \ln x - 1$

$\dfrac{8}{3} = \ln x$

$x = e^{8/3} \approx 14.39$

which confirms the graphical solution obtained above.

81. (a)

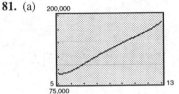

(b) $S(11) \approx \$160,316.21$ million

(c) $S'(t) = -8,001,622\left[\dfrac{t^2(1/t) - 2t \ln t}{t^4}\right] - 2(9,205,046.5)t^{-3} + 0.03e^t$

$S'(11) \approx \$10,783.74$ million per year

Section 4.5   Exponential Growth and Decay

1. Since $y = 2$ when $t = 0$, it follows that $C = 2$. Moreover, since $y = 3$ when $t = 4$, we have $3 = 2e^{4k}$ and

$k = \dfrac{\ln(3/2)}{4} \approx 0.1014.$

Thus, $y = 2e^{0.1014t}$.

3. Since $y = 4$ when $t = 0$, it follows that $C = 4$. Moreover, since $y = \frac{1}{2}$ when $t = 5$, we have $\frac{1}{2} = 4e^{5k}$ and

$k = \dfrac{\ln(1/8)}{5} \approx -0.4159.$

Thus, $y = 4e^{-0.4159t}$.

5. Using the fact that $y = 1$ when $t = 1$ and $y = 5$ when $t = 5$, we have $1 = Ce^k$ and $5 = Ce^{5k}$. From these two equations, we have $Ce^k = \left(\frac{1}{5}\right)Ce^{5k}$. Thus,

$k = \dfrac{\ln 5}{4} \approx 0.4024$

and we have $y = Ce^{0.4024t}$. Since $1 - Ce^{0.4024}$, it follows that $C \approx 0.6687$ and $y = 0.6687e^{0.4024t}$

7. $\dfrac{dy}{dt} = 2y, \ y = 10$ when $t = 0$.

$y = 10e^{2t}$

$\dfrac{dy}{dt} = 10(2)e^{2t} = 2(10e^{2t}) = 2y$

Exponential growth

9. $\dfrac{dy}{dt} = -4y$, $y = 30$ when $t = 0$.

$$y = 30e^{-4t}$$

$$\dfrac{dy}{dt} = 30(-4)e^{-4t} = -4(30e^{-4t}) = -4y$$

Exponential decay

11. From Example 1, we have $y = 10e^{[\ln(1/2)/1620]t}$.

(a) When $t = 1000$,

$$y = 10e^{[\ln(1/2)/1620](1000)} \approx 6.519 \text{ grams.}$$

(b) When $t = 10{,}000$,

$$y = 10e^{[\ln(1/2)/1620](10{,}000)} \approx 0.139 \text{ gram.}$$

13. Since $y = Ce^{[\ln(1/2)/5730]t}$, we have

$$2 = Ce^{[\ln(1/2)/5730](10{,}000)} \Longrightarrow C \approx 6.705$$

which implies that the initial quantity is 6.705 grams. When $t = 1000$, we have

$$y = 6.705e^{[\ln(1/2)/5730](1000)} \approx 5.941 \text{ grams.}$$

15. Since $y = Ce^{[\ln(1/2)/24,360]t}$, we have

$$2.1 = Ce^{[\ln(1/2)/24,360](1000)} \Longrightarrow C \approx 2.161.$$

Thus, the initial quantity is 2.161 grams. When $t = 10{,}000$,

$$y = 2.161e^{[\ln(1/2)/24,360](10{,}000)} \approx 1.626 \text{ grams.}$$

17. $y = Ce^{[\ln(1/2)/1620t]}$ (See Example 1.)

When $t = 900$, $y = Ce^{[\ln(1/2)/1620 \cdot 900]} \approx 0.68C$. After 900 years, approximately 68% of the radioactive radium will remain.

19.

$$0.15C = Ce^{[\ln(1/2)/5730]t}$$

$$\ln 0.15 = \left[\dfrac{\ln(1/2)}{5730}\right]t$$

$$\dfrac{5730 \ln 0.15}{\ln 0.5} = t$$

$$t \approx 15{,}683 \text{ years}$$

21. The model is $y = Ce^{kt}$. Since $y = 150$ when $t = 0$, we have $C = 150$. Furthermore,

$$450 = 150e^{k5}$$

$$3 = e^{5k}$$

$$k = \dfrac{\ln 3}{5}.$$

Therefore, $y = 150e^{[(\ln 3)/5]t} \approx 150e^{0.2197t}$.

(a) When $t = 10$, $y = 150^{[(\ln 3)/5]10} = 1350$ bacteria.

(b) To find the time required for the population to double, solve for t.

$$300 = 150e^{[(\ln 3)/5]t}$$

$$2 = e^{[(\ln 3)/5]t}$$

$$\ln 2 = \dfrac{\ln 3}{5}t$$

$$t = \dfrac{5 \ln 2}{\ln 3} \approx 3.15 \text{ hours}$$

(c) No, the doubling time is always 3.15 hours.

23. Since $A = 1000e^{0.12t}$, the time to double is given by $2000 = 1000e^{0.12t}$ and we have

$$t = \dfrac{\ln 2}{0.12} \approx 5.776 \text{ years.}$$

Amount after 10 years: $A = 1000e^{1.2} \approx \$3320.12$

Amount after 25 years: $A = 1000e^{0.12(25)} \approx \$20{,}085.54$

25. Since $A = 750e^{rt}$ and $A = 1500$ when $t = 7.75$, we have

$$1500 = 750e^{7.75r}$$

$$r = \dfrac{\ln 2}{7.75} \approx 0.0894 = 8.94\%.$$

Amount after 10 years: $A = 750e^{0.0894(10)} \approx \$1833.67$

Amount after 25 years: $A = 750e^{0.0894(25)} \approx \$7009.86$

27. Since $A = 500e^{rt}$ and $A = 1292.85$ when $t = 10$, we have

$$1292.85 = 500e^{10r}$$

$$r = \dfrac{\ln(1292.85/500)}{10} \approx 0.095 = 9.5\%.$$

The time to double is given by

$$1000 = 500e^{0.095t}$$

$$t = \dfrac{\ln 2}{0.095} \approx 7.296 \text{ years.}$$

Amount after 25 years: $A = 500e^{0.095(25)} \approx \$5375.51$

29. (a) $P(1 + i)^t = P\left(1 + \dfrac{r}{n}\right)^{nt}$

$\sqrt[t]{(1 + i)^t} = \sqrt[t]{\left(1 + \dfrac{r}{n}\right)^{nt}}$

$1 + i = \left(1 + \dfrac{r}{n}\right)^n$

$i = \left(1 + \dfrac{r}{n}\right)^n - 1$

(b) If $r = 0.06$ and $n = 12$, then

$i = \left(1 + \dfrac{0.06}{12}\right)^{12} - 1 \approx 0.0617$ or 6.17%.

31.

Number of compoundings per year	4	12	365	Continuous
Effective yield	5.095%	5.116%	5.127%	5.127%

$n = 4$: $i = \left(1 + \dfrac{0.05}{4}\right)^4 - 1 \approx 0.05095 \approx 5.095\%$

$n = 12$: $i = \left(1 + \dfrac{0.05}{12}\right)^{12} - 1 \approx 0.05116 \approx 5.116\%$

$n = 365$: $i = \left(1 + \dfrac{0.05}{365}\right)^{365} - 1 \approx 0.05127 \approx 5.127\%$

Continuous: $i = e^{0.05} - 1 \approx 0.05127 \approx 5.127\%$

33. $2P = Pe^{rt}$

$2 = e^{rt}$

$\ln 2 = rt$

$t = \dfrac{\ln 2}{4} \approx \dfrac{0.6931}{r} \approx \dfrac{0.70}{r}$

If r is entered as a percentage and not as a decimal, then

$t \approx 100\left(\dfrac{0.70}{r}\right) = \dfrac{70}{r}$.

35. Let $t = 0$ correspond to 1987: $(0, 486.8)$, $(9, 1005.8)$

(a) $y = 486.8e^{kt}$

$1005.8 = 486.8e^{k(9)}$

$k = \dfrac{1}{9}\ln\left(\dfrac{1005.8}{486.8}\right) \approx 0.08063$

$y = 486.8e^{0.08063t}$

For 2000, $t = 13$ and $y \approx 1388.6$ million.

(b) $y - 486.8 = \dfrac{1005.8 - 486.8}{9 - 0}(t - 0)$

$y = 57.67t + 486.8$

For 2000, $t = 13$ and $y \approx 1236.5$ million.

(c)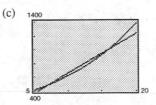

37. $S = Ce^{k/t}$

(a) Since $S = 5$ when $t = 1$, we have $5 = Ce^k$ and

$$\lim_{t \to \infty} Ce^{k/t} = C = 30.$$

Therefore, $5 = 30e^k$, $k = \ln(1/6) \approx -1.7918$, and $S = 30e^{-1.7918/t}$.

(b) $S(5) = 30e^{-1.7918/5} \approx 20.9646 \approx 20,965$ units

(c)

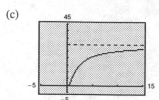

39. $N = 30(1 - e^{kt})$

Since $19 = 30(1 - e^{20k})$, it follows that

$$30e^{20k} = 11$$

$$k = \frac{\ln(11/30)}{20} \approx -0.0502$$

$$N = 30(1 - e^{-0.0502t})$$

$$25 = 30(1 - e^{-0.0502t})$$

$$e^{-0.0502t} = \frac{1}{6}$$

$$t = \frac{\ln 6}{0.0502} \approx 36 \text{ days.}$$

41. (a) Since $p = Ce^{kx}$ where $p = 45$ when $x = 1000$ and $p = 40$ when $x = 1200$, we have the following.

$$45 = Ce^{1000k} \text{ and } 40 = Ce^{1200k}$$

$$\ln 45 = \ln C + 1000k$$

$$\ln 40 = \ln C + 1200k$$

$$\ln 45 - \ln 40 = -200k$$

$$k = \frac{\ln(45/40)}{-200} \approx -0.0005889$$

Therefore, we have $45 = Ce^{1000(-0.0005889)}$ which implies that $C \approx 81.0915$ and $p = 81.0915e^{-0.0005889x}$.

(b) Since $R = xp = 81.0915xe^{-0.0005889x}$, we have the following.

$$R' = 81.0915[-0.0005889xe^{-0.0005889x} + e^{-0.0005889x}]$$

$$= 81.0915e^{-0.0005889x}[1 - 0.0005889x]$$

$$= 0$$

Since $R' = 0$ when $x = 1/0.0005889 \approx 1698$ units, we have $p = 81.0915e^{-0.0005889(1698)} \approx \$29.83$.

43. $A = Ve^{-0.04t}$

$$= 100,000e^{0.75\sqrt{t}}e^{-0.04t}$$

$$= 100,000e^{(0.75\sqrt{t} - 0.04t)}$$

$$A'(t) = 100,000\left(\frac{0.75}{2\sqrt{t}} - 0.04\right)e^{(0.75\sqrt{t} - 0.04t)} = 0$$

$$\frac{0.75}{2\sqrt{t}} = 0.04$$

$$\sqrt{t} = \frac{0.75}{(0.04)(2)} = 9.375$$

$$t = 87.89 \approx 88$$

The timber should be harvested in 2078 to maximize the present value.

45. (a) If $I_0 = 1$, then we have

$$R = \frac{\ln I}{\ln 10} \text{ and } 8.3 = \frac{\ln I}{\ln 10}.$$

Therefore, $I = e^{8.3 \ln 10} \approx 199,526,231.5$.

(b) $2R = \dfrac{\ln I}{\ln 10}$ implies that $I = e^{2R \ln 10} = (e^{R \ln 10})^2$. The intensity is squared if R is doubled.

(c) $\dfrac{dR}{dI} = \dfrac{1}{\ln 10}\left(\dfrac{1}{I}\right) = \dfrac{1}{I \ln 10}$

Review Exercises for Chapter 4

1. $4(4^4) = 4(256) = 1024$

3. $\left(\frac{1}{5}\right)^4 = \frac{1}{625}$

5. $\left(\frac{25}{4}\right)^0 = 1$

7. $\frac{6^3}{36^2} = \frac{6 \cdot 6^2}{36 \cdot 6^2} = \frac{6}{36} = \frac{1}{6}$

9. $5^x = 625$
$x = 4$

11. $e^{-1/2} = e^{x-1}$
$-\frac{1}{2} = x - 1$
$x = \frac{1}{2}$

13. $y = 978.665(1.0954)^t, \quad 5 \le t \le 55$

For 1950, $t = 10$ and $y = 2434.2$ million shares.

For 1970, $t = 30$ and $y = 15{,}059.7$ million shares.

For 1990, $t = 50$ and $y = 93{,}168.9$ million shares.

15. $f(x) = 9^{x/2}$

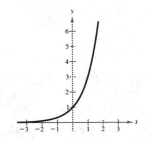

17. $f(x) = \left(\frac{1}{2}\right)^{2x} + 4$

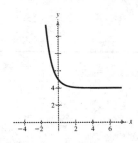

19. $p = 12{,}500 - \dfrac{10{,}000}{2 + e^{-0.001x}}$

$\lim_{x \to \infty} p = 12{,}500 - \dfrac{10{,}000}{2 + 0} - 7500$

21. $f(x) = 2e^{x-1}$
$f(2) = 2e^{2-1} = 2e \approx 5.4366$

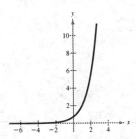

23. $g(t) = 12e^{-0.2t}$
$g(17) = 12e^{-0.2(17)} \approx 0.4005$

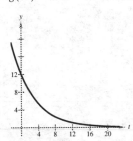

25. (a) $P = \dfrac{10{,}000}{1 + 19e^{-t/5}}, \quad t \ge 0$

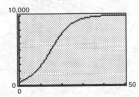

(b) When $t = 4$, $P \approx 1049$ fish.

(c) Yes, P approaches 10,000 fish as $t \to \infty$.

(d) The population is increasing most rapidly at the inflection point, around $t = 15$ months ($P = 5000$).

27.

n	1	2	4	12	365	Continuous
A	1216.65	1218.99	1220.19	1221.00	1221.39	1221.40

$$A = P\left(1 + \frac{r}{n}\right)^{nt} = 1000\left(1 + \frac{0.04}{n}\right)^{5n}$$

$$A = Pe^{rt} = 1000e^{(0.04)5} \approx 1221.50 \text{ (Continuous)}$$

29. (a) $A = Pe^{rt} = 2000e^{0.05(10)} \approx \$3297.44$

(b) $A = P\left(1 + \frac{r}{n}\right)^{nt} = 2000\left(1 + \frac{0.06}{4}\right)^{4(10)} \approx \$3628.04$

Account (b) will be greater.

31. $A = 22.56 + \dfrac{1}{0.299 + 55.56e^{-0.4115t}}, \quad 0 \le t \le 20$

1970 $(t = 0)$: $A \approx 22.58$ years

1980 $(t = 10)$: $A \approx 23.39$ years

1990 $(t = 20)$: $A \approx 25.75$ years

33. $y = 4e^{x^2}$

$y' = 4e^{x^2}(2x) = 8xe^{x^2}$

35. $y = \dfrac{x}{e^{2x}}$

$y' = \dfrac{e^{2x}(1) - x2e^{2x}}{(e^{2x})^2} = \dfrac{1 - 2x}{e^{2x}}$

37. $y = \sqrt{4e^{4x}} = (4e^{4x})^{1/2} = 2e^{2x}$

$y' = 4e^{2x}$

39. $y = \dfrac{5}{1 + e^{2x}} = 5(1 + e^{2x})^{-1}$

$y' = -5(1 + e^{2x})^{-2}(2e^{2x}) = \dfrac{-10e^{2x}}{(1 + e^{2x})^2}$

41. $f(x) = 4e^{-x}$

No relative extrema

No inflection point

Horizontal asymptote: $y = 0$

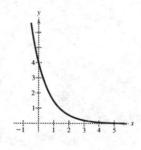

43. $f(x) = \dfrac{e^x}{x^2}$

$f'(x) = (x - 2)\dfrac{e^x}{x^3}$

$f''(x) = (x^2 - 4x + 6)\dfrac{e^x}{x^4}$

Relative minimum: $(2, 1.847)$

Horizontal asymptote: $y = 0$

Vertical asymptote: $x = 0$

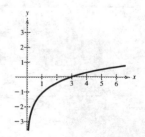

45. $\ln 12 \approx 2.4849$

$e^{2.4849} \approx 12$

47. $e^{1.5} \approx 4.4817$

$\ln 4.4817 \approx 1.5$

49. $y = \ln(4 - x)$

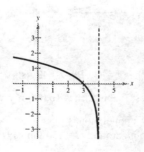

51. $y = \ln\dfrac{x}{3} = \ln x - \ln 3$

53. $\ln\sqrt{x^2(x-1)} = \frac{1}{2}\ln[x^2(x-1)]$

$$= \frac{1}{2}[\ln x^2 + \ln(x-1)]$$

$$= \ln x + \frac{1}{2}\ln(x-1)$$

55. $\ln\left(\dfrac{1-x}{3x}\right)^3 = 3\ln\left(\dfrac{1-x}{3x}\right)$

$$= 3[\ln(1-x) - \ln 3x]$$

$$= 3[\ln(1-x) - \ln 3 - \ln x]$$

57. $e^{\ln x} = 3$

   $x = 3$

59. $\ln 2x - \ln(3x-1) = 0$

$$\ln 2x = \ln(3x-1)$$

$$2x = 3x - 1$$

$$x = 1$$

61. $\ln x + \ln(x-3) = 0$

$$\ln[x(x-3)] = 0$$

$$x(x-3) = 1$$

$$x^2 - 3x - 1 = 0$$

$$x = \frac{3 \pm \sqrt{13}}{2}$$

$x = \dfrac{3 + \sqrt{13}}{2} \approx 3.3028$ is the only solution in the domain.

63. $9^{6x} - 27 = 0$

$$3^{12x} = 3^3$$

$$12x = 3$$

$$x = \frac{1}{4}$$

65. (a) $M = 100{,}000\left(\dfrac{\dfrac{0.08}{12}}{1 - \left(\dfrac{1}{(0.08/12)+1}\right)^{12t}}\right)$

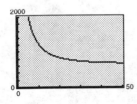

(b) A 30-year term has a smaller monthly payment, but takes more time to pay off than a 20-year term.

67. $f(x) = \ln 3x^2 = \ln 3 + 2\ln x$

$$f'(x) = \frac{2}{x}$$

69. $y = x\sqrt{\ln x}$

$$y' = \sqrt{\ln x} + \frac{1}{2}x(\ln x)^{-1/2}\frac{1}{x} = \sqrt{\ln x} + \frac{1}{2\sqrt{\ln x}}$$

71. $y = \dfrac{\ln x}{x^3}$

$$y' = \frac{x^3(1/x) - 3x^2 \cdot \ln x}{x^6} = \frac{1 - 3\ln x}{x^4}$$

73. $f(x) = \ln e^{-x} = -x^2$

$$f'(x) = -2x$$

75. $y = \ln(x + 3)$

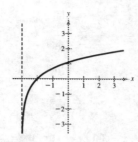

77. $y = \dfrac{\ln 10}{x + 2}$

$= \ln 10 - \ln(x + 2)$

No relative extrema or inflection points

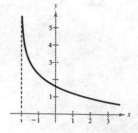

79. $\log_7 49 = \log_7 y^2 = 2\log_7 7 = 2$

81. $\log_{10} 1 = 0$

83. $\log_5 10 = \dfrac{\ln 10}{\ln 5} \approx 1.4307$

85. $\log_{16} 64 = \dfrac{\ln 64}{\ln 16} = 1.5$

87. $y = \log_3(2x - 1)$

$y' = \dfrac{1}{\ln 3} \cdot \dfrac{2}{2x - 1} = \dfrac{2}{(2x - 1)\ln 3}$

89. $y = \log_2 \dfrac{1}{x^2} = \log_2 1 - \log_2 x^2 = -2\log_2 x$

$y' = -2 \dfrac{1}{\ln 2} \cdot \dfrac{1}{x} = \dfrac{-2}{x \ln 2}$

91. $V = 20,000(0.75)^t$

(a) $V(2) = 11,250$

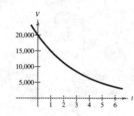

(b) $V'(t) = 20,000 \cdot \ln\left(\dfrac{3}{4}\right)(0.75)^t$

$V'(1) = -4315.23$ dollars/year

$V'(40) = -1820.49$ dollars/year

(c) $V = 5000 = 20,000(0.75)^t$

$\dfrac{1}{4} = (0.75)^t$

$t = \dfrac{\ln(1/4)}{\ln(3/4)} \approx 4.8$ years

93. $A = Ce^{kt} = 500e^{kt}$

$A = 500$ when $t = 0$.

$A = 300$ after 40 days.

$300 = 500e^{40k}$

$\dfrac{3}{5} = e^{40k}$

$40k = \ln\left(\dfrac{3}{5}\right)$

$k = \dfrac{\ln(3/5)}{40} \approx -0.01277$

$y = 500e^{0.01277t}$

95. $y = 50e^{kt}$

$42.031 = 50e^{7k}$

$k = \dfrac{1}{7}\ln\left(\dfrac{42.031}{50}\right) \approx -0.0248$

$25 = 50e^{-0.0248t}$

$\dfrac{1}{2} = e^{-0.0248t}$

$t = \dfrac{\ln(1/2)}{-0.0248} \approx 27.9$ years

97. Let $t = 0$ correspond to 1987.

$y = 78.1e^{kt}$

$2176.0 = 78.1e^{k(9)}$

$k = \dfrac{1}{9}\ln\left(\dfrac{2176.0}{78.1}\right) \approx 0.3697$

For 2000, $t = 13$ and $y \approx \$9547$ million.

Practice Test for Chapter 4

1. Evaluate each of the following expressions.

   (a) $27^{4/3}$ (b) $4^{-5/2}$ (c) $(8^{2/3})(64^{-1/3})$

2. Solve for x.

   (a) $4^{x+1} = 64$ (b) $x^{6/5} = 64$ (c) $(2x + 3)^{10} = 13^{10}$

3. Sketch the graph of (a) $f(x) = 3^x$, and (b) $g(x) = \left(\frac{4}{9}\right)^x$.

4. Find the amount in an account in which $2000 is invested for 7 years at 8.5% if the interest is compounded (a) annually, (b) monthly, and (c) continuously.

5. Differentiate $y = e^{3x^2}$.

6. Differentiate $y = e^{\sqrt[3]{x}}$.

7. Differentiate $y = \sqrt{e^x + e^{-x}}$.

8. Differentiate $y = x^3 e^{2x}$.

9. Differentiate $y = \dfrac{e^x + 3}{4x}$.

10. Write $\ln 5 = 1.6094 \ldots$ as an exponential equation.

11. Sketch the graph of (a) $y = \ln(x + 2)$, and (b) $y = \ln x + 2$.

12. Write the given expression as a single logarithm.

    (a) $\ln(3x + 1) - \ln(2x - 5)$ (b) $4 \ln x - 3 \ln y - \frac{1}{2} \ln z$

13. Solve for x.

    (a) $\ln x = 17$ (b) $5^{3x} = 2$

14. Differentiate $y = \ln(6x - 7)$.

15. Differentiate $y = \ln\left(\dfrac{x^3}{4x + 10}\right)$.

16. Differentiate $y = \ln \sqrt[3]{\dfrac{x}{x + 3}}$.

17. Differentiate $y = x^4 \ln x$.

18. Differentiate $y = \sqrt{\ln x + 1}$.

19. Find the exponential function $y = Ce^{kt}$ that passes through the following points.

   (a) $(0, 7), \left(4, \frac{1}{3}\right)$ (b) $\left(3, \frac{2}{3}\right), (8, 8)$

20. If \$5000 is invested in an account in which the interest rate of 12% is compounded continuously, find the time required for the investment to double.

Graphing Calculator Required

21. Use a graphing calculator to graph both $y = \ln\left[x^3 \sqrt{x + 3}\right]$ and $y = 3 \ln x + \frac{1}{2} \ln(x + 3)$ on the same set of axes. What do you notice about the graphs?

22. Graph the function

$$f(t) = \frac{4200}{7 + e^{-0.9t}}$$

and use the graph to find $\lim\limits_{t \to \infty} f(t)$ and $\lim\limits_{t \to -\infty} f(t)$.

C H A P T E R 5
Integration and Its Applications

CHAPTER 5
Integration and Its Applications

Section 5.1 Antiderivatives and Indefinite Integrals

Solutions to Odd-Numbered Exercises

1. $\dfrac{d}{dx}\left(\dfrac{3}{x^3} + C\right) = \dfrac{d}{dx}(3x^{-3} + C) = -9x^{-4} = \dfrac{-9}{x^4}$

3. $\dfrac{d}{dx}\left(x^4 + \dfrac{1}{x} + C\right) = 4x^3 - \dfrac{1}{x^2}$

5. $\dfrac{d}{dx}\left(\dfrac{4}{9}x^{9/2} + C\right) = \dfrac{4}{9} \cdot \dfrac{9}{2}x^{7/2} = 2x^3\sqrt{x}$

7. $\dfrac{d}{dx}\left(\dfrac{2(x^2 + 3)}{3\sqrt{x}} + C\right) = \dfrac{d}{dx}\left(\dfrac{2}{3}x^{2/3} + 2x^{-1/2} + C\right)$

$$= x^{1/2} - x^{-3/2} = \dfrac{x^2 - 1}{x^{3/2}}$$

9. $\displaystyle\int 6\,dx = 6x + C$

$\dfrac{d}{dx}[6x + C] = 6$

11. $\displaystyle\int 3t^2\,dt = t^3 + C$

$\dfrac{d}{dt}[t^3 + C] = 3t^2$

13. $\displaystyle\int 5x^{-3}\,dx = \dfrac{5x^{-2}}{-2} + C = \dfrac{-5}{2x^2} + C$

$\dfrac{d}{dx}\left[-\dfrac{5}{2}x^{-2} + C\right] = 5x^{-3}$

15. $\displaystyle\int du = u + C$

$\dfrac{d}{du}[u + C] = 1$

17. $\displaystyle\int x^{3/2}\,dx = \dfrac{2}{5}x^{5/2} + C$

$\dfrac{d}{dx}\left[\dfrac{2}{5}x^{5/2} + C\right] = x^{3/2}$

19.

Given	Rewrite	Integrate	Simplify
$\displaystyle\int \sqrt[3]{x}\,dx$	$\displaystyle\int x^{1/3}\,dx$	$\dfrac{x^{4/3}}{4/3} + C$	$\dfrac{3}{4}x^{4/3} + C$

21.

Given	Rewrite	Integrate	Simplify
$\displaystyle\int \dfrac{1}{x\sqrt{x}}\,dx$	$\displaystyle\int x^{-3/2}\,dx$	$\dfrac{x^{-1/2}}{-1/2} + C$	$-\dfrac{2}{\sqrt{x}} + C$

23.

Given	Rewrite	Integrate	Simplify
$\displaystyle\int \dfrac{1}{2x^3}\,dx$	$\dfrac{1}{2}\displaystyle\int x^{-3}\,dx$	$\dfrac{1}{2}\left(\dfrac{x^{-2}}{-2}\right) + C$	$-\dfrac{1}{4x^2} + C$

25. If $f'(x) = 2$, then $f(x) = 2x + C$. For example, $f(x) = 2x$ or $f(x) = 2x + 1$.

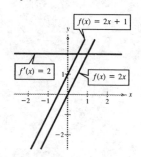

27. $\displaystyle\int (x^3 + 2)\,dx = \dfrac{x^4}{4} + 2x + C$

$\dfrac{d}{dx}\left[\dfrac{x^4}{4} + 2x + C\right] = x^3 + 2$

29. $\displaystyle\int (2x^{4/3} + 3x - 1)\, dx = \frac{6}{7}x^{7/3} + \frac{3}{2}x^2 - x + C$

$\displaystyle\frac{d}{dx}\left[\frac{6}{7}x^{7/3} + \frac{3}{2}x^2 - x + C\right] = 2x^{4/3} - 3x - 1$

31. $\displaystyle\int \sqrt[3]{x^2}\, dx = \int x^{2/3}\, dx = \frac{3}{5}x^{5/3} + C$

$\displaystyle\frac{d}{dx}\left[\frac{3}{5}x^{5/3} + C\right] = \sqrt[3]{x^2}$

33. $\displaystyle\int \frac{1}{x^4}\, dx = \int x^{-4}\, dx = \frac{x^{-3}}{-3} + C = \frac{-1}{3x^3} + C$

$\displaystyle\frac{d}{dx}\left[\frac{-1}{3x^3} + C\right] = \frac{1}{x^4}$

35. $\displaystyle\int (2x + x^{-1/2})\, dx = x^2 + \frac{x^{1/2}}{1/2} + C = x^2 + 2\sqrt{x} + C$

$\displaystyle\frac{d}{dx}\left[x^2 + 2\sqrt{x} + C\right] = 2x + x^{-1/2}$

37. $\displaystyle\int u(3u^2 + 1)\, du = \int (3u^3 + u)\, du$

$\displaystyle\qquad\qquad = \frac{3}{4}u^4 + \frac{1}{2}u^2 + C$

$\displaystyle\frac{d}{du}\left[\frac{3}{4}u^4 + \frac{1}{2}u^2 + C\right] = 3u^3 + u = u(3u^2 + 1)$

39. $\displaystyle\int (x - 1)(6x - 5)\, dx = \int (6x^2 - 11x + 5)\, dx$

$\displaystyle\qquad\qquad = 2x^3 - \frac{11}{2}x^2 + 5x + C$

$\displaystyle\frac{d}{dx}\left[2x^3 - \frac{11}{2}x^2 + 5x + C\right] = 6x^2 - 11x + 5$

41. $\displaystyle\int y^2\sqrt{y}\, dy = \int y^{5/2}\, dy = \frac{2}{7}y^{7/2} + C$

$\displaystyle\frac{d}{dy}\left[\frac{2}{7}y^{7/2} + C\right] = y^{5/2} = y^2\sqrt{y}$

43. $f(x) = \displaystyle\int (3x^{1/2} + 3)\, dx = 2x^{3/2} + 3x + C$

$f(1) = 4 = 2(1) + 3(1) + C = 5 + C \Rightarrow C = -1$

$f(x) = 2x^{3/2} + 3x - 1$

45. $f(x) = \displaystyle\int 6x(x - 1)\, dx$

$\qquad = \displaystyle\int (6x^2 - 6x)\, dx$

$\qquad = 2x^3 - 3x^2 + C$

$f(1) = -1 = 2 - 3 + C = -1 + C \Rightarrow C = 0$

$f(x) = 2x^3 - 3x^2$

47. $f(x) = \displaystyle\int \frac{2 - x}{x^3}\, dx = \int (2x^{-3} - x^{-2})\, dx$

$\qquad = -x^{-2} + x^{-1} + C = \frac{-1}{x^2} + \frac{1}{x} + C$

$f(2) = \dfrac{3}{4} = -\dfrac{1}{4} + \dfrac{1}{2} + C = \dfrac{1}{4} + C \Rightarrow C = \dfrac{1}{2}$

$f(x) = -\dfrac{1}{x^2} + \dfrac{1}{x} + \dfrac{1}{2}$

49. $y = \displaystyle\int (-5x - 2)\, dx = -\frac{5}{2}x^2 - 2x + C$

At $(0, 2)$, $2 = C$. Thus, $y = -\frac{5}{2}x^2 - 2x + 2$.

51. $f(x) = \displaystyle\int (6\sqrt{x} - 10)\, dx$

$\qquad = \displaystyle\int (6x^{1/2} - 10)\, dx$

$\qquad = 4x^{3/2} - 10x + C$

$\qquad = 4x\sqrt{x} - 10x + C$

At $(4, 2)$, $2 = 4(4)\sqrt{4} - 10(4) + C$ which implies that $C = 10$. Thus, $f(x) = 4x\sqrt{x} - 10x + 10$.

53. $f'(x) = \displaystyle\int 2\, dx = 2x + C_1$

Since $f'(2) = 4 + C_1 = 5$, we know that $C_1 = 1$. Thus, $f'(x) = 2x + 1$.

$f(x) = \displaystyle\int (2x + 1)\, dx = x^2 + x + C_2$

Since $f(2) = 4 + 2 + C_2 = 10$, we know that $C_2 = 4$. Thus, $f(x) = x^2 + x + 4$.

55. $f'(x) = \displaystyle\int x^{-2/3}\, dx = 3x^{1/3} + C_1$

Since $f'(8) = 3(2) + C_1 = 6$, we have $C_1 = 0$. Thus, $f'(x) = 3x^{1/3}$.

$f(x) = \displaystyle\int 3x^{1/3}\, dx = \frac{9}{4}x^{4/3} + C_2$

Since $f(0) = 0 = C_2$, $f(x) = \frac{9}{4}x^{4/3}$.

57. $C = \int 85 \, dx = 85x + k$

When $x = 0$, $C = 5500 = k$. Thus, $C = 85x + 5500$.

59. $C = \int \left(\frac{1}{20}x^{-1/2} + 4\right) dx = \frac{1}{10}x^{1/2} + 4x + k$

When $x = 0$,

$$C = \frac{1}{10}\left(\sqrt{0}\right) + 4(0) + k = 750 \Rightarrow k = 750.$$

Thus,

$$C = \frac{\sqrt{x}}{10} + 4x + 750.$$

61. $R = \int (225 - 3x) \, dx = 225x - \frac{3}{2}x^2 + C$

Since $R = 0$ when $x = 0$, it follows that $C = 0$. Thus, $R = 225x - \frac{3}{2}x^2$ and the demand function is

$$p = \frac{R}{x} = 225 - \frac{3}{2}x.$$

63. $P = \int (-18x + 1650) \, dx = -9x^2 + 1650x + C$

$P(15) = 22{,}725 = -9(15)^2 + 1650(15) + C \Rightarrow C = 0$

$P = -9 \text{ (as in } -9x^2) + 1650x$

65. $v(t) = \int -32 \, dt = -32t + C_1$

Since $v(0) = 60$, it follows that $C_1 = 60$. Thus, we have $v(t) = -32t + 60$.

$$s(t) = \int (-32t + 60) \, dt = -16t^2 + 60t + C_2$$

Since $s(0) = 0$, it follows that $C_2 = 0$. Therefore, the position function is $s(t) = -16t^2 + 60t$. Now, since $v(t) = 0$ when $t = \frac{60}{32} = 1.875$ seconds, the maximum height of the ball is $s(1.875) = 56.25$ feet.

67. $v(t) = \int -32 \, dt = -32t + C_1$

Letting v_0 be the initial velocity, we have $C_1 = v_0$.

$$s(t) = \int (-32t + v_0) \, dt = -16t^2 + v_0 t + C_2$$

Since $s(0) = 0$, we have $C_2 = 0$. Therefore, the position function is $s(t) = -16t^2 + v_0 t$. At the highest point, the velocity is zero. Therefore, we have $v(t) = -32t + v_0 = 0$, and $t = v_0/32$ seconds. Finally, substituting this value into the position function, we have

$$s\left(\frac{v_0}{32}\right) = -16\left(\frac{v_0}{32}\right)^2 + v_0\left(\frac{v_0}{32}\right) = 550$$

which implies that $v_0^2 = 35{,}200$ and the initial velocity should be $v_0 = 40\sqrt{22} \approx 187.617$ ft/sec.

69. (a) $C(x) = \int (2x - 12) \, dx = x^2 - 12x + C_1$

Since $C(0) = 125$, it follows that $C_1 = 125$. Thus, $C(x) = x^2 - 12x + 125$ and the average cost is $C/x = x - 12 + (125/x)$.

(b) $C(50) = 50^2 - 12(50) + 125 = \$2025$

(c) \$125 is fixed, \$1900 is variable.

71. $S = \int (0.012t^2 - 0.182t + 1.045) \, dt$

$$= 0.004t^3 - 0.091t^2 + 1.045t + C$$

To find C, use $S(5) = 17.8$.

$$17.8 = 0.004(5)^3 - 0.091(5)^2 + 1.045(5) + C$$

$$14.35 = C$$

$$S = 0.004t^3 - 0.091t^2 + 1.045t + 14.35$$

$S(14) = 22.12$ quadrillion Btu's

Section 5.2   The General Power Rule

$\int u^n \dfrac{du}{dx}\, dx$	u	$\dfrac{du}{dx}$
1. $\int (5x^2 + 1)^2(10x)\, dx$	$5x^2 + 1$	$10x$

$\int u^n \dfrac{du}{dx}\, dx$	u	$\dfrac{du}{dx}$
3. $\int \sqrt{1 - x^2}\,(-2x)\, dx$	$1 - x^2$	$-2x$

$\int u^n \dfrac{du}{dx}\, dx$	u	$\dfrac{du}{dx}$
5. $\int \left(4 + \dfrac{1}{x^2}\right)\left(\dfrac{-2}{x^3}\right) dx$	$4 + \dfrac{1}{x^2}$	$-\dfrac{2}{x^3}$

7. $\int (1 + 2x)^4(2)\, dx = \dfrac{(1 + 2x)^5}{5} + C$

9. $\int \sqrt{5x^2 - 4}\,(10x)\, dx = \int (5x^2 - 4)^{1/2}(10x)\, dx$

$\qquad = \dfrac{2}{3}(5x^2 - 4)^{3/2} + C$

11. $\int (x - 1)^4\, dx = \int (x - 1)^4(1)\, dx$

$\qquad = \dfrac{(x - 1)^5}{5} + C$

13. $\int x(x^2 - 1)^7\, dx = \dfrac{1}{2}\int (x^2 - 1)^7(2x)\, dx$

$\qquad = \dfrac{1}{2}\dfrac{(x^2 - 1)^8}{8} + C$

$\qquad = \dfrac{(x^2 - 1)^8}{16} + C$

15. $\int \dfrac{x^2}{(1 + x^3)^2}\, dx = \dfrac{1}{3}\int (1 + x^3)^{-2}(3x^2)\, dx$

$\qquad = \dfrac{1}{3}\dfrac{(1 + x^3)^{-1}}{-1} + C$

$\qquad = -\dfrac{1}{3(1 + x^3)} + C$

17. $\int \dfrac{x + 1}{(x^2 + 2x - 3)^2}\, dx = \dfrac{1}{2}\int (x^2 + 2x - 3)^{-2}2(x + 1)\, dx$

$\qquad = \dfrac{1}{2}\dfrac{(x^2 + 2x - 3)^{-1}}{-1} + C$

$\qquad = -\dfrac{1}{2(x^2 + 2x - 3)} + C$

19. $\int \dfrac{x - 2}{\sqrt{x^2 - 4x + 3}}\, dx = \dfrac{1}{2}\int (x^2 - 4x + 3)^{-1/2}(2x - 4)\, dx$

$\qquad = \dfrac{1}{2}(2)(x^2 - 4x + 3)^{1/2} + C$

$\qquad = \sqrt{x^2 - 4x + 3} + C$

21. $\int 5x\sqrt[3]{1 - x^2}\, dx = 5\left(-\dfrac{1}{2}\right)\int (1 - x^2)^{1/3}(-2x)\, dx$

$\qquad = -\dfrac{5}{2}\left(\dfrac{3}{4}\right)(1 - x^2)^{4/3} + C$

$\qquad = \dfrac{-15(1 - x^2)^{4/3}}{8} + C$

23. $\int \dfrac{4x}{\sqrt{1 + x^2}}\, dx = 4\left(\dfrac{1}{2}\right)\int (1 + x^2)^{-1/2}(2x)\, dx$

$\qquad = 2(2)(1 + x^2)^{1/2} + C$

$\qquad = 4\sqrt{1 + x^2} + C$

25. $\int \dfrac{-3}{\sqrt{2x + 3}}\, dx = -\dfrac{3}{2}\int (2x + 3)^{-1/2}(2)\, dx$

$\qquad = -\dfrac{3}{2}(2)(2x + 3)^{1/2} + C$

$\qquad = -3\sqrt{2x + 3} + C$

27. $\int \dfrac{x^3}{\sqrt{1 - x^4}}\, dx = -\dfrac{1}{4}\int (1 - x^4)^{-1/2}(-4x^3)\, dx$

$\qquad = -\dfrac{1}{4}(2)(1 - x^4)^{1/2} + C$

$\qquad = -\dfrac{\sqrt{1 - x^4}}{2} + C$

29. $\displaystyle\int \frac{1}{\sqrt{2x}}\, dx = \frac{1}{\sqrt{2}}\int x^{-1/2}\, dx$

$\qquad = \dfrac{1}{\sqrt{2}}(2x^{1/2}) + C$

$\qquad = \sqrt{2x} + C$

31. $\displaystyle\int (x^3 + 3x)(x^2 + 1)\, dx = \frac{1}{3}\int (x^3 + 3x)^1(3x^2 + 3)\, dx$

$\qquad = \dfrac{1}{3}\dfrac{(x^3 + 3x)^2}{2} + C$

$\qquad = \dfrac{1}{6}(x^3 + 3x)^2 + C$

33. Let $u = 6x^2 - 1$. Then $du = 12x\, dx$ and $x\, dx = \frac{1}{12}\, du$.

$\displaystyle\int x(6x^2 - 1)^3\, dx = \int (6x^2 - 1)^3(x\, dx)$

$\qquad = \displaystyle\int u^3 \frac{1}{12}\, du$

$\qquad = \dfrac{1}{12}\dfrac{u^4}{4} + C$

$\qquad = \dfrac{1}{48}(6x^2 - 1)^4 + C$

35. Let $u = 2 - 3x^3$, then $du = -9x^2\, dx$ which implies that $x^2\, dx = -\frac{1}{9}\, du$.

$\displaystyle\int x^2(2 - 3x^3)^{3/2}\, dx = \int (2 - 3x^3)^{3/2}(x^2)\, dx$

$\qquad = \displaystyle\int u^{3/2}\left(-\frac{1}{9}\right) du$

$\qquad = -\dfrac{1}{9}\left(\dfrac{2}{5}\right)u^{5/2} + C$

$\qquad = -\dfrac{2}{45}(2 - 3x^3)^{5/2} + C$

37. Let $u = x^2 + 25$, then $du = 2x\, dx$ which implies that $x\, dx = \frac{1}{2}\, du$.

$\displaystyle\int \frac{x}{\sqrt{x^2 + 25}}\, dx = \int (x^2 + 25)^{-1/2}(x)\, dx$

$\qquad = \displaystyle\int u^{-1/2}\left(\frac{1}{2}\right) du$

$\qquad = \dfrac{1}{2}(2u^{1/2}) + C$

$\qquad = \sqrt{u} + C$

$\qquad = \sqrt{x^2 + 25} + C$

39. Let $u = x^3 + 3x + 4$, then $du = (3x^2 + 3)\, dx = 3(x^2 + 1)\, dx$ and $(x^2 + 1)\, dx = \frac{1}{3}\, du$.

$\displaystyle\int \frac{x^2 + 1}{\sqrt{x^3 + 3x + 4}}\, dx = \int (x^3 + 3x + 4)^{-1/2}(x^2 + 1)\, dx$

$\qquad = \displaystyle\int u^{-1/2}\left(\frac{1}{3}\right) du$

$\qquad = \left(\dfrac{1}{3}\right)2u^{1/2} + C$

$\qquad = \dfrac{2}{3}\sqrt{u} + C$

$\qquad = \dfrac{2}{3}\sqrt{x^3 + 3x + 4} + C$

41. $\displaystyle\int (2x - 1)^2\, dx = \frac{1}{2}\int (2x - 1)^2(2)\, dx$

$\qquad = \dfrac{1}{2}\dfrac{(2x - 1)^3}{3} + C_1$

$\qquad = \dfrac{1}{6}(2x - 1)^3 + C_1$

$\qquad = \dfrac{1}{6}(8x^3 - 12x^2 + 6x - 1) + C_1$

$\qquad = \dfrac{4}{3}x^3 - 2x^2 + x - \dfrac{1}{6} + C_1$

$\displaystyle\int (2x - 1)^2\, dx = \int (4x^2 - 4x + 1)\, dx$

$\qquad = \dfrac{4}{3}x^3 - 2x^2 + x + C_2$

The two answers differ by a constant.

43. $\displaystyle\int x(x^2 - 1)^2\, dx = \frac{1}{2}\int (x^2 - 1)^2(2x)\, dx$

$\qquad = \dfrac{1}{2}\dfrac{(x^2 - 1)^3}{3} + C_1$

$\qquad = \dfrac{1}{6}(x^6 - 3x^4 + 3x^2 - 1) + C_1$

$\qquad = \dfrac{1}{6}x^6 - \dfrac{1}{2}x^4 + \dfrac{1}{2}x^2 - \dfrac{1}{6} + C_1$

$\displaystyle\int x(x^2 - 1)^2\, dx = \int (x^5 - 2x^3 + x)\, dx$

$\qquad = \dfrac{1}{6}x^6 - \dfrac{1}{2}x^4 + \dfrac{1}{2}x^2 + C_2$

The two answers differ by a constant.

45. $f(x) = \int x\sqrt{1 - x^2} \, dx$

$$= -\frac{1}{2} \int (1 - x^2)^{1/2}(-2x) \, dx$$

$$= -\frac{1}{2}\left(\frac{2}{3}\right)(1 - x^2)^{3/2} + C$$

$$= -\frac{1}{3}(1 - x^2)^{3/2} + C$$

Since $f(0) = \frac{4}{3}$, it follows that $C = \frac{5}{3}$ and we have

$$f(x) = -\frac{1}{3}(1 - x^2)^{3/2} + \frac{5}{3}$$

$$= \frac{1}{3}[5 - (1 - x^2)^{3/2}].$$

47. (a) $C = \int \frac{4}{\sqrt{x + 1}} \, dx$

$$= 4 \int (x + 1)^{-1/2} \, dx$$

$$= 4(2)(x + 1)^{1/2} + K$$

$$= 8\sqrt{x + 1} + K$$

Since $C(15) = 50$, it follows that $K = 18$, and we have $C = 8\sqrt{x + 1} + 18$.

(b)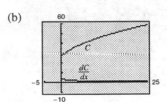

49. $x = \int p\sqrt{p^2 - 25} \, dp$

$$= \frac{1}{2} \int (p^2 - 25)^{1/2}(2p) \, dp$$

$$= \frac{1}{2} \cdot \frac{2}{3}(p^2 - 25)^{3/2} + C$$

$$= \frac{1}{3}(p^2 - 25)^{3/2} + C$$

Since $x = 600$ when $p = 13$, it follows that

$$x = 600 = \frac{1}{3}(13^2 - 25)^{3/2} + C = 576 + C$$

and $C = 24$. Therefore, $x = \frac{1}{3}(p^2 - 25)^{3/2} + 24$.

51. $x = \int -\frac{6000p}{(p^2 - 16)^{3/2}} \, dp$

$$= -\frac{6000}{2} \int (p^2 - 16)^{-3/2}(2p) \, dp$$

$$= -3000(-2)(p^2 - 16)^{-1/2} + C$$

$$= \frac{6000}{\sqrt{p^2 - 16}} + C$$

Since $x = 5000$ when $p = 5$, it follows that $C = 3000$, and we have

$$x = \frac{6000}{\sqrt{p^2 - 16}} + 3000.$$

53. (a) $h = \int \frac{17.6t}{\sqrt{17.6t^2 + 1}} \, dt$

$$= \frac{1}{2} \int (17.6t^2 + 1)^{-1/2}(35.2t) \, dt$$

$$= (17.6t^2 + 1)^{1/2} + C$$

$$h(0) = 6 \Rightarrow 6 = 1 + C \Rightarrow C = 5$$

$$h(t) = \sqrt{17.6t^2 + 1} + 5$$

(b) $h(5) = 26$ inches

55. (a) $Q = \int \frac{0.95}{(x - 19,999)^{0.05}} \, dx = (x - 19,999)^{0.95} + C$

Since $Q = 20,000$ when $x = 20,000$,

$$20,000 = (20,000 - 19,999)^{0.95} + C = 1 + C$$

$$C = 19,999.$$

Thus, $Q = (x - 19,999)^{0.95} + 19,999.$

(b)

x	20,000	50,000	100,000	150,000
Q	20,000	37,916.56	65,491.59	92,151.16
$X - Q$	0	12,083.44	34,508.41	57,848.84

(c)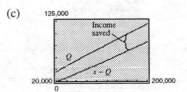

57. $\int \dfrac{1}{\sqrt{x} + \sqrt{x+1}} \, dx = -\dfrac{2}{3}x^{3/2} + \dfrac{2}{3}(x+1)^{3/2} + C$

Section 5.3   Exponential and Logarithmic Integrals

1. $\int e^{2x}(2) \, dx = e^{2x} + C$

3. $\int e^{4x} \, dx = \dfrac{1}{4}\int e^{4x}(4) \, dx = \dfrac{1}{4}e^{4x} + C$

5. $\int 9xe^{-x^2} \, dx = -\dfrac{9}{2}\int e^{-x^2}(-2x) \, dx = -\dfrac{9}{2}e^{-x^2} + C$

7. $\int 5x^2e^{x^3} \, dx = \dfrac{5}{3}\int e^{x^3}(3x^2) \, dx = \dfrac{5}{3}e^{x^3} + C$

9. $\dfrac{1}{3}\int e^{x^3 + 3x^2 - 1}(3x^2 + 6x) \, dx = \dfrac{1}{3}e^{x^3 + 3x^2 - 1} + C$

11. $\int 5e^{2-x} \, dx = -5\int e^{2-x}(-1) \, dx = -5e^{2-x} + C$

13. $\int \dfrac{1}{x+1} \, dx = \ln|x+1| + C$

15. $-\dfrac{1}{2}\int \dfrac{1(-2)}{3 - 2x} \, dx = -\dfrac{1}{2}\ln|3 - 2x| + C$

17. $\dfrac{1}{2}\int \dfrac{x(2)}{x^2 + 1} \, dx = \dfrac{1}{2}\ln(x^2 + 1) + C = \ln\sqrt{x^2 + 1} + C$

19. $\int \dfrac{x^2}{x^3 + 1} \, dx = \dfrac{1}{3}\int \dfrac{3x^2}{x^3 + 1} \, dx = \dfrac{1}{3}\ln|x^3 + 1| + C$

21. $\int \dfrac{x+3}{x^2 + 6x + 7} \, dx = \dfrac{1}{2}\int \dfrac{2(x+3)}{x^2 + 6x + 7} \, dx$
$$= \dfrac{1}{2}\ln|x^2 + 6x + 7| + C$$

23. $\int \dfrac{1}{x \ln x} \, dx = \int \dfrac{1}{\ln x}\left(\dfrac{1}{x}\right) \, dx = \ln|\ln x| + C$

25. $-\dfrac{1}{2}\int e^{2/x}(-2x^{-2}) \, dx = -\dfrac{1}{2}e^{2/x} + C$

27. $2\int e^{\sqrt{x}}\dfrac{1}{2\sqrt{x}} \, dx = 2e^{\sqrt{x}} + C$

29. $-\int \dfrac{1}{1 + e^{-x}}(-e^{-x}) \, dx = -\ln(1 + e^{-x}) + C$

31. $2\int \dfrac{1}{5 - e^{2x}}2e^{2x} \, dx = 2\ln|5 - e^{2x}| + C$

33. $\int \dfrac{e^{2x} + 2e^x + 1}{e^x} \, dx = \int (e^x + 2 + e^{-x}) \, dx$
$$= e^x + 2x - e^{-x} + C$$

35. $\int e^x\sqrt{1 - e^x} \, dx = -\int (1 - e^x)^{1/2}(-e^x) \, dx$
$$= -\dfrac{2}{3}(1 - e^x)^{3/2} + C$$

37. $\int (x-1)^{-2} \, dx = \dfrac{(x-1)^{-1}}{-1} = \dfrac{-1}{x-1} + C$

39. $\int \dfrac{x^2 - 4}{x} \, dx = \int \left[x - 4\left(\dfrac{1}{x}\right)\right] \, dx = \dfrac{x^2}{2} - 4\ln|x| + C$

41. $\int \dfrac{x^3 - 8x}{2x^2} \, dx = \int \left(\dfrac{x}{2} - \dfrac{4}{x}\right) \, dx = \dfrac{x^2}{4} - 4\ln|x| + C$

43. $\int \dfrac{2}{1 + e^{-x}} \, dx = 2\int \dfrac{1}{e^x + 1}e^x \, dx = 2\ln(e^x + 1) + C$

45. $\int \dfrac{x^2 + 2x + 5}{x - 1} \, dx = \int \left(x + 3 + \dfrac{8}{x-1}\right) \, dx$
$$= \dfrac{1}{2}x^2 + 3x + 8\ln|x - 1| + C$$

47. $\int \dfrac{1 + e^{-x}}{1 + xe^{-x}} \, dx = \int \dfrac{e^x + 1}{e^x + x} \, dx = \ln|e^x + x| + C$

49. Dividing, you obtain

$$\frac{x^2 + 4x + 3}{x - 1} = x + 5 + \frac{8}{x - 1}.$$

Hence,

$$f(x) = \int \left(x + 5 + \frac{8}{x - 1} \right) dx$$

$$= \frac{x^2}{2} + 5x + 8 \ln|x - 1| + C.$$

$$f(2) = 4 \Rightarrow$$

$$4 = \frac{2^2}{2} + 5(2) + 8 \ln|2 - 1| + C$$

$$-8 = C$$

Hence,

$$f(x) = \frac{x^2}{2} + 5x + 8 \ln|x - 1| - 8.$$

53. (a) $p = \displaystyle\int 0.1e^{-x/500} \, dx = -50e^{-x/500} + C$

Since $x = 600$ when $p = 30$, we have:

$$30 = -50e^{600/500} + C \Rightarrow C \approx 45.06$$

$$p = -50e^{-x/500} + 45.06$$

(b)

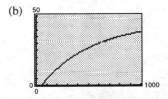

Price increases as demand increases.

(c) When $p = 22, x \approx 387$.

51. (a) $P = \displaystyle\int \frac{3000}{1 + 0.25t} \, dt$

$$= \frac{3000}{0.25} \int \frac{0.25}{1 + 0.25t} \, dt$$

$$= 12{,}000 \ln|1 + 0.25t| + C$$

Since $P(0) = 12{,}000 \ln 1 + C = 1000$, it follows that $C = 1000$. Therefore,

$$P(t) = 12{,}000 \ln|1 + 0.25t| + 1000$$

$$= 1000[12 \ln|1 + 0.25t| + 1]$$

$$= 1000[1 + \ln(1 + 0.25t)^{12}].$$

(b) The population when $t = 3$ is

$$P(3) = 1000[1 + \ln(1 + 0.75)^{12}]$$

$$\approx 7715 \text{ bacteria.}$$

(c) Using a graphing utility,

$$12{,}000 = 1000[1 + \ln(1 + 0.25t)^{12}]$$

$$\Rightarrow t \approx 6 \text{ days.}$$

55. (a) $T = \displaystyle\int 189.167e^{0.11t} \, dt$

$$= \frac{(189.167)e^{0.11t}}{0.11} + C$$

$$= 1719.7e^{0.11t} + C$$

1993: $T(8) = 7705 = 1719.7e^{0.11(8)} + C$

$$C = 3558.976$$

$$T(t) = 1719.7e^{0.11t} + 3558.976$$

(b) 1990: $T(5) \approx 6540$ million transactions

Section 5.4   Area and the Fundamental Theorem of Calculus

1. $\displaystyle\int_0^2 3 \, dx$

Area = (base)(height)

   $= (2)(3) = 6$

3. $\displaystyle\int_0^5 (x + 1) \, dx$

Area = $\frac{1}{2}$(base)(height)

   $= \frac{1}{2}(5)(1 + 6) = \frac{35}{2}$

5. $A = \displaystyle\int_0^1 (x - x^2)\, dx = \left[\dfrac{x^2}{2} - \dfrac{x^3}{3}\right]_0^1 = \dfrac{1}{6}$

7. $A = \displaystyle\int_{-1}^1 (1 - x^4)\, dx$

$\quad = 2\displaystyle\int_0^1 (1 - x^4)\, dx$

$\quad = 2\left(x - \dfrac{x^5}{5}\right)\Big]_0^1$

$\quad = \dfrac{8}{5}$ square units

9. $A = \displaystyle\int_0^4 \sqrt[3]{2x}\, dx$

$\quad = \dfrac{1}{2}\displaystyle\int_0^4 (2x)^{1/3}(2)\, dx$

$\quad = \dfrac{1}{2}\left(\dfrac{3}{4}\right)(2x)^{4/3}\Big]_0^4$

$\quad = \dfrac{3}{8}(\sqrt[3]{2x})^4\Big]_0^4$

$\quad = \dfrac{3}{8}(16) - \dfrac{3}{8}(0)$

$\quad = 6$ square units

11. $\displaystyle\int_0^1 2x\, dx = x^2\Big]_0^1 = 1 - 0 = 1$

13. $\displaystyle\int_{-1}^0 (2x + 1)\, dx = \left[x^2 + x\right]_{-1}^0 = 0 - 0 = 0$

15. $\displaystyle\int_{-1}^1 (2t - 1)^2\, dt = \dfrac{1}{6}(2t - 1)^3\Big]_{-1}^1$

$\qquad\qquad = \dfrac{1}{6} - \left(\dfrac{-27}{6}\right)$

$\qquad\qquad = \dfrac{14}{3}$

17. $\displaystyle\int_0^1 (2t - 1)^2\, dt = \dfrac{1}{2}\displaystyle\int_0^1 (2t - 1)^2(2)\, dt$

$\qquad\qquad = \dfrac{1}{6}(2t - 1)^3\Big]_0^1$

$\qquad\qquad = \dfrac{1}{6} - \left(-\dfrac{1}{6}\right)$

$\qquad\qquad = \dfrac{1}{3}$

19. $\displaystyle\int_{-1}^1 \left(\sqrt[3]{t} - 2\right) dt = \left[\dfrac{3}{4}t^{4/3} - 2t\right]_{-1}^1$

$\qquad\qquad = -\dfrac{5}{4} - \dfrac{11}{4}$

$\qquad\qquad = -4$

21. $\displaystyle\int_1^4 \dfrac{2u - 1}{\sqrt{u}}\, du = \displaystyle\int_1^4 (2u^{1/2} - u^{-1/2})\, du$

$\qquad\qquad = \left[\dfrac{4}{3}u^{3/2} - 2u^{1/2}\right]_1^4$

$\qquad\qquad = \left(\dfrac{32}{3} - 4\right) - \left(\dfrac{4}{3} - 2\right)$

$\qquad\qquad = \dfrac{22}{3}$

23. $\displaystyle\int_{-1}^0 (t^{1/3} - t^{2/3})\, dt = \left[\dfrac{3}{4}t^{4/3} - \dfrac{3}{5}t^{5/3}\right]_{-1}^0$

$\qquad\qquad = 0 - \left(\dfrac{3}{4} + \dfrac{3}{5}\right)$

$\qquad\qquad = -\dfrac{27}{20}$

25. $\displaystyle\int_0^4 \dfrac{1}{\sqrt{2x + 1}}\, dx = \dfrac{1}{2}\displaystyle\int_0^4 (2x + 1)^{-1/2}(2)\, dx$

$\qquad\qquad = \left[\dfrac{1}{2}(2)(2x + 1)^{1/2}\right]_0^4$

$\qquad\qquad = \sqrt{2x + 1}\Big]_0^4$

$\qquad\qquad = 3 - 1$

$\qquad\qquad = 2$

27. $\displaystyle\int_0^1 e^{-2x}\,dx = -\frac{1}{2}e^{-2x}\Big]_0^1$

$\qquad = -\frac{e^{-2}}{2} + \frac{1}{2}$

$\qquad = \frac{1}{2}(1 - e^{-2})$

$\qquad \approx 0.432$

29. $\displaystyle\int_1^3 \frac{e^{3/x}}{x^2}\,dx = -\frac{1}{3}\int_1^3 e^{3/x}\left(-\frac{3}{x^2}\right)dx$

$\qquad = -\frac{1}{3}e^{3/x}\Big]_1^3$

$\qquad = -\frac{1}{3}(e - e^3)$

$\qquad = \frac{e^3 - e}{3}$

$\qquad \approx 5.789$

31. $\displaystyle\int_{-1}^1 |4x|\,dx = \int_{-1}^0 -4x\,dx + \int_0^1 4x\,dx$

$\qquad = \left[-2x^2\right]_{-1}^0$

$\qquad = (0 + 2) + (2 - 0)$

$\qquad = 4$

33. $\displaystyle\int_0^4 (2 - |x - 2|)\,dx = \int_0^2 \{2 - [-(x - 2)]\}\,dx + \int_2^4 [2 - (x - 2)]\,dx$

$\qquad = \int_0^2 x\,dx + \int_2^4 (4 - x)\,dx$

$\qquad = \frac{x^2}{2}\Big]_0^2 + \left[4x - \frac{x^2}{2}\right]_2^4$

$\qquad = (2 - 0) + (8 - 6)$

$\qquad = 4$

35. $\displaystyle\int_{-1}^2 \frac{x}{x^2 - 9}\,dx = \left[\frac{1}{2}\ln|x^2 - 9|\right]_{-1}^2$

$\qquad = \frac{1}{2}\ln 5 - \frac{1}{2}\ln 8$

$\qquad \approx -0.235$

37. $\displaystyle\int_0^3 \frac{2e^x}{2 + e^x}\,dx = \left[2\ln(2 + e^x)\right]_0^3$

$\qquad = 2\ln(2 + e^3) - 2\ln 3$

$\qquad \approx 3.993$

39. $\displaystyle\int_1^3 (4x - 3)\,dx = \left[2x^2 - 3\right]_1^3$

$\qquad = 9 - (-1)$

$\qquad = 10$

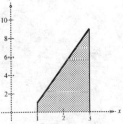

41. $\displaystyle\int_0^1 (x - x^3)\,dx = \left[\frac{x^2}{2} - \frac{x^4}{4}\right]_0^1$

$\qquad = \left(\frac{1}{2} - \frac{1}{4}\right) - 0$

$\qquad = \frac{1}{4}$

43. $\int_2^4 \frac{3x^2}{x^3 - 1} \, dx = \left[\ln(x^3 - 1) \right]_2^4$

$$= \ln 63 - \ln 7$$

$$= \ln 9$$

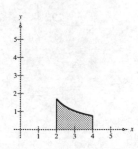

45. $A = \int_0^2 (3x^2 + 1) \, dx$

$$= \left[x^3 + x \right]_0^2$$

$$= 10 \text{ square units}$$

47. $A = \int_1^5 \frac{x + 5}{x} \, dx$

$$= \int_1^5 \left[1 + 5\left(\frac{1}{x}\right) \right] dx$$

$$= \left[x + 5 \ln|x| \right]_1^5$$

$$= (5 + 5 \ln 5) - (1 + 0)$$

$$= 4 + 5 \ln 5$$

$$\approx 12.047 \text{ square units}$$

49. (a) $\int_0^5 [f(x) + g(x)] \, dx = \int_0^5 f(x) \, dx + \int_0^5 g(x) \, dx = 8 + 3 = 11$

(b) $\int_0^5 [f(x) - g(x)] \, dx = \int_0^5 f(x) \, dx - \int_0^5 g(x) \, dx = 8 - 3 = 5$

(c) $\int_0^5 -4f(x) \, dx = -4 \int_0^5 f(x) \, dx = -4(8) = -32$

(d) $\int_0^5 [f(x) - 3g(x)] \, dx = \int_0^5 f(x) \, dx - 3 \int_0^5 g(x) \, dx = 8 - 3(3) = -1$

51. Average value $= \dfrac{1}{2 - (-2)} \displaystyle\int_{-2}^2 (6 - x^2) \, dx$

$$= \frac{1}{4} \left[6x - \frac{x^3}{3} \right]_{-2}^2$$

$$= \frac{1}{4} \left(\frac{28}{3} + \frac{28}{3} \right)$$

$$= \frac{14}{3}$$

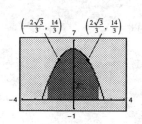

To find the x-values for which $f(x) = \frac{14}{3}$, we let $6 - x^2 = \frac{14}{3}$ and solve for x.

$$x^2 = \frac{4}{3}$$

$$x = \pm 2\sqrt{3} \approx \pm 1.155$$

53. Average value $= \dfrac{1}{2-0} \displaystyle\int_0^2 x\sqrt{4-x^2}\,dx$

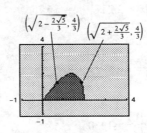

$$= \frac{1}{2}\left(-\frac{1}{2}\right)\int_0^2 (4-x^2)^{1/2}(-2x)\,dx$$

$$= \frac{1}{2}\left(-\frac{1}{2}\right)\left(\frac{2}{3}\right)(4-x^2)^{3/2}\bigg]_0^2$$

$$= -\frac{1}{6}(4-x^2)^{3/2}\bigg]_0^2$$

$$= 0 + \frac{4}{3}$$

$$= \frac{4}{3}$$

To find the x-values for which $f(x) = \frac{4}{3}$, we let $x\sqrt{4-x^2} = \frac{4}{3}$ and solve for x to obtain the following.

$$x^2(4-x^2) = \frac{16}{9}$$

$$36x^2 - 9x^4 = 16$$

$$9x^4 - 36x^2 + 16 = 0$$

$$x^2 = \frac{36 \pm \sqrt{720}}{18} = 2 \pm \frac{2\sqrt{5}}{3}$$

$$x = \sqrt{2 \pm \frac{2\sqrt{5}}{3}} \text{ in the interval } (0, 2).$$

55. Average value $= \dfrac{1}{4-0} \displaystyle\int_0^4 \left(x - 2\sqrt{x}\right) dx = \frac{1}{4}\left(\frac{x^2}{2} - \frac{4}{3}x^{3/2}\right)\bigg]_0^4 = \frac{1}{4}\left(8 - \frac{32}{3}\right) = -\frac{2}{3}$

To find the x-values for which $f(x) = -\frac{2}{3}$, we let $x - 2\sqrt{x} = -\frac{2}{3}$ and solve for x to obtain the following.

$$x + \frac{2}{3} = 2\sqrt{x}$$

$$x^2 + \frac{4}{3}x + \frac{4}{9} = 4x$$

$$x^2 - \frac{8}{3}x + \frac{4}{9} = 0$$

$$9x^2 - 24x + 4 = 0$$

$$x = \frac{24 \pm \sqrt{432}}{18} = \frac{4 \pm 2\sqrt{3}}{3}$$

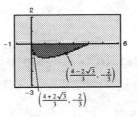

57. Since $f(-x) = 3(-x)^4 = 3x^4 = f(x)$, the function is even.

59. Since $g(-t) \neq g(t)$ nor $g(-t) = -g(t)$, the function is neither even nor odd.

61. Since $f(x) = x^2$ is an **even** function, we have the following.

(a) $\displaystyle\int_{-2}^0 x^2\,dx - \int_0^2 x^2\,dx = \frac{8}{3}$ (b) $\displaystyle\int_{-2}^2 x^2\,dx = 2\int_0^2 x^2\,dx = \frac{16}{3}$ (c) $\displaystyle\int_0^2 -x^2\,dx = -\int_0^2 x^2\,dx = -\frac{8}{3}$

63. $\Delta C = \displaystyle\int_{100}^{103} 2.25 \, dx = 2.25x \Big]_{100}^{103} = \$6.75$

65. $\Delta R = \displaystyle\int_{12}^{15} (48 - 3x) \, dx = \left[48x - \frac{3}{2}x^2 \right]_{12}^{15} = \$22.50$

67. $\Delta P = \displaystyle\int_{200}^{203} \frac{400 - x}{150} \, dx = \frac{1}{150} \left[400x - \frac{x^2}{2} \right]_{200}^{203} = \$3.97$

69. $\displaystyle\int_{0}^{5} 500 \, dt = 500t \Big]_{0}^{5} = \$2500$

71. $\displaystyle\int_{0}^{5} 100t \, dt = 50t^2 \Big]_{0}^{5} = \$1250$

73. $C(x) = 5000 \left(25 + 3 \displaystyle\int_{0}^{x} t^{1/4} \, dt \right)$

$\qquad = 5000 \left(25 + \left[\frac{12}{5} t^{5/4} \right]_{0}^{x} \right)$

$\qquad = 5000 \left(25 + \frac{12}{5} t^{5/4} \right)$

(a) $C(1) = 5000 \left[25 + \left(\frac{12}{5} \right) \right] = \$137,000.00$

(b) $C(5) = 5000 \left[25 + \left(\frac{12}{5} \right)(5)^{5/4} \right] \approx \$214,720.93$

(c) $C(10) = 5000 \left[25 + \left(\frac{12}{5} \right)(10)^{5/4} \right] \approx \$338,393.53$

75. Average balance $= \dfrac{1}{5 - 0} \displaystyle\int_{0}^{5} 2250 e^{0.12t} \, dt$

$\qquad = 450 \displaystyle\int_{0}^{5} e^{0.12t} \, dt$

$\qquad = 450 \left(\frac{1}{0.12} \right) e^{0.12t} \Big]_{0}^{5}$

$\qquad = 3750(e^{0.6} - 1)$

$\qquad \approx \$3082.95$

77. (a) $R(t) = \displaystyle\int (1.62t + 21.14e^{-t}) \, dt = 0.81t^2 - 21.14e^{-t} + C$

Since $R(0) = 45.1 = C$, $R(t) = 0.81t^2 - 21.14e^{-t} + 66.24$.

(b) Average revenue $= \dfrac{1}{9 - 0} \displaystyle\int_{0}^{9} (0.81t^2 - 21.14e^{-t} + 66.24) \, dt \approx \$85.76$ billion

79. $\displaystyle\int_{3}^{6} \frac{x}{3\sqrt{x^2 - 8}} \, dx \approx 1.4305 = \frac{2}{3}\sqrt{7} - \frac{1}{3}$

81. $\displaystyle\int_{2}^{5} \left(\frac{1}{x^2} - \frac{1}{x^3} \right) dx = 0.195 = \frac{39}{200}$

Section 5.5   The Area of a Region Bounded by Two Graphs

1. $\displaystyle\int_{0}^{6} [0 - (x^2 - 6x)] \, dx = -\left(\frac{x^3}{3} - 3x^2 \right) \Big]_{0}^{6} = 36$

3. $\displaystyle\int_{0}^{3} [(-x^2 + 2x + 3) - (x^2 - 4x + 3)] \, dx = \displaystyle\int_{0}^{3} (-2x^2 + 6x) \, dx$

$\qquad\qquad\qquad\qquad\qquad\qquad = \left[\frac{-2x^3}{3} + 3x^2 \right]_{0}^{3}$

$\qquad\qquad\qquad\qquad\qquad\qquad = 9$

5. $A = 2 \displaystyle\int_{0}^{1} [0 - 3(x^3 - x)] \, dx$

$\qquad = -6 \left(\frac{x^4}{4} - \frac{x^2}{2} \right) \Big]_{0}^{1}$

$\qquad = \dfrac{3}{2}$ (by symmetry)

7. $A = \displaystyle\int_{0}^{1} [(e^x - 1) - 0] \, dx$

$\qquad = e^x - x \Big]_{0}^{1}$

$\qquad = (e - 1) - 1$

$\qquad = e - 2$

9. The region is bounded by the graphs of $y = x + 1$, $y = x/2$, $x = 0$, and $x = 4$, as shown in the figure.

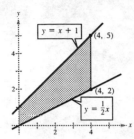

11.

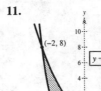

13.

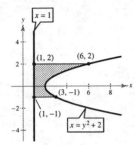

15. To find the points of intersection, we solve:

$$2x = 4 - 2x$$

$$4x = 4$$

$$x = 1$$

$$A = \int_0^1 2x \, dx + \int_1^2 (4 - 2x) \, dx$$

$$= x^2 \Big]_0^1 + 4x - x^2 \Big]_1^2$$

$$= 1 + (4 - 3)$$

$$= 2$$

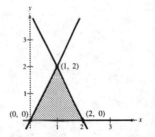

17. $x^3 - 3x^2 + 3x = x^2$

$x^3 - 4x^2 + 3x = 0$

$x(x - 1)(x - 3) = 0$

$\qquad x = 0, 1, 3$

$$A = \int_0^1 \left[(x^3 - 3x^2 + 3x) - x^2 \right] dx + \int_1^3 \left[x^2 - (x^3 - 3x^2 + 3x) \right] dx$$

$$= \int_0^1 (x^3 - 4x^2 + 3x) \, dx + \int_1^3 (-x^3 + 4x^2 - 3x) \, dx$$

$$= \left[\frac{x^4}{4} - \frac{4x^3}{3} + \frac{3x^2}{2} \right]_0^1 + \left[\frac{-x^4}{4} + \frac{4x^3}{3} - \frac{3x^2}{2} \right]_1^3$$

$$= \frac{5}{12} + \frac{8}{3}$$

$$= \frac{37}{12}$$

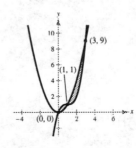

19. $A = 2 \int_0^1 \left(\sqrt[3]{x} - x \right) dx$

$\qquad = 2 \left[\dfrac{3}{4} x^{4/3} - \dfrac{x^2}{2} \right]_0^1$

$\qquad = \dfrac{1}{2}$

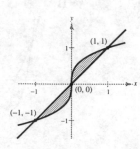

21. The points of intersection of the two graphs are found by equating y-values and solving for x.

$\qquad x^2 - 4x + 3 = 3 + 4x - x^2$

$\qquad\qquad 2x^2 - 8x = 0$

$\qquad\qquad 2x(x - 4) = 0$

$\qquad\qquad\qquad x = 0, 4$

$\qquad A = \int_0^4 \left[(3 + 4x - x^2) - (x^2 - 4x + 3) \right] dx$

$\qquad = \int_0^4 (-2x^2 + 8x)\, dx$

$\qquad = \left[-\dfrac{2x^3}{3} + 4x^2 \right]_0^4$

$\qquad = \dfrac{64}{3}$

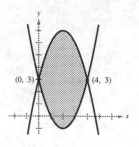

23. $2y - y^2 = -y$

$\qquad 0 = y^2 - 3y = y(y - 3)$

$\qquad y = 0, 3$

$\qquad A = \int_0^3 \left[(2y - y^2) - (-y) \right] dy$

$\qquad = \int_0^3 (3y - y^2)\, dy$

$\qquad = \left[\dfrac{3y^2}{2} - \dfrac{y^3}{3} \right]_0^3$

$\qquad = \dfrac{9}{2}$

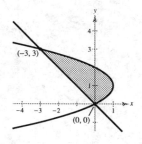

25. $A = \int_1^2 \left[e^{1/2x} - \left(-\dfrac{1}{x} \right) \right] dx$

$\qquad = 2e^{1/2x} + \ln x \Big]_1^2$

$\qquad = (2e + \ln 2) - 2e^{1/2}$

$\qquad \approx 3.832$

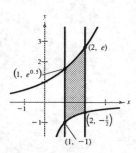

27. $A = \int_1^2 \left(\frac{4}{x} - x\right) dx + \int_2^4 \left(x - \frac{4}{x}\right) dx$

$= \left[4 \ln x - \frac{x^2}{2}\right]_1^2 + \left[\frac{x^2}{2} - 4 \ln x\right]_2^4$

$= \left(4 \ln 2 - \frac{3}{2}\right) + (6 - 4 \ln 2)$

$= \frac{9}{2}$

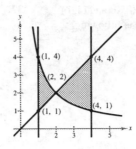

29. $A = \int_0^1 xe^{-x^2} dx$

$= -\frac{1}{2} e^{-x^2} \Big]_0^1$

$= -\frac{1}{2} e^{-1} + \frac{1}{2}$

≈ 0.316

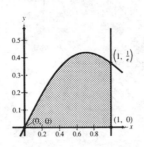

31. The points of intersection of f and g are found by setting $f(x) = g(x)$ and solving for x.

$x^2 - 4x = 0$

$x(x - 4) = 0$

$x = 0, 4$

$A = \int_0^4 [0 - (x^2 - 4x)] dx$

$= -\left(\frac{x^3}{3} - 2x^2\right)\Big]_0^4$

$= \frac{32}{3}$

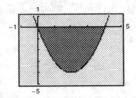

33. The points of intersection are found by setting $f(x) = g(x)$.

$x^2 + 2x + 1 = x + 1$

$x^2 + x = 0$

$x(x + 1) = 0$

$x = 0, -1$

$A = \int_{-1}^0 [(x + 1) - (x^2 - 2x + 1)] dx$

$= \int_{-1}^0 (-x^2 - x) dx$

$= \frac{-x^3}{3} - \frac{x^2}{2}\Big]_{-1}^0$

$= 0 - \left(\frac{1}{3} - \frac{1}{2}\right)$

$= \frac{1}{6}$

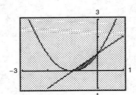

35. The equation of the line passing through $(0, 0)$ and $(4, 4)$ is $y = x$. Therefore, the area is given by the following.

$$A = \int_0^4 x \, dx = \frac{x^2}{2}\Big]_0^4 = 8$$

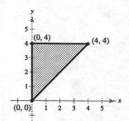

37. The point of equilibrium is found by equating $50 - 0.5x = 0.125x$ to obtain $x = 80$ and $p = 10$.

$$CS = \int_0^{80} [(50 - 0.5x) - 10] \, dx$$

$$= \left[-\frac{0.5x^2}{2} + 40x\right]_0^{80} = 1600$$

$$PS = \int_0^{80} (10 - 0.125x) \, dx$$

$$= \left[10x - \frac{0.125x^2}{2}\right]_0^{80} = 400$$

39. The point of equilibrium is found by equating the supply and demand functions.

$$200 - 0.02x^2 = 100 + x$$

$$0.02x^2 + x - 100 = 0$$

$$x^2 + 50x - 5000 = 0$$

$$(x + 100)(x - 50) = 0$$

$$x = 50 \text{ and } p = 150$$

$$CS = \int_0^{50} [(200 - 0.02x^2) - 150] \, dx = \left[50x - \frac{0.02x^3}{3}\right]_0^{50} \approx 1666.67$$

$$PS = \int_0^{50} [150 - (100 + x)] \, dx = \left[50x - \frac{x^2}{2}\right]_0^{50} = \$1250.00$$

41. The point of equilibrium is found by equating the demand and supply functions.

$$\frac{10,000}{\sqrt{x + 100}} = 100\sqrt{0.05x + 10}$$

$$100 = \sqrt{(x + 100)(0.05x + 10)}$$

$$10,000 = 0.05x^2 + 15x + 1000$$

$$5x^2 + 1500x - 900,000 = 0$$

$$5(x^2 + 300x - 180,000) = 0$$

$$5(x + 600)(x - 300) = 0$$

$$x = 300 \text{ and } p = 500$$

$$CS = \int_0^{300} \left(\frac{10,000}{\sqrt{x + 100}} - 500\right) dx$$

$$= \left[20,000\sqrt{x + 100} - 500x\right]_0^{300}$$

$$= 250,000 - 200,000$$

$$= 50,000$$

$$PS = \int_0^{300} \left(500 - 100\sqrt{0.05x + 10}\right) dx$$

$$= \left[500x - \frac{4000}{3}(0.05x + 10)^{3/2}\right]_0^{300}$$

$$= -\frac{50,000}{3} + \frac{40,000\sqrt{10}}{3}$$

$$= \frac{10,000}{3}\left(4\sqrt{10} - 5\right)$$

$$\approx 25,497$$

43. A typical demand function is decreasing, while a typical supply function is increasing.

45. The model R_1 projects greater revenue than R_2. The difference in total revenue is

$$\int_6^{10} (R_1 - R_2)\, dt = \int_6^{10} [(7.21 - 0.58t) - (7.21 + 0.45t)]\, dt$$

$$= \int_6^{10} 0.13t\, dt$$

$$= \frac{0.13t^2}{2}\Bigg]_6^{10}$$

$$= \$4.16 \text{ billion.}$$

47. The total savings is given by

$$\int_5^{15} (C_1 - C_2)\, dt = \int_5^{15} [(568.50 + 7.15t) - (525.60 + 6.43t)]\, dt$$

$$= \int_5^{15} (42.90 + 0.72t)\, dt$$

$$= \Big[42.90t + 0.36t^2\Big]_5^{15}$$

$$= \$501 \text{ million.}$$

49. (a)

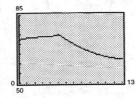

(b) The difference in consumption is given by

$$\int_5^{13} \Big[(72.1 + 0.98\sqrt{t}) - (94.6 - 4.71t + 0.17t^2)\Big]\, dt \approx 65.03 \text{ pounds.}$$

51. To find the demand function we use the points (6000, 325) and (8000, 300) to obtain

$$p - 325 = \frac{300 - 325}{8000 - 6000}(x - 6000)$$

$$p = -\frac{1}{80}x + 400.$$

The point of equilibrium is obtained by equating $-\frac{1}{80}x + 400 = \frac{11}{400}x$ to obtain $x = 10,000$ and $p = 275$.

$$\text{Consumer Surplus} = \int_0^{10,000} \left(-\frac{1}{80}x + 400 - 275\right) dx = \$625,000$$

$$\text{Producer Surplus} = \int_0^{10,000} \left(275 - \frac{11}{400}x\right) dx = \$1,375,000$$

53. $P = \int_0^{10} (R - C)\, dt$

$$= \int_0^{10} [(100 + 0.08t) - (60 + 0.2t^2)]\, dt$$

$$= \int_0^{10} (-0.2t^2 + 0.08t + 40)\, dt$$

$$= \left[-\frac{0.2t^3}{3} + 0.04t^2 + 40t\right]_0^{10}$$

$$\approx \$337.33 \text{ million}$$

55.

Quintile	Lowest	2nd	3rd	4th	Highest
Percent	13.19	17.77	20.74	23.13	25.17

$y(20) - y(0) \approx 13.19 - 0 = 13.19\%$

$y(40) - y(20) \approx 30.96 - 13.19 = 17.77\%$

$y(60) - y(40) \approx 51.70 - 30.96 = 20.74\%$

$y(80) - y(60) \approx 74.83 - 51.70 = 23.13\%$

$y(100) - y(80) \approx 100 - 74.83 = 25.17\%$

Section 5.6 The Definite Integral as the Limit of a Sum

1. The midpoints of the four intervals are $\frac{1}{8}, \frac{3}{8}, \frac{5}{8}$, and $\frac{7}{8}$. The approximate area is

$$A \approx \frac{1-0}{4}\left[f\left(\frac{1}{8}\right) + f\left(\frac{3}{8}\right) + f\left(\frac{5}{8}\right) + f\left(\frac{7}{8}\right) \right] = \frac{1}{4}\left[\frac{11}{4} + \frac{9}{4} + \frac{7}{4} + \frac{5}{4} \right] = 2.$$

The exact area is

$$A = \int_0^1 (-2x + 3)\, dx = \left[-x^2 + 3x \right]_0^1 = 2.$$

3. The midpoints of the four intervals are $\frac{1}{8}, \frac{3}{8}, \frac{5}{8}$, and $\frac{7}{8}$. The approximate area is

$$A \approx \frac{1-0}{4}\left[\sqrt{\frac{1}{8}} + \sqrt{\frac{3}{8}} + \sqrt{\frac{5}{8}} + \sqrt{\frac{7}{8}} \right] = \frac{1}{4}\left[\frac{\sqrt{2}}{4} + \frac{\sqrt{6}}{4} + \frac{\sqrt{10}}{4} + \frac{\sqrt{14}}{4} \right] \approx 0.6730.$$

The exact area is

$$A = \int_0^1 \sqrt{x}\, dx = \frac{2}{3}x^{3/2}\Big]_0^1 = \frac{2}{3} \approx 0.6667.$$

5. The midpoints of the four intervals are $-\frac{3}{4}, -\frac{1}{4}, \frac{1}{4}$, and $\frac{3}{4}$. The approximate area is

$$A \approx \frac{1-(-1)}{4}\left[\frac{41}{16} + \frac{33}{16} + \frac{33}{16} + \frac{41}{16} \right] = \frac{1}{2}\left[\frac{148}{16} \right] = \frac{37}{8} = 4.625.$$

The exact area is

$$A = \int_{-1}^1 (x^2 + 2)\, dx = \left[\frac{x^3}{3} + 2x \right]_{-1}^1 = \frac{14}{3} \approx 4.6667.$$

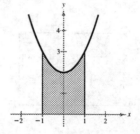

7. The midpoints of the four intervals are $\frac{5}{4}, \frac{7}{4}, \frac{9}{4}$, and $\frac{11}{4}$. The approximate area is

$$A \approx \frac{3-1}{4}\left[f\left(\frac{5}{4}\right) + f\left(\frac{7}{4}\right) + f\left(\frac{9}{4}\right) + f\left(\frac{11}{4}\right) \right]$$

$$= \frac{1}{2}\left[\frac{25}{8} + \frac{49}{8} + \frac{81}{8} + \frac{121}{8} \right]$$

$$= \frac{69}{4}$$

$$= 17.25.$$

The exact area is

$$A = \int_1^3 2x^2\, dx = \frac{2x^3}{3}\Big]_1^3 = \frac{52}{3} \approx 17.3333.$$

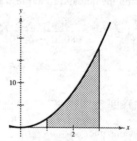

9. The midpoints of the four intervals are $\frac{1}{8}, \frac{3}{8}, \frac{5}{8}$, and $\frac{7}{8}$. The approximate area is

$$A \approx \frac{1-0}{4}\left[\frac{7}{512} + \frac{45}{512} + \frac{75}{512} + \frac{49}{512} \right] \approx 0.0859.$$

The exact area is

$$A = \int_0^1 (x^2 - x^3)\, dx = \left[\frac{x^3}{3} - \frac{x^4}{4} \right]_0^1 = \frac{1}{12} \approx 0.0833.$$

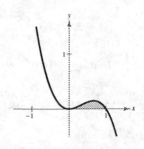

11. The midpoints of the four intervals are $\frac{1}{8}, \frac{3}{8}, \frac{5}{8},$ and $\frac{7}{8}$. The approximate area is

$$A \approx \frac{1-0}{4}\left[\frac{49}{512} + \frac{75}{512} + \frac{45}{512} + \frac{7}{512}\right] = \frac{11}{128} \approx 0.0859.$$

The exact area is

$$A = \int_0^1 x(1-x)^2\, dx = \int_0^1 (x^3 - 2x^2 + x)\, dx = \left[\frac{x^4}{4} - \frac{2x^3}{3} + \frac{x^2}{2}\right]_0^1 = \frac{1}{12} \approx 0.0833.$$

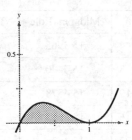

13. $\displaystyle\int_0^4 (2x^2 + 3)\, dx = \frac{164}{3} \approx 54.6667$

Using the Midpoint Rule with $n = 34$, we get 54.657.

15. $\displaystyle\int_1^2 (2x^2 - x + 1)\, dx = \left[\frac{2x^3}{3} - \frac{x^2}{2} + x\right]_1^2$

$$= \left(\frac{16}{3} - 2 + 2\right) - \left(\frac{2}{3} - \frac{1}{2} + 1\right)$$

$$= \frac{25}{6}$$

$$\approx 4.16\overline{6}$$

Using the Midpoint Rule with $n = 5$, we get 4.16.

17. The midpoints of the four intervals are $\frac{9}{4}, \frac{11}{4}, \frac{13}{4},$ and $\frac{15}{4}$. The approximate area is

$$A \approx \frac{4-2}{4}\left[f\left(\frac{9}{4}\right) + f\left(\frac{11}{4}\right) + f\left(\frac{13}{4}\right) + f\left(\frac{15}{4}\right)\right] = \frac{1}{2}(3) = 1.5.$$

The exact area is

$$A = \int_2^4 \frac{1}{4}y\, dy = \left.\frac{y^2}{8}\right]_2^4 = 2 - \frac{1}{2} = 1.5.$$

(The Midpoint Rule is exact for lines.)

19. Exact: $\displaystyle\int_0^2 x^3\, dx = \left.\frac{x^4}{4}\right]_0^2 = 4$

Trapezoidal: $\displaystyle\int_0^2 x^3\, dx \approx \frac{2-0}{2(8)}\left[(0)^3 + 2\left(\frac{1}{4}\right)^3 + 2\left(\frac{1}{2}\right)^3 + 2\left(\frac{3}{4}\right)^3 + 2(1)^3 + 2\left(\frac{5}{4}\right)^3 + 2\left(\frac{3}{2}\right)^3 + 2\left(\frac{7}{4}\right)^3 + (2)^3\right] = 4.0625$

Midpoint: 3.9688

The Midpoint Rule is a better approximation in this example.

21. $\displaystyle\int_0^2 \frac{1}{x+1}\, dx \approx \frac{2-0}{2(4)}\left[1 + 2\left(\frac{2}{3}\right) + 2\left(\frac{1}{2}\right) + 2\left(\frac{2}{5}\right) + \frac{1}{3}\right] \approx 1.1167$

(Exact answer is $\ln 3 \approx 1.0986$.)

23. $\displaystyle\int_{-1}^1 \frac{1}{x^2 + 1}\, dx \approx \frac{1-(-1)}{2(4)}\left[\frac{1}{(-1)^2} + 1 + \frac{2}{(-1/2)^2 + 1} + \frac{2}{(0)^2 + 1} + \frac{2}{(1/2)^2 + 1} + \frac{1}{(1)^2 + 1}\right] = 1.55$

(Exact value is $\pi/2$.)

25.

n	Midpoint Rule	Trapezoidal Rule
4	15.3965	15.6055
8	15.4480	15.5010
12	15.4578	15.4814
16	15.4613	15.4745
20	15.4628	15.4713

27. $A = \displaystyle\int_0^3 \sqrt{\dfrac{x^3}{4-x}}\, dx$

$\approx \dfrac{3-0}{2(10)}\left[0 + 2f\left(\dfrac{3}{10}\right) + 2f\left(\dfrac{3}{5}\right) + 2f\left(\dfrac{9}{10}\right) + 2f\left(\dfrac{6}{5}\right) + 2f\left(\dfrac{3}{2}\right) + 2f\left(\dfrac{9}{5}\right) + 2f\left(\dfrac{21}{10}\right) + 2f\left(\dfrac{12}{5}\right) + 2f\left(\dfrac{27}{10}\right) + f(3)\right]$

≈ 4.81

29. $s = \displaystyle\int_0^{20} v\, dt$

$\approx \dfrac{20-0}{2(4)}[v(0) + 2v(5) + 2v(10) + 2v(15) + v(20)]$

$\approx \dfrac{5}{2}[0 + 58.6 + 102.6 + 132 + 73.3]$

$= \dfrac{5}{2}(366.5)$

$= 916.25$ feet

31. Midpoint Rule: 3.1468

Trapezoidal Rule: 3.1312

Graphing utility: 3.141593

Section 5.7   Volumes of Solids of Revolution

1. $V = \pi \displaystyle\int_0^2 \left(\sqrt{4-x^2}\right)^2 dx$

$= \pi \displaystyle\int_0^2 (4 - x^2)\, dx$

$= \pi\left(4x - \dfrac{x^3}{3}\right)\Big]_0^2$

$= \dfrac{16\pi}{3}$

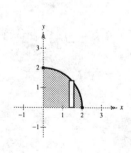

3. $V = \pi \displaystyle\int_1^4 \left(\sqrt{x}\right)^2 dx$

$= \pi \displaystyle\int_1^4 x\, dx$

$= \pi\left(\dfrac{x^2}{2}\right)\Big]_1^4$

$= 8\pi - \dfrac{\pi}{2}$

$= \dfrac{15\pi}{2}$

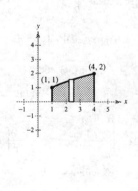

5. $V = \pi \displaystyle\int_{-2}^2 (4 - x^2)^2 dx$

$= \pi \displaystyle\int_{-2}^2 (16 - 8x^2 + x^4)\, dx$

$= \pi\left[16x - \dfrac{8}{3}x^3 + \dfrac{x^5}{5}\right]_{-2}^2$

$= \dfrac{512\pi}{15}$

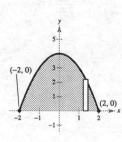

7. $V = \pi \displaystyle\int_{-2}^2 \left(1 - \dfrac{x^2}{4}\right)^2 dx$

$= 2\pi \displaystyle\int_0^2 \left(1 - \dfrac{x^2}{2} + \dfrac{x^4}{16}\right) dx$

$= 2\pi\left[x - \dfrac{1}{6}x^3 + \dfrac{1}{80}x^5\right]_0^2$

$= \dfrac{32\pi}{15}$

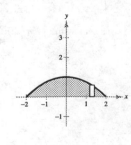

9. $V = \pi \int_0^1 (-x + 1)^2 \, dx$

$= \pi \int_0^1 (x^2 - 2x + 1) \, dx$

$= \pi \left(\dfrac{x^3}{3} - x^2 + x \right) \Big]_0^1$

$= \dfrac{\pi}{3}$

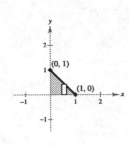

11. $V = \pi \int_1^2 \left[\left(-\dfrac{1}{2}x + 1 \right)^2 - \left(\dfrac{1}{x} - \dfrac{1}{2} \right)^2 \right] dx$

$= \pi \int_1^2 \left[\left(\dfrac{1}{4}x^2 - x + 1 \right) - \left(\dfrac{1}{x^2} + \dfrac{1}{4} - \dfrac{1}{x} \right) \right] dx$

$= \left[\dfrac{x^3}{12} - \dfrac{1}{2}x^2 + \dfrac{3}{4}x + \ln x + \dfrac{1}{x} \right]_1^2$

$= \pi \left[\ln 2 - \dfrac{2}{3} \right]$

≈ 0.0832

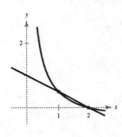

13. $V = \pi \int_0^2 (2x^2)^2 \, dx$

$= \pi \int_0^2 4x^4 \, dx$

$= \pi \left(\dfrac{4x^5}{5} \right) \Big]_0^2$

$= \dfrac{128\pi}{5}$

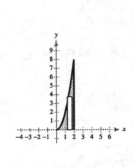

15. $V = \pi \int_0^1 (e^x)^2 \, dx$

$= \pi \dfrac{1}{2} e^{2x} \Big]_0^1$

$= \dfrac{\pi}{2}(e^2 - 1)$

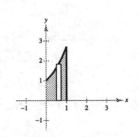

17. The points of intersection of the three graphs occur when $y = 0$ and $y = 4$.

$V = \pi \int_0^4 (\sqrt{y})^2 \, dy$

$= \pi \int_0^4 y \, dy$

$= \pi \left(\dfrac{y^2}{2} \right) \Big]_0^4$

$= 8\pi$

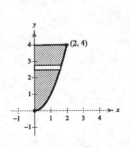

19. $V = \pi \int_0^2 \left(1 - \dfrac{1}{2}y \right)^2 dy$

$= \pi \int_0^2 \left[1 - y + \dfrac{1}{4}y^2 \right] dy$

$= \pi \left[y - \dfrac{y^2}{2} + \dfrac{1}{12}y^3 \right]_0^2$

$= \pi \left[2 - 2 + \dfrac{1}{12} \cdot 8 \right]$

$= \dfrac{2}{3}\pi$

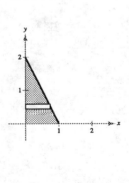

21. The points of intersection of the three graphs occur when $y = 0$ and $y = 1$.

$V = \pi \int_0^1 (y^{3/2})^2 \, dy$

$= \pi \int_0^1 y^3 \, dy$

$= \pi \left(\dfrac{y^4}{4} \right) \Big]_0^1$

$= \dfrac{\pi}{4}$

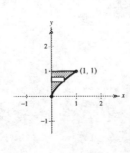

23. The points of intersection of the three graphs occur when $y = 0$ and $y = 2$.

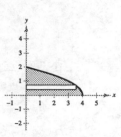

$$V = \pi \int_0^2 (4 - y^2)^2 \, dy$$

$$= \pi \int_0^2 (16 - 8y^2 + y^4) \, dy$$

$$= \pi \left[16y - \frac{8y^3}{3} + \frac{y^5}{5} \right]_0^2$$

$$= \pi \left[32 - \frac{64}{3} + \frac{32}{5} \right]$$

$$= \frac{256\pi}{15}$$

25. $V = \pi \int_0^6 \left(\frac{1}{2}x \right)^2 \, dx$

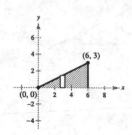

$$= \frac{\pi}{4} \int_0^6 x^2 \, dx$$

$$= \frac{\pi}{4} \left(\frac{x^3}{3} \right) \Big]_0^6$$

$$= 18\pi$$

27. A sphere of radius r can be formed by revolving the graph of $y = \sqrt{r^2 - x^2}$ about the x-axis.

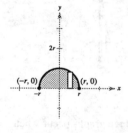

$$V = \pi \int_{-r}^{r} \left(\sqrt{r^2 - x^2} \right)^2 \, dx$$

$$= \pi \int_{-r}^{r} (r^2 - x^2) \, dx$$

$$= \pi \left[r^2 x - \frac{x^3}{3} \right]_{-r}^{r}$$

$$= \pi \left[\left(r^3 - \frac{r^3}{3} \right) - \left(-r^3 + \frac{r^3}{3} \right) \right]$$

$$= \frac{4\pi r^3}{3}$$

29. The upper half of the ellipse is given by

$$y = \sqrt{9 - \frac{9x^2}{16}}.$$

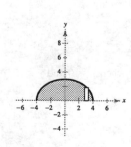

$$V = \pi \int_{-4}^{4} \left(9 - \frac{9}{16}x^2 \right) \, dx$$

$$= 18\pi \int_0^4 \left(1 - \frac{x^2}{16} \right) \, dx$$

$$= 18\pi \left[x - \frac{x^3}{48} \right]_0^4$$

$$= 18\pi \left[4 - \frac{4}{3} \right]$$

$$= 48\pi$$

31. (a) $y = 20[(0.005x)^2 - 1]$

Solving for x, we obtain

$$x = 200\sqrt{\frac{y}{20} + 1}.$$

$$V = \pi \int_{-20}^{0} \left(200\sqrt{\frac{y}{20} + 1}\right)^2 dy$$

$$= 40,000\pi \int_{-20}^{0} \left(\frac{y}{20} + 1\right) dy$$

$$= 40,000\pi \left[\frac{y^2}{40} + y\right]_{-20}^{0}$$

$$= 40,000\pi(10)$$

$$\approx 1,256,637 \text{ cubic feet}$$

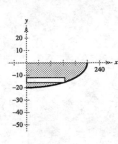

(b) The maximum number of fish that can be supported is $\frac{1,256,637}{500} \approx 2513$ fish.

33. Using the program on page 362 in the text, and $n = 100$,

$$V = \pi \int_{0}^{7} (\sqrt[3]{x + 1})^2 dx = \pi \int_{0}^{7} (x + 1)^{2/3} dx \approx 58.434 \text{ cubic units.}$$

Review Exercises for Chapter 5

1. $\displaystyle\int 16\, dx = 16x + C$

3. $\displaystyle\int (2x^2 + 5x)\, dx = \frac{2}{3}x^3 + \frac{5}{2}x^2 + C$

5. $\displaystyle\int \frac{2}{3\sqrt[3]{x}}\, dx = \int \frac{2}{3}x^{-1/3}\, dx = x^{2/3} + C$

7. $\displaystyle\int \left(\sqrt[3]{x^4} + 3x\right) dx = \int (x^{4/3} + 3x)\, dx$

$$= \frac{3}{7}x^{7/3} + \frac{3}{2}x^2 + C$$

9. $f(x) = \displaystyle\int (3x + 1)\, dx = \frac{3}{2}x^2 + x + C$

$f(2) = 6 = \frac{3}{2}(2)^2 + 2C = 8 + C \Rightarrow C = -2$

Hence, $f(x) = \frac{3}{2}x^2 + x - 2$.

11. $f'(x) = \displaystyle\int 2x^2\, dx = \frac{2}{3}x^3 + C$

$f'(3) = 10 = 18 + C_1 \Rightarrow C_1 = -8$

$f'(x) = \frac{2}{3}x^3 - 8$

$f(x) = \displaystyle\int \left(\frac{2}{3}x^3 - 8\right) dx = \frac{1}{6}x^4 - 8x + C_2$

$f(3) = 6 = \frac{81}{6} - 24 + C_2 \Rightarrow C_2 = 30 - \frac{81}{6} = \frac{33}{2}$

13. (a) $s(t) = -16t^2 + 80t$

$v(t) = -32t + 80 = 0$

$32t = 80$

$t = 2.5$ seconds

(b) $s(2.5) = 100$ feet

(c) $v(t) = -32t + 80 = 40$

$32t = 40$

$t = 1.25$ seconds

(d) $s(1.25) = 75$ feet

15. $\displaystyle\int (1 + 5x)^2 \, dx = \int (1 + 10x + 25x^2) \, dx$

$$= x + 5x^2 + \frac{25}{3}x^3 + C$$

or

$$\int (1 + 5x)^2 \, dx = \frac{1}{5}\frac{(1 + 5x)^3}{3} + C_1$$

$$= \frac{1}{15}(1 + 5x)^3 + C_1$$

17. $\displaystyle\int \frac{1}{\sqrt{5x - 1}} \, dx = \frac{1}{5}\int (5x - 1)^{-1/2}(5) \, dx$

$$= \frac{1}{5}(2)(5x - 1)^{1/2} + C$$

$$= \frac{2}{5}\sqrt{5x - 1} + C$$

19. $\displaystyle\int x(1 - 4x^2) \, dx = \int (x - 4x^3) \, dx$

$$= \frac{x^2}{2} - x^4 + C$$

21. $\displaystyle\int (x^4 - 2x)(2x^3 - 1) \, dx = \frac{1}{2}\int (x^4 - 2x)(4x^3 - 2) \, dx$

$$= \frac{1}{2} \cdot \frac{1}{2}(x^4 - 2x)^2 + C$$

$$= \frac{1}{4}(x^4 - 2x)^2 + C$$

23. $\displaystyle P = \int 2t(0.001t^2 + 0.5)^{1/4} \, dt$

$$= 800(0.001t^2 + 0.5)^{5/4} + C$$

$$P(0) = 0 = 800(0.5)^{5/4} + C \Rightarrow C \approx -336.36$$

$$P = 800(0.001t^2 + 0.5)^{5/4} - 336.36$$

(a) $P(6) \approx 30.5$ board feet

(b) $P(12) \approx 125.2$ board feet

25. $\displaystyle\int 3e^{-3x} \, dx = -e^{-3x} + C$

27. $\displaystyle\int (x - 1)e^{x^2 - 2x} \, dx = \frac{1}{2}e^{x^2 - 2x} + C$

29. $\displaystyle\int \frac{x^2}{1 - x^3} \, dx = -\frac{1}{3}\int \frac{1}{1 - x^3}(-3x^2) \, dx$

$$= -\frac{1}{3}\ln|1 - x^3| + C$$

31. $\displaystyle\int \frac{(\sqrt{x} + 1)^2}{\sqrt{x}} \, dx = \int \frac{x + 2\sqrt{x} + 1}{\sqrt{x}} \, dx$

$$= \int (x^{1/2} + 2 + x^{-1/2}) \, dx$$

$$= \frac{2}{3}x^{3/2} + 2x + 2x^{1/2} + C$$

33. $\displaystyle A = \int_0^2 (4 - 2x) \, dx$

$$= \left[4x - x^2 \right]_0^2$$

$$= 4$$

35. $\displaystyle A = \int_0^2 (y - 2)^2 \, dy$

$$= \left[\frac{y^3}{3} - 2y^2 + 4y - \frac{8}{3} \right]_0^2$$

$$= \frac{8}{3}$$

37. $\displaystyle A = \int_0^1 \frac{2}{x + 1} \, dx$

$$= 2\ln(x + 1) \Big]_0^1$$

$$= 2\ln 2$$

39. $\displaystyle\int_0^4 (2 + x) \, dx = \left[2x + \frac{x^2}{2} \right]_0^4 = 16$

41. $\displaystyle\int_{-1}^1 (4t^3 - 2t) \, dt = \left[t^4 - t^2 \right]_{-1}^1 = 0$

43. $\displaystyle\int_0^3 \frac{1}{\sqrt{1+x}}\,dx = \int_0^3 (1+x)^{-1/2}\,dx$

$$= 2\sqrt{1+x}\,\Big]_0^3$$

$$= 2$$

45. $\displaystyle\int_1^2 \left(\frac{1}{x^2} - \frac{1}{x^3}\right)dx = \int_1^2 (x^{-2} - x^{-3})\,dx$

$$= \left[\frac{x^{-1}}{-1} - \frac{x^{-2}}{-2}\right]_1^2$$

$$= \left[-\frac{1}{x} + \frac{1}{2x^2}\right]_1^2$$

$$= -\frac{3}{8} - \left(-\frac{1}{2}\right)$$

$$= \frac{1}{8}$$

47. $\displaystyle\int_1^3 \frac{3\ln x}{x}\,dx = \frac{1}{2}(3 + \ln x)^2\,\Big]_1^3$

$$= \frac{1}{2}[(3 + \ln 3)^2 - 3^2]$$

$$= \frac{1}{2}[6\ln 3 + (\ln 3)^2]$$

$$\approx 3.899$$

49. $\displaystyle\int_{-1}^1 3xe^{x^2-1}\,dx = \frac{3}{2}e^{x^2-1}\,\Big]_{-1}^1$

$$= \frac{3}{2}(1 - 1)$$

$$= 0$$

51. $\displaystyle C = \int (675 + 0.5x)\,dx = 675x + \frac{1}{4}x^2 + C_1$

$C(51) - C(50) = \$700.25$

53. Average value $\displaystyle = \frac{1}{10 - 5}\int_5^{10} \frac{4}{\sqrt{x-1}}\,dx$

$$= \frac{4}{5}\int_5^{10} (x-1)^{-1/2}\,dx$$

$$= \frac{8}{5}(x-1)^{1/2}\,\Big]_5^{10}$$

$$= \frac{8}{5}(3 - 2)$$

$$= \frac{8}{5}$$

To find the values for which $f(x) = \frac{8}{5}$, solve for x in the equation.

$$\frac{4}{\sqrt{x-1}} = \frac{8}{5}$$

$$\frac{5}{2} = \sqrt{x-1}$$

$$x - 1 = \frac{25}{4}$$

$$x = \frac{29}{4}$$

55. Average value $\displaystyle = \frac{1}{5 - 2}\int_2^5 e^{5-x}\,dx$

$$= \frac{1}{3}\left[-e^{5-x}\right]_2^5$$

$$= \frac{1}{3}(-1 + e^3)$$

To find the value of x for which $f(x) = \frac{1}{3}(-1 + e^3)$, we solve $e^{5-x} = \frac{1}{3}(-1 + e^3)$. Using a graphing utility, we obtain $x \approx 3.150$.

57. Average value $= \dfrac{1}{b-a} \displaystyle\int_a^b f(t)\, dt$

$\qquad = \dfrac{1}{2-0} \displaystyle\int_0^2 500 e^{0.04t}\, dt$

$\qquad = 250 \left[\dfrac{e^{0.04t}}{0.04} \right]_0^2$

$\qquad = 6250[e^{0.08} - 1]$

$\qquad \approx \$520.54$

59. (a) $S = 145.3t - 39.9t^2 + 4.6t^3 - 0.1t^4 + C$

$\qquad S(0) = 144 \Rightarrow C = 144$

$\qquad S = 145.3t - 39.9t^2 + 4.6t^3 - 0.1t^4 + 144$

(b) $S = 10{,}000$ (\$10,000,000 salary) when $t \approx 22$ (2002).

61. $\displaystyle\int_{-2}^{2} 6x^5\, dx = 0$ (odd function)

63. $\displaystyle\int_{-2}^{-1} \dfrac{4}{x^2}\, dx = \int_{1}^{2} \dfrac{4}{x^2}\, dx = 2$ (symmetric about y-axis)

65. Area $= \displaystyle\int_{1/2}^{3} \left(4 - \dfrac{1}{x^2} \right) dx$

$\qquad = \left[4x + \dfrac{1}{x} \right]_{1/2}^{3}$

$\qquad = \left(12 + \dfrac{1}{3} \right) - (2 + 2)$

$\qquad = \dfrac{25}{3}$

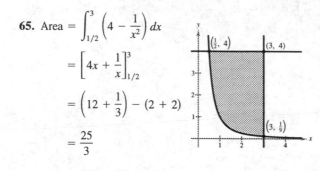

67. $A = \displaystyle\int_0^8 \dfrac{4}{\sqrt{x+1}}\, dx$

$\qquad = 4 \displaystyle\int_0^8 (x+1)^{-1/2}\, dx$

$\qquad = \left[4(2)(x+1)^{1/2} \right]_0^8$

$\qquad = \left[8\sqrt{x+1} \right]_0^8$

$\qquad = 8(3 - 1)$

$\qquad = 16$

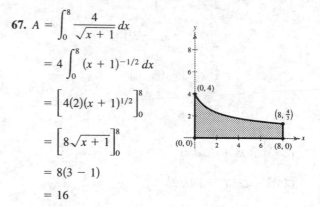

69. $\quad (x-3)^2 = 8 - (x-3)^2$

$\qquad 2(x-3)^2 = 8$

$\qquad (x-3)^2 = 4$

$\qquad x - 3 = \pm 2$

$\qquad x = 1, 5$

$A = \displaystyle\int_1^5 \{[8 - (x-3)^2] - (x-3)^2\}\, dx$

$\qquad = \displaystyle\int_1^5 [8 - 2(x-3)^2]\, dx$

$\qquad = \left[8x - \dfrac{2}{3}(x-3)^3 \right]_1^5$

$\qquad = \left(40 - \dfrac{16}{3} \right) - \left(8 + \dfrac{16}{3} \right)$

$\qquad = \dfrac{64}{3}$

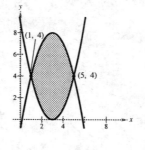

71. $\qquad\qquad x = 2 - x^2$

$\qquad\quad x^2 + x - 2 = 0$

$\quad (x+2)(x-1) = 0$

$\qquad\qquad\quad x = -2, 1$

$A = \displaystyle\int_{-2}^{1} (2 - x^2 - x)\, dx = \left[2x - \dfrac{x^3}{3} - \dfrac{x^2}{2} \right]_{-2}^{1} = \dfrac{9}{2}$

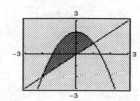

73. Demand function = Supply function

$$500 - x = 1.25x + 162.5$$

$$2.25x = 337.5$$

$$x = 150 \Rightarrow \text{price} = 350$$

Consumer surplus $= \displaystyle\int_0^{150} (\text{demand function} - \text{price})\, dx$

$$= \int_0^{150} [(500 - x) - 350]\, dx$$

$$= \left[150x - \frac{x^2}{2} \right]_0^{150}$$

$$= 11{,}250$$

Producer surplus $= \displaystyle\int_0^{150} (\text{price} - \text{supply function})\, dx$

$$= \int_0^{150} [350 - (1.25x + 162.5)]\, dx$$

$$= \left[187.5x - 1.25\frac{x^2}{2} \right]_0^{150}$$

$$= 14{,}062.5$$

75. Notice that the curves intersect at $t \approx 11$ ($t = 10.887$).

$$\int_{11}^{24} (y_1 - y_2)\, dt \approx 66.1 \text{ gallons more}$$

77. $\displaystyle\int_{12}^{16} [(-509.32 + 173.92t - 19.06t^2 + 0.75t^3) - (-1067.22 + 251.47t - 17.83t^2 + 0.43t^3)]\, dt \approx 501.92 \text{ million more}$

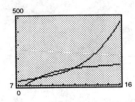

79. $\displaystyle\int_0^2 (x^2 + 1)^2\, dx$

$n = 4$: 13.3203

$n = 20$: 13.7167 (exact $13.7\overline{3}$)

81. $\displaystyle\int_0^1 \frac{1}{x^2 + 1}\, dx$

$n = 4$: 0.7867

$n = 20$: 0.7855

83. $V = \pi \displaystyle\int_1^4 \left(\frac{1}{\sqrt{x}}\right)^2 dx$

$$= \pi \int_1^4 \frac{1}{x}\, dx$$

$$= \pi \ln x \Big]_1^4$$

$$= \pi \ln 4$$

$$\approx 4.355$$

85. $V = \pi \displaystyle\int_0^2 (e^{1-x})^2\, dx$

$$= \pi \int_0^2 e^{2-2x}\, dx$$

$$= -\frac{\pi}{2}\left[e^{2-2x} \right]_0^2$$

$$= -\frac{\pi}{2}[e^{-2} - e^2]$$

$$= \frac{\pi}{2}(e^2 - e^{-2})$$

$$\approx 11.394$$

87. $V = \pi \int_0^2 [(2x + 1)^2 - 1^2] \, dx$

$= \pi \int_0^2 (4x^2 + 4x) \, dx$

$= \pi \left[\frac{4}{3}x^3 + 2x^2 \right]_0^2$

$= \pi \left(\frac{32}{3} + 8 \right)$

$= \frac{56}{3} \pi$

89. $V = \pi \int_0^1 [(x^2)^2 - (x^3)^2] \, dx$

$= \pi \left[\frac{x^5}{5} - \frac{x^7}{7} \right]_0^1$

$= \frac{2}{35} \pi$

91. $x^2 + y^2 = 1, \; y = \frac{1}{4}$

$x = \sqrt{1 - \left(\frac{1}{4}\right)^2} = \frac{\sqrt{15}}{4}$

$V = \pi \int_{-\sqrt{15}/4}^{\sqrt{15}/4} \left[(\sqrt{1 - x^2})^2 - \left(\frac{1}{4}\right)^2 \right] dx$

$= 2\pi \int_0^{\sqrt{15}/4} \left(1 - x^2 - \frac{1}{16} \right) dx$

$= \frac{5}{16} \pi \sqrt{15}$

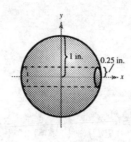

Practice Test for Chapter 5

1. Evaluate $\int (3x^2 - 8x + 5) \, dx$.

2. Evaluate $\int (x + 7)(x^2 - 4) \, dx$.

3. Evaluate $\int \dfrac{x^3 - 9x^2 + 1}{x^2} \, dx$.

4. Evaluate $\int x^3 \sqrt[4]{1 - x^4} \, dx$.

5. Evaluate $\int \dfrac{3}{\sqrt[3]{7x}} \, dx$.

6. Evaluate $\int \sqrt{6 - 11x} \, dx$.

7. Evaluate $\int \left(\sqrt[4]{x} + \sqrt[6]{x} \right) dx$.

8. Evaluate $\int \left(\dfrac{1}{x^4} - \dfrac{1}{x^5} \right) dx$.

9. Evaluate $\int (1 - x^2)^3 \, dx$.

10. Evaluate $\int \dfrac{5x}{(1 + 3x^2)^3} \, dx$.

11. Evaluate $\int e^{7x} \, dx$.

12. Evaluate $\int xe^{4x^2} \, dx$.

13. Evaluate $\int e^x(1 + 4e^x)^3 \, dx$.

14. Evaluate $\int (e^x + 2)^2 \, dx$.

15. Evaluate $\int \dfrac{e^{3x} - 4e^x + 1}{e^x} \, dx$.

16. Evaluate $\int \dfrac{1}{x + 6} \, dx$.

17. Evaluate $\int \dfrac{x^2}{8 - x^3} \, dx$.

18. Evaluate $\int \dfrac{e^x}{1 + 3e^x} \, dx$.

19. Evaluate $\int \dfrac{(\ln x)^6}{x} \, dx$.

20. Evaluate $\int \dfrac{x^2 + 5}{x - 1} \, dx$.

21. Evaluate $\int_0^3 (x^2 - 4x + 2) \, dx$.

22. Evaluate $\int_1^8 x \sqrt[3]{x} \, dx$.

23. Evaluate $\int_{\sqrt{5}}^{\sqrt{13}} \dfrac{x}{\sqrt{x^2 - 4}} \, dx$.

24. Sketch the region bounded by the graphs of $f(x) = x^2 - 6x$ and $g(x) = 0$ and find the area of the region.

25. Sketch the region bounded by the graphs of $f(x) = x^3 + 1$ and $g(x) = x + 1$ and find the area of the region.

26. Sketch the region bounded by the graphs of $f(y) = 1/y^2$, $x = 0$, $y = 1$, and $y = 3$ and find the area of the region.

27. Approximate the definite integral by the Midpoint Rule using $n = 4$.

$$\int_0^1 \sqrt{x^3 + 2}\, dx$$

28. Approximate the definite integral by the Midpoint Rule using $n = 4$.

$$\int_3^4 \frac{1}{x^2 - 5}\, dx$$

29. Find the volume of the solid generated by revolving the region bounded by the graphs of $f(x) = 1/\sqrt[3]{x}$, $x = 1$, $x = 8$, and $y = 0$ about the x-axis.

30. Find the volume of the solid generated by revolving the region bounded by the graphs of $y = \sqrt{25 - x}$, $y = 0$, and $x = 0$ about the y-axis.

Graphing Calculator Required

31. Use a program similar to that on page 362 of the textbook to approximate the following integral for $n = 50$ and $n = 100$.

$$\int_0^4 \sqrt{1 + x^4}\, dx$$

32. Use a graphing calculator to sketch the region bounded by $f(x) = 3 - \sqrt{x}$ and $g(x) = 3 - \frac{1}{3}x$. Based on the graph alone (do no calculations), determine which value best approximates the bounded area.

(a) 13   (b) 3   (c) 5   (d) 6

CHAPTER 6
Techniques of Integration

CHAPTER 6
Techniques of Integration

Section 6.1   Integration by Substitution

Solutions to Odd-Numbered Exercises

1. $\int (x-2)^4\,dx = \dfrac{(x-2)^5}{5} + C$

3. $\int \dfrac{2}{(t-9)^2}\,dt = 2\int (t-9)^{-2}\,dt$

$\qquad\qquad\qquad = (2)\dfrac{(t-9)^{-1}}{-1} + C$

$\qquad\qquad\qquad = -\dfrac{2}{t-9} + C$

$\qquad\qquad\qquad = \dfrac{2}{9-t} + C$

5. $\int \sqrt{1+x}\,dx = \int (1+x)^{1/2}\,dx$

$\qquad\qquad = \dfrac{2}{3}(1+x)^{3/2} + C$

7. $\int \dfrac{12x+2}{3x^2+x}\,dx = 2\int \dfrac{6x+1}{3x^2+x}\,dx$

$\qquad\qquad\qquad = 2\ln|3x^2+x| + C$

$\qquad\qquad\qquad = \ln(3x^2+x)^2 + C$

9. $\int \dfrac{1}{(5x+1)^3}\,dx = \dfrac{1}{5}\int (5x+1)^{-3}(5)\,dx$

$\qquad\qquad = \left(\dfrac{1}{5}\right)\dfrac{(5x+1)^{-2}}{-2} + C$

$\qquad\qquad = -\dfrac{1}{10(5x+1)^2} + C$

11. $\int \dfrac{1}{\sqrt{x+1}}\,dx = \int (x+1)^{-1/2}\,dx$

$\qquad\qquad = 2(x+1)^{1/2} + C$

13. $\int \dfrac{e^{3x}}{1-e^{3x}}\,dx = -\dfrac{1}{3}\int \dfrac{-3e^{3x}}{1-e^{3x}}\,dx$

$\qquad\qquad = -\dfrac{1}{3}\ln|1-e^{3x}| + C$

15. $\int \dfrac{x^2}{x-1}\,dx = \int \left(x+1+\dfrac{1}{x-1}\right)\,dx$

$\qquad\qquad = \dfrac{x^2}{2} + x + \ln|x-1| + C$

17. $\int x\sqrt{x^2+4}\,dx = \dfrac{1}{2}\int (x^2+4)^{1/2}\,2x\,dx$

$\qquad\qquad = \dfrac{1}{3}(x^2+4)^{3/2} + C$

19. $\int e^{5x}\,dx = \dfrac{1}{5}\int e^{5x}(5)\,dx = \dfrac{1}{5}e^{5x} + C$

21. Let $u = x+1, du = dx, x = u-1.$

$\qquad \int \dfrac{x}{(x+1)^4}\,dx = \int \dfrac{u-1}{u^4}\,du = \int u^{-3}\,du - \int u^{-4}\,du$

$\qquad\qquad\qquad = \dfrac{u^{-2}}{-2} - \dfrac{u^{-3}}{-3} + C = \dfrac{-1}{2u^2} + \dfrac{1}{3u^3} + C = \dfrac{-1}{2(x+1)^2} + \dfrac{1}{3(x+1)^3} + C$

23. Let $u = 3x - 1$, then $x = (u + 1)/3$ and $dx = (1/3) \, du$.

$$\int \frac{x}{(3x - 1)^2} \, dx = \int \frac{(u + 1)/3}{u^2} \left(\frac{1}{3}\right) du$$

$$= \frac{1}{9} \int \left(\frac{1}{u} + \frac{1}{u^2}\right) du$$

$$= \frac{1}{9}\left[\ln|u| - \frac{1}{u}\right] + C$$

$$= \frac{1}{9}\left[\ln|3x - 1| - \frac{1}{3x - 1}\right] + C$$

25. Let $u = \sqrt{t} - 1$, $t = (u + 1)^2$, and $dt = 2(u + 1) \, du$.

$$\int \frac{1}{\sqrt{t} - 1} \, dt = \int \frac{2(u + 1)}{u} du$$

$$= \int \left(2 + \frac{2}{u}\right) du$$

$$= 2u + 2 \ln|u| + C$$

$$= 2(\sqrt{t} - 1) + 2 \ln\left|\sqrt{t} - 1\right| + C$$

27. $\displaystyle\int \frac{2\sqrt{t} + 1}{t} \, dt = \int \left(2t^{-1/2} + \frac{1}{t}\right) dt$

$$= 4t^{1/2} + \ln|t| + C$$

$$= 4\sqrt{t} + \ln|t| + C$$

29. $\displaystyle\int_0^4 \sqrt{2x + 1} \, dx = \frac{1}{2}\int_0^4 (2x + 1)^{1/2}(2) \, dx$

$$= \frac{1}{2}\left(\frac{2}{3}\right)(2x + 1)^{3/2}\Big]_0^4$$

$$= \frac{1}{3}(9)^{3/2} - \frac{1}{3}(1)^{3/2}$$

$$= \frac{26}{3}$$

31. $\displaystyle\int_0^1 3xe^{x^2} \, dx = \frac{3}{2}\int_0^1 2xe^{x^2} \, dx$

$$= \frac{3}{2}e^{x^2}\Big]_0^1$$

$$= \frac{3}{2}(e - 1)$$

$$\approx 2.577$$

33. Let $u = x + 4$, $x = u - 4$, and $du = dx$. $u = 4$ when $x = 0$, $u = 8$ when $x = 4$.

$$\int_0^4 \frac{x}{(x + 4)^2} \, dx = \int_4^8 \frac{u - 4}{u^2} \, du$$

$$= \int_4^8 \left(\frac{1}{u} - 4u^{-2}\right) du$$

$$= \left[\ln u + 4\frac{1}{u}\right]_4^8$$

$$= \left(\ln 8 + \frac{1}{2}\right) - (\ln 4 + 1)$$

$$= \ln 2 - \frac{1}{2}$$

$$\approx 0.193$$

35. Let $u = 1 - x$, then $x = 1 - u$ and $dx = -du$. $u = 1$ when $x = 0$, and $u = 0.5$ when $x = 0.5$.

$$\int_0^{0.5} x(1 - x)^3 \, dx = \int_1^{0.5} (1 - u)u^3(-du)$$

$$= \int_1^{0.5} (u^4 - u^3) \, du$$

$$= \left(\frac{u^5}{5} - \frac{u^4}{4}\right)\Big]_1^{0.5}$$

$$= \left(\frac{1}{160} - \frac{1}{64}\right) - \left(\frac{1}{5} - \frac{1}{4}\right)$$

$$= \frac{13}{320}$$

37. Let $u = \sqrt{x - 3}$, then $x = u^2 + 3$ and $dx = 2u\,du$. $u = 0$ when $x = 3$, and $u = 2$ when $x = 7$.

$$\int_3^7 x\sqrt{x - 3}\,dx = \int_0^2 (u^2 + 3)u(2u\,du)$$

$$= 2\int_0^2 (u^4 + 3u^2)\,du$$

$$= 2\left(\frac{u^5}{5} + u^3\right)\Bigg]_0^2$$

$$= 2\left(\frac{32}{5} + 8\right)$$

$$= \frac{144}{5}$$

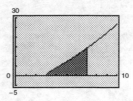

39. Let $u = \sqrt{1 - x}$, then $x = 1 - u^2$ and $dx = -2u\,du$. $u = 2$ when $x = -3$, and $u = 1$ when $x = 0$.

$$\text{Area} = \int_{-3}^0 x^2\sqrt{1 - x}\,dx$$

$$= \int_2^1 (1 - u^2)^2 u(-2u)\,du$$

$$= -2\int_2^1 (u^6 - 2u^4 + u^2)\,du$$

$$= -2\left[\frac{u^7}{7} - 2\frac{u^5}{5} + \frac{u^3}{3}\right]_2^1$$

$$= -2\left[\left(\frac{1}{7} - \frac{2}{5} + \frac{1}{3}\right) - \left(\frac{128}{7} - \frac{64}{5} + \frac{8}{3}\right)\right]$$

$$= -2\left(\frac{8}{105} - \frac{856}{105}\right)$$

$$= \frac{1696}{105}$$

$$\approx 16.1524$$

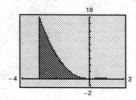

41. Let $u = \sqrt{2x - 1}$, then $x = \frac{1}{2}(u^2 + 1)$, and $dx = u\,du$. $u = 1$ when $x = 1$, and $u = 3$ when $x = 5$.

$$\text{Area} = \int_1^5 \frac{x^2 - 1}{\sqrt{2x - 1}}\,dx$$

$$= \int_1^3 \frac{\left(\frac{u^2 + 1}{2}\right)^2 - 1}{u}u\,du$$

$$= \frac{1}{4}\int_1^3 (u^4 + 2u^2 - 3)\,du$$

$$= \frac{1}{4}\left[\frac{u^5}{5} + \frac{2u^3}{3} - 3u\right]_1^3$$

$$= \frac{224}{15}$$

$$= 14.9\overline{3}$$

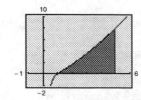

43. Let $u = \sqrt[3]{x + 1}$, then $x = u^3 - 1$ and $dx = 3u^2\,du$. $u = 1$ when $x = 0$, and $u = 2$ when $x = 7$.

$$\int_0^7 x\sqrt[3]{x + 1}\,dx = 3\int_1^2 (u^3 - 1)u^3\,du$$

$$= 3\int_1^2 (u^6 - u^3)\,du$$

$$= 3\left(\frac{u^7}{7} - \frac{u^4}{4}\right)\Big]_1^2$$

$$= 3\left(\frac{128}{7} - 4\right) - 3\left(\frac{1}{7} - \frac{1}{4}\right)$$

$$= \frac{1209}{28}$$

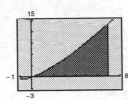

45. Let $u = x + 2$, $x = u - 2$, and $du = dx$. $u = 0$ when $x = -2$, and $u = 2$ when $x = 0$.

$$\int_{-2}^0 -x\sqrt{x + 2}\,dx = \int_0^2 -(u - 2)\sqrt{u}\,du$$

$$= \int_0^2 (-u^{3/2} + 2u^{1/2})\,du$$

$$= \left[-\frac{2}{5}u^{5/2} + \frac{4}{3}u^{3/2}\right]_0^2$$

$$= -\frac{2}{5}\cdot 2^{3/2} + \frac{4}{3}\cdot 2^{3/2}$$

$$= \sqrt{2}\left(\frac{8}{3} - \frac{8}{5}\right)$$

$$= \frac{16\sqrt{2}}{15}$$

47. Use $y = x\sqrt{1 - x^2}$ and $y = 0$ and multiply by 4.

$$A = 4\int_0^1 x\sqrt{1 - x^2}\,dx$$

$$= \frac{4}{-2}\int_0^1 (1 - x^2)^{1/2}(-2x)\,dx$$

$$= -2\left(\frac{2}{3}\right)(1 - x^2)^{3/2}\Big]_0^1$$

$$= -\frac{4}{3}(0 - 1)$$

$$= \frac{4}{3}\text{ square units}$$

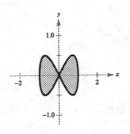

49. $V = 2\pi\int_0^1 \left(x\sqrt{1 - x^2}\right)^2 dx$

$$= 2\pi\int_0^1 x^2(1 - x^2)\,dx$$

$$= 2\pi\left[\frac{x^3}{3} - \frac{x^5}{5}\right]_0^1$$

$$= 2\pi\left(\frac{1}{3} - \frac{1}{5}\right)$$

$$= \frac{4\pi}{15}\text{ cubic unit}$$

51. $\dfrac{1}{1 - 0}\displaystyle\int_0^1 [f(x) - g(x)]\,dx = \int_0^1 \left[\frac{1}{x + 1} - \frac{x}{(x + 1)^2}\right] dx$

$$= \int_0^1 \frac{1}{(x + 1)^2}\,dx$$

$$= -\frac{1}{x + 1}\Big]_0^1$$

$$= -\frac{1}{2} + 1$$

$$= \frac{1}{2}$$

53. Let $u = \sqrt{1 - x}$, then $x = 1 - u^2$ and $dx = -2u\,du$.

$$\int \frac{15}{4} x\sqrt{1 - x}\,dx = \frac{1}{2}(1 - x)^{3/2}(-3x - 2) + C$$

(a) $P(0.40 \leq x \leq 0.80) = \frac{1}{2}(1 - x)^{3/2}(-3x - 2)\Big]_{0.40}^{0.80} = 0.547$

(b) $P(0 \leq x \leq b) = \frac{1}{2}(1 - x)^{3/2}(-3x - 2)\Big]_{0}^{b}$

$$= \frac{1}{2}[(1 - b)^{3/2}(-3b - 2) + 2]$$

$$= 0.5$$

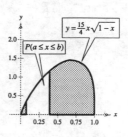

Solving this equation for b produces

$$(1 - b)^{3/2}(-3b - 2) + 2 = 1$$

$$(1 - b)^{3/2}(-3b - 2) = -1$$

$$(1 - b)^{3/2}(3b + 2) = 1$$

$$(1 - b)^3(3b + 2)^2 = 1$$

$$b \approx 0.586.$$

55. Average $= \dfrac{1}{14 - 0} \displaystyle\int_0^{14} t\sqrt{14 - t}\,dt$

Let $u = 14 - t$, $t = 14 - u$, and $dt = -du$. When $t = 0$, $u = 14$ and when $t = 14$, $u = 0$.

$$\text{Average} = \frac{-1}{14} \int_{14}^{0} (14 - u)\sqrt{u}\,du$$

$$= \frac{-1}{14} \int_{14}^{0} (14u^{1/2} - u^{3/2})\,du$$

$$= \frac{-1}{14} \left[\frac{28}{3} u^{3/2} - \frac{2}{5} u^{5/2} \right]_{14}^{0}$$

$$= \frac{-1}{14} \left[-\frac{28}{3}(14)^{3/2} + \frac{2}{5}(14)^{5/2} \right]$$

$$\approx 13.97 \text{ inches}$$

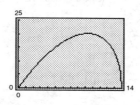

57. Certain fruit and vegetable sales resemble this pattern.

59. $A = 4 \displaystyle\int_0^1 x\sqrt{1 - x^2}\,dx \approx 1.346$

Section 6.2   Integration by Parts and Present Value

1. Let $u = x$ and $dv = e^{3x}\,dx$, then $du = dx$ and $v = \frac{1}{3}e^{3x}$.

$$\int xe^{3x}\,dx = \frac{1}{3}xe^{3x} - \int \frac{1}{3}e^{3x}\,dx$$

$$= \frac{1}{3}xe^{3x} - \frac{1}{9}e^{3x} + C$$

$$= \frac{1}{9}e^{3x}(3x - 1) + C$$

3. Let $u = x^2$ and $dv = e^{-x}\,dx$, then $du = 2x\,dx$ and $v = -e^{-x}$.

$$\int x^2 e^{-x}\,dx = -x^2 e^{-x} + 2\int xe^{-x}\,dx$$

Let $u = x$ and $dv = e^{-x}\,dx$, then $du = dx$ and $v = -e^{-x}$.

$$\int x^2 e^{-x}\,dx = -x^2 e^{-x} + 2\left[-xe^{-x} + \int e^{-x}\,dx \right]$$

$$= -x^2 e^{-x} - 2xe^{-x} - 2e^{-x} + C$$

$$= -e^{-x}(x^2 + 2x + 2) + C$$

5. Let $u = \ln 2x$ and $dv = dx$, then $du = (1/x)\, dx$ and $v = x$.

$$\int \ln 2x\, dx = x \ln 2x - \int x\left(\frac{1}{x}\right) dx$$

$$= x \ln 2x - \int dx$$

$$= x \ln 2x - x + C$$

$$= x(\ln 2x - 1) + C$$

7. $\displaystyle\int e^{4x}\, dx = \frac{1}{4} \int e^{4x}(4)\, dx = \frac{1}{4} e^{4x} + C$

9. Let $u = x$ and $dv = e^{4x}\, dx$, then $du = dx$ and $v = \frac{1}{4} e^{4x}$.

$$\int xe^{4x}\, dx = \frac{1}{4} xe^{4x} - \frac{1}{4} \int e^{4x}\, dx$$

$$= \frac{1}{4} xe^{4x} - \frac{1}{16} e^{4x} + C$$

$$= \frac{e^{4x}}{16}(4x - 1) + C$$

11. $\displaystyle\int xe^{x^2}\, dx = \frac{1}{2} e^{x^2} + C$

13. Let $u = x^2$ and $dv = e^x\, dx$, then $du = 2x\, dx$ and $v = e^x$.

$$\int x^2 e^x\, dx = x^2 e^x - \int 2xe^x\, dx$$

Let $u = 2x$ and $dv = e^x\, dx$, then $du = 2\, dx$ and $v = e^x$.

$$\int x^2 e^x\, dx = x^2 e^x - \left[2xe^x - \int e^x(2\, dx) \right]$$

$$= (x^2 - 2x + 2)e^x + C$$

15. Let $u = 1/t$, $du = (-1/t^2)\, dt$.

$$\int \frac{e^{1/t}}{t^2}\, dt = -\int e^u\, du$$

$$= -e^u + C$$

$$= -e^{1/t} + C$$

17. Let $u = \ln(t + 1)$ and $dv = t\, dt$, then $du = [1/(t + 1)]\, dt$ and $v = t^2/2$.

$$\int t \ln(t + 1)\, dt = \frac{t^2}{2} \ln(t + 1) - \frac{1}{2} \int \frac{t^2}{t + 1}\, dt$$

$$= \frac{t^2}{2} \ln(t + 1) - \frac{1}{2} \int \left(t - 1 + \frac{1}{t + 1} \right) dt$$

$$= \frac{t^2}{2} \ln(t + 1) - \frac{1}{2} \left[\frac{t^2}{2} - t + \ln(t + 1) \right] + C$$

$$= \frac{1}{4} [2(t^2 - 1) \ln(t + 1) + t(2 - t)] + C$$

19. Let $u = (\ln x)^2$ and $dv = x\, dx$, then $du = [(2 \ln x)/x]\, dx$ and $v = x^2/2$.

$$\int x(\ln x)^2\, dx = \frac{x^2}{2} (\ln x)^2 - \int x \ln x\, dx$$

Let $u = \ln x$ and $v = x\, dx$, then $du = (1/x)\, dx$ and $v = x^2/2$.

$$\int x(\ln x)^2\, dx = \frac{x^2}{2} (\ln x)^2 - \left[\frac{x^2}{2} \ln x - \int \frac{x}{2}\, dx \right]$$

$$= \frac{x^2}{2} (\ln x)^2 - \frac{x^2}{2} \ln x + \frac{x^2}{4} + C$$

21. $\displaystyle\int \frac{(\ln x)^2}{x}\, dx = \int (\ln x)^2 \left(\frac{1}{x}\right) dx$

$$= \frac{(\ln x)^3}{3} + C$$

23. $\displaystyle\int x(x+1)^2\,dx = \int x(x^2+2x+1)\,dx$

$\displaystyle\qquad = \int (x^3+2x^2+x)\,dx$

$\displaystyle\qquad = \frac{x^4}{4} + \frac{2x^3}{3} + \frac{x^2}{2} + C$

25. Let $u = xe^{2x}$ and $dv = (2x+1)^{-2}\,dx$, then
$du = e^{2x}(2x+1)\,dx$ and $v = -1/[2(2x+1)]$.

$\displaystyle\int \frac{xe^{2x}}{(2x+1)^2}\,dx = -\frac{xe^{2x}}{2(2x+1)} + \frac{1}{2}\int e^{2x}\,dx$

$\displaystyle\qquad = -\frac{xe^{2x}}{2(2x+1)} + \frac{1}{4}e^{2x} + C$

$\displaystyle\qquad = \frac{e^{2x}}{4(2x+1)} + C$

27. Area $= \displaystyle\int_0^2 x^3 e^x\,dx$

First use integration by parts three times to evaluate the indefinite integral. Let $u = x^3$, $dv = e^x\,dx$, $du = 3x^2\,dx$, $v = e^x$.

$\displaystyle\int x^3 e^x\,dx = x^3 e^x - \int 3x^2 e^x\,dx = x^3 e^x - 3\int x^2 e^x\,dx$

Let $u = x^2$, $dv = e^x\,dx$, $du = 2x\,dx$, $v = e^x$.

$\displaystyle\int x^3 e^x\,dx = x^3 e^x - 3\left[x^2 e^x - \int 2xe^x\,dx\right] = x^3 e^x - 3x^2 e^x + 6\int xe^x\,dx$

Let $u = x$, $dv = e^x\,dx$, $du = dx$, $v = e^x$.

$\displaystyle\int x^3 e^x\,dx = x^3 e^x - 3x^2 e^x + 6\left[xe^x - \int e^x\,dx\right] = (x^3 - 3x^2 + 6x - 6)e^x + C$

Area $= \left[(x^3 - 3x^2 + 6x - 6)e^x\right]_0^2 = 2e^2 + 6 \approx 20.778$

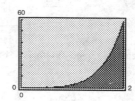

29. $A = \displaystyle\int_1^e x^2 \ln x\,dx$

Use integration by parts: $u = \ln x$, $dv = x^2$, $du = (1/x)\,dx$, $v = x^3/3$.

$\displaystyle\int x^2 \ln x\,dx = \frac{x^3}{3}\ln x - \int \frac{x^3}{3}\left(\frac{1}{x}\right)\,dx = \frac{x^3}{3}\ln x - \frac{x^3}{9} + C$

$A = \left[\dfrac{x^3}{3}\ln x - \dfrac{x^3}{9}\right]_1^e = \left(\dfrac{e^3}{3} - \dfrac{e^3}{9}\right) - \left(-\dfrac{1}{9}\right) = \dfrac{1}{9} + \dfrac{2e^3}{9} = \dfrac{1}{9}(1 + 2e^3)$

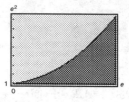

31. Let $u = x^2$ and $dv = e^x\,dx$, then $du = 2x\,dx$ and $v = e^x$.

$\displaystyle\int_0^1 x^2 e^x\,dx = x^2 e^x\Big]_0^1 - 2\int_0^1 xe^x\,dx$

Let $u = x$ and $dv = e^x\,dx$, then $du = dx$ and $v = e^x$.

$\displaystyle\int_0^1 x^2 e^x\,dx = x^2 e^x\Big]_0^1 - 2\left(xe^x\Big]_0^1 - \int_0^1 e^x\,dx\right)$

$\qquad = e - 2\left(e - \left[e^x\right]_0^1\right)$

$\qquad = e - 2e + 2(e-1)$

$\qquad = e - 2$

$\qquad \approx 0.718$

33. Let $u = \ln x$ and $dv = x^5\,dx$, then $du = (1/x)\,dx$ and $v = x^6/6$.

$\displaystyle\int_1^e x^5 \ln x\,dx = \frac{x^6}{6}\ln x\Big]_1^e - \int_1^e \frac{x^5}{6}\,dx$

$\qquad = \dfrac{e^6}{6} - \left[\dfrac{x^6}{36}\right]_1^e$

$\qquad = \dfrac{e^6}{6} - \dfrac{e^6}{36} + \dfrac{1}{36}$

$\qquad = \dfrac{5}{36}e^6 + \dfrac{1}{36}$

$\qquad \approx 56.060$

35. (a) Let $u = 2x$ and $dv = \sqrt{2x - 3}\, dx$, then $du = 2\, dx$ and $v - \frac{1}{3}(2x - 3)^{3/2}$.

$$\int 2x\sqrt{2x - 3}\, dx = \frac{2}{3}x(2x - 3)^{3/2} - \frac{2}{3}\int (2x - 3)^{3/2}\, dx$$

$$= \frac{2}{3}x(2x - 3)^{3/2} - \frac{2}{15}(2x - 3)^{5/2} + C$$

$$= \frac{2}{15}(2x - 3)^{3/2}(3x + 3) + C$$

$$= \frac{2}{5}(2x - 3)^{3/2}(x + 1) + C$$

(b) Let $u = \sqrt{2x - 3}$, then $x = (u^2 + 3)/2$ and $dx = u\, du$.

$$\int 2x\sqrt{2x - 3}\, dx = \int (u^2 + 3)u^2\, du$$

$$= \frac{u^5}{5} + u^3 + C$$

$$= \frac{u^3}{5}(u^2 + 5) + C$$

$$= \frac{2}{5}(2x - 3)^{3/2}(x + 1) + C$$

37. (a) Let $u = x$ and $dv = \left(1/\sqrt{4 + 5x}\right) dx$, then $du = dx$ and $v = (2/5)\sqrt{4 + 5x}$.

$$\int \frac{x}{\sqrt{4 + 5x}}\, dx = \frac{2}{5}x\sqrt{4 + 5x} - \frac{2}{5}\int \sqrt{4 + 5x}\, dx$$

$$= \frac{2}{5}x\sqrt{4 + 5x} - \frac{4}{75}(4 + 5x)^{3/2} + C$$

$$= \frac{2}{75}\sqrt{4 + 5x}(5x - 8) + C$$

(b) Let $u = \sqrt{4 + 5x}$, then $x = (u^2 - 4)/5$ and $dx = (2u/5)\, du$.

$$\int \frac{x}{\sqrt{4 + 5x}}\, dx = \int \frac{(u^2 - 4)/5}{u}\left(\frac{2u}{5}\right) du$$

$$= \frac{2}{25}\int (u^2 - 4)\, du$$

$$= \frac{2}{25}\left(\frac{u^3}{3} - 4u\right) + C$$

$$= \frac{2}{75}\sqrt{4 + 5x}(5x - 8) + C$$

39. Let $u = \ln x$ and $dv = x^n\, dx$, then $du = (1/x)\, dx$ and $v = x^{n+1}/(n + 1)$.

$$\int x^n \ln x\, dx = \frac{x^{n+1}}{n + 1}\ln x - \int \frac{1}{x}\cdot\frac{x^{n+1}}{n + 1}\, dx$$

$$= \frac{x^{n+1}}{n + 1}\ln x - \frac{1}{n + 1}\int x^n\, dx$$

$$= \frac{x^{n+1}}{n + 1}\ln x - \frac{1}{n + 1}\cdot\frac{x^{n+1}}{n + 1} + C$$

$$= \frac{x^{n+1}}{(n + 1)^2}[-1 + (n + 1)\ln x] + C$$

41. Using $n = 2$ and $a = 5$, we have

$$\int x^2 e^{5x}\, dx = \frac{x^2 e^{5x}}{5} - \frac{2}{5}\int x e^{5x}\, dx.$$

Now, using $n = 1$ and $a = 5$, we have the following.

$$\int x^2 e^{5x}\, dx = \frac{x^2 e^{5x}}{5} - \frac{2}{5}\left[\frac{x e^{5x}}{5} - \frac{1}{5}\int e^{5x}\, dx\right]$$

$$= \frac{x^2 e^{5x}}{5} - \frac{2x e^{5x}}{25} + \frac{2e^{5x}}{125} + C$$

$$= \frac{e^{5x}}{125}(25x^2 - 10x + 2) + C$$

43. Using $n = -2$, Exercise 39 yields

$$\int x^{-2} \ln x \, dx = \frac{x^{-2+1}}{(-2+1)^2}[-1 + (-2+1)\ln x] + C$$

$$= \frac{1}{x}(-1 - \ln x) + C$$

$$= -\frac{1}{x} - \frac{\ln x}{x} + C.$$

45. Letting $u = x$ and $dv = e^{-x} \, dx$, we have $du = dx$ and $v = -e^{-x}$, and the area is

$$A = \int_0^4 xe^{-x} \, dx$$

$$= -xe^{-x} \Big]_0^4 + \int_0^4 e^{-x} \, dx$$

$$= -4e^{-4} - \Big[e^{-x}\Big]_0^4$$

$$= -4e^{-4} - e^{-4} + 1$$

$$= 1 - 5e^{-4}$$

$$\approx 0.908.$$

47. (a) Letting $u = 2 \ln x$ and $dv = dx$, we have $du = (2/x) \, dx$ and $v = x$.

$$\text{Area} = \int_1^e 2 \ln x \, dx$$

$$= 2x \ln x \Big]_1^e - \int_1^e 2 \, dx$$

$$= 2e - \Big[2x\Big]_1^e$$

$$= 2e - 2e + 2$$

$$= 2$$

(b) Letting $u = (2 \ln x)^2$ and $dv = dx$, we have $du = 8 \ln x (1/x) \, dx$ and $v = x$.

$$\text{Volume} = \pi \int_1^e (2 \ln x)^2 \, dx$$

$$= \pi \left(\Big[x(2 \ln x)^2\Big]_1^e - \int_1^e 8 \ln x \, dx \right)$$

$$= \pi \left(4e - \Big[8(x \ln x - x)\Big]_1^e \right)$$

$$= \pi(4e - [8])$$

$$= 4\pi(e - 2)$$

$$\approx 9.026$$

49. $\displaystyle\int_0^2 t^3 e^{-4t} \, dt = \frac{3}{128} - \frac{379}{128}e^{-8} \approx 0.022$

51. $\displaystyle\int_0^5 x^4(25 - x^2)^{3/2} \, dx = \frac{1{,}171{,}875}{256}\pi \approx 14{,}381{,}070$

53. (a) $x = 500(20 + te^{-0.1t})$, $0 \le t \le 10$

$$\frac{dx}{dt} = 500(-0.1te^{-0.1t} + e^{-0.1t})$$

$$= 500e^{-0.1t}(-0.1t + 1)$$

On the interval $(0, 10)$, $dx/dt > 0$. Thus, x is increasing. The corporate officers are forecasting an increase in demand over the next decade.

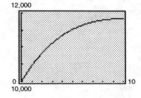

(b) $\displaystyle\int_0^{10} 500(20 + te^{-0.1t}) \, dt = 500\left[\int_0^{10} 20 \, dt + \int_0^{10} te^{-0.1t} \, dt\right]$

$$= 500(200) + 500\int te^{-0.1t} \, dt$$

Let $u = t$ and $dv = e^{-0.1t} \, dt$, then $du = dt$ and $v = -10e^{-0.1t}$.

$$= 100{,}000 + 500\left[-10te^{-0.1t} + 10\int e^{-0.1t} \, dt\right]_0^{10}$$

$$= 100{,}000 + 500\left\{-100e^{-1} - \Big[100e^{-0.1t}\Big]_0^{10}\right\}$$

$$= 100{,}000 + 500(-100e^{-1} - 100e^{-1} + 100)$$

$$= 100{,}000 + 50{,}000(1 - 2e^{-1})$$

$$\approx 113{,}212 \text{ units}$$

(c) Average $= \dfrac{113{,}212}{10}$

$$\approx 11{,}321 \text{ per year}$$

55. (a) Average $= \displaystyle\int_1^2 (1.6t \ln t + 1)\, dt$

$= (0.8t^2 \ln t - 0.4t^2 + t)\Big]_1^2$

$= 3.2(\ln 2) - 0.2$

≈ 2.0181

(b) Average $= \displaystyle\int_3^4 (1.6t \ln t + 1)\, dt$

$= (0.8t^2 \ln t - 0.4t^2 + t)\Big]_3^4$

$= 12.8(\ln 4) - 7.2(\ln 3) - 1.8$

≈ 8.0346

57. $V = \displaystyle\int_0^{t_1} c(t)e^{-rt}\, dt = \int_0^4 5000e^{-0.05t}\, dt$

$= \dfrac{5000}{-0.05}e^{-0.05t}\Big]_0^4$

$\approx \$18{,}126.92$

59. $V = \displaystyle\int_0^{t_1} c(t)e^{-rt} = \int_0^{10} (150{,}000 + 2500t)e^{-0.04t}\, dt$

$\approx \$1{,}332{,}474.72$

61. $V = \displaystyle\int_0^{t_1} c(t)e^{-rt}\, dt = \int_0^4 (1000 + 50e^{t/2})e^{-0.06t}\, dt$

$\approx \$4103.07$

63. $c = 150{,}000 + 75{,}000t$

(a) $\displaystyle\int_0^4 (150{,}000 + 75{,}000t)\, dt = 150{,}000t + 37{,}500t^2\Big]_0^4 = \$1{,}200{,}000$

(b) $V = \displaystyle\int_0^{t_1} c(t)e^{-rt}\, dt = \int_0^4 (150{,}000 + 75{,}000t)e^{-0.04t}\, dt \approx \$1{,}094{,}142.26$

65. $\displaystyle\int_1^4 \dfrac{4}{\sqrt{x} + \sqrt[3]{x}}\, dx = 4\int_1^4 \dfrac{1}{\sqrt{x} + \sqrt[3]{x}}\, dx$

$\approx 4\left(\dfrac{4-1}{10}\right)\left[f\!\left(\dfrac{23}{20}\right) + f\!\left(\dfrac{29}{20}\right) + f\!\left(\dfrac{35}{20}\right) + f\!\left(\dfrac{41}{20}\right) + f\!\left(\dfrac{47}{20}\right) + f\!\left(\dfrac{53}{20}\right) + f\!\left(\dfrac{59}{20}\right) + f\!\left(\dfrac{65}{20}\right) + f\!\left(\dfrac{71}{20}\right) + f\!\left(\dfrac{77}{20}\right)\right]$

≈ 4.254

67. Future value $= e^{rt_1}\displaystyle\int_0^{t_1} f(t)e^{-rt}\, dt$

$= e^{(0.08)10}\displaystyle\int_0^{10} 3000e^{-0.08t}\, dt$

$= e^{0.8}\left[\dfrac{3000}{-0.08}e^{-0.08t}\right]_0^{10}$

$\approx \$45{,}957.78$

69. (a) Future value $= e^{(0.07)(10)}\displaystyle\int_0^{10} 1200e^{-(0.07)t}\, dt = e^{0.7}\int_0^{10} 1200e^{-0.07t}\, dt \approx \$17{,}378.62$

(b) Difference $= \left[e^{(0.10)(15)}\displaystyle\int_0^{15} 1200e^{-0.10t}\, dt\right] - \left[e^{(0.09)(15)}\int_0^{15} 1200e^{-0.09t}\, dt\right]$

$= \left[e^{1.5}\displaystyle\int_0^{15} 1200e^{-0.1t}\, dt\right] - \left[e^{1.35}\int_0^{15} 1200e^{-0.09t}\, dt\right]$

$\approx \$41{,}780.27 - \$38{,}099.01$

$= \$3681.26$

Section 6.3   Partial Fractions and Logistics Growth

1. $\dfrac{2x + 40}{(x - 5)(x + 5)} = \dfrac{A}{x - 5} + \dfrac{B}{x + 5}$

Basic equation: $2x + 40 = A(x + 5) + B(x - 5)$

When $x = 5$: $50 = 10A, A = 5$

When $x = -5$: $30 = -10B, B = -3$

$\dfrac{2(x + 20)}{x^2 - 25} = \dfrac{5}{x - 5} - \dfrac{3}{x + 5}$

3. $\dfrac{8x + 3}{x(x - 3)} = \dfrac{A}{x} + \dfrac{B}{x - 3}$

Basic equation: $8x + 3 = A(x - 3) + Bx$

When $x = 0$: $3 = -3A, A = -1$

When $x = 3$: $27 = 3B, B = 9$

$\dfrac{8x + 3}{x^2 - 3x} = \dfrac{9}{x - 3} - \dfrac{1}{x}$

5. $\dfrac{4x - 13}{(x - 5)(x + 2)} = \dfrac{A}{x - 5} + \dfrac{B}{x + 2}$

Basic equation: $4x - 13 = A(x + 2) + B(x - 5)$

When $x = 5$: $7 = 7A, A = 1$

When $x = -2$: $-21 = -7B, B = 3$

$\dfrac{4x - 13}{x^2 - 3x - 10} = \dfrac{1}{x - 5} + \dfrac{3}{x + 2}$

7. $\dfrac{3x^2 - 2x - 5}{x^2(x + 1)} = \dfrac{A}{x} + \dfrac{B}{x^2} + \dfrac{C}{x + 1}$

Basic equation:

$$3x^2 - 2x - 5 = Ax(x + 1) + B(x + 1) + Cx^2$$

When $x = 0$: $-5 = B$

When $x = 1$: $0 = C$

When $x = 1$: $-4 = 2A - 10 \Rightarrow A = 3$

$\dfrac{3x^2 - 2x - 5}{x^2(x + 1)} = \dfrac{3}{x} - \dfrac{5}{x^2}$

9. $\dfrac{1}{3}\left[\dfrac{x + 1}{(x - 2)^2}\right] = \dfrac{1}{3}\left[\dfrac{A}{x - 2} + \dfrac{B}{(x - 2)^2}\right]$

Basic equation: $x + 1 = A(x - 2) + B$

When $x = 2$: $3 = B$

When $x = 3$: $4 = A + B \Rightarrow A = 1$

$\dfrac{x + 1}{3(x - 2)^2} = \dfrac{1}{3}\left[\dfrac{1}{x - 2} + \dfrac{3}{(x - 2)^2}\right]$

$\phantom{\dfrac{x + 1}{3(x - 2)^2}} = \dfrac{1}{3(x - 2)} + \dfrac{1}{(x - 2)^2}$

11. $\dfrac{8x^2 + 15x + 9}{(x + 1)^3} = \dfrac{A}{x + 1} + \dfrac{B}{(x + 1)^2} + \dfrac{C}{(x + 1)^3}$

Basic equation:

$$8x^2 + 15x + 9 = A(x + 1)^2 + B(x + 1) + C$$
$$= Ax^2 + (2A + B)x + (A + B + C)$$

Therefore, $A = 8$, $2A + B = 15$, and $A + B + C = 9$. Solving these equations yields $A = 8$, $B = -1$, and $C = 2$.

$\dfrac{8x^2 + 15x + 9}{(x + 1)^3} = \dfrac{8}{x + 1} - \dfrac{1}{(x + 1)^2} + \dfrac{2}{(x + 1)^3}$

13. $\dfrac{1}{(x + 1)(x - 1)} = \dfrac{A}{x + 1} + \dfrac{B}{x - 1}$

Basic equation: $1 = A(x - 1) + B(x + 1)$

When $x = -1$: $1 = -2A, A = -\dfrac{1}{2}$

When $x = 1$: $1 = 2B, B = \dfrac{1}{2}$

$\displaystyle\int \dfrac{1}{x^2 - 1}\, dx = \dfrac{-1}{2}\int \dfrac{1}{x + 1}\, dx + \dfrac{1}{2}\int \dfrac{1}{x - 1}\, dx$

$\phantom{\int \dfrac{1}{x^2 - 1}\, dx} = \dfrac{-1}{2}\ln|x + 1| + \dfrac{1}{2}\ln|x - 1| + C$

$\phantom{\int \dfrac{1}{x^2 - 1}\, dx} = \dfrac{1}{2}\ln\left|\dfrac{x - 1}{x + 1}\right| + C$

15. $\dfrac{-2}{(x + 4)(x - 4)} = \dfrac{A}{x + 4} + \dfrac{B}{x - 4}$

Basic equation: $-2 = A(x - 4) + B(x + 4)$

When $x = -4$: $-2 = -8A, A = \dfrac{1}{4}$

When $x = 4$: $-2 = 8B, B = -\dfrac{1}{4}$

$\displaystyle\int \dfrac{-2}{x^2 - 16}\, dx = \dfrac{1}{4}\int \dfrac{1}{x + 4}\, dx - \dfrac{1}{4}\int \dfrac{1}{x - 4}\, dx$

$\phantom{\int \dfrac{-2}{x^2 - 16}\, dx} = \dfrac{1}{4}\ln|x + 4| - \dfrac{1}{4}\ln|x - 4| + C$

$\phantom{\int \dfrac{-2}{x^2 - 16}\, dx} = \dfrac{1}{4}\ln\left|\dfrac{x + 4}{x - 4}\right| + C$

17. $\dfrac{1}{3x^2 - x} = \dfrac{1}{x(3x - 1)} = \dfrac{A}{x} + \dfrac{B}{3x - 1}$

Basic equation: $1 = A(3x - 1) + Bx$

When $x = 0$: $1 = -A \Longrightarrow A = -1$

When $x = \dfrac{1}{3}$: $1 = B\left(\dfrac{1}{3}\right) \Longrightarrow B = 3$

$\displaystyle \int \dfrac{1}{3x^2 - x}\, dx = \int \dfrac{-1}{x}\, dx + \int \dfrac{3}{3x - 1}\, dx$

$\qquad = -\ln|x| + \ln|3x - 1| + C$

19. $\dfrac{1}{x(2x + 1)} = \dfrac{A}{x} + \dfrac{B}{2x + 1}$

Basic equation: $1 = A(2x + 1) + Bx$

When $x = 0$: $A = 1$

When $x = -\dfrac{1}{2}$: $B = -2$

$\displaystyle \int \dfrac{1}{2x^2 + x}\, dx = \int \dfrac{1}{x}\, dx - \int \dfrac{2}{2x + 1}\, dx$

$\qquad = \ln|x| - \ln|2x + 1| + C$

$\qquad = \ln\left|\dfrac{x}{2x + 1}\right| + C$

21. $\dfrac{3}{(x - 1)(x + 2)} = \dfrac{A}{x - 1} + \dfrac{B}{x + 2}$

Basic equation: $3 = A(x + 2) + B(x - 1)$

When $x = 1$: $3 = 3A, A = 1$

When $x = -2$: $3 = -3B, B = -1$

$\displaystyle \int \dfrac{3}{x^2 + x - 2}\, dx = \int \dfrac{1}{x - 1}\, dx - \int \dfrac{1}{x + 2}\, dx$

$\qquad = \ln|x - 1| - \ln|x + 2| + C$

$\qquad = \ln\left|\dfrac{x - 1}{x + 2}\right| + C$

23. $\dfrac{5 - x}{(2x - 1)(x + 1)} = \dfrac{A}{2x - 1} + \dfrac{B}{x + 1}$

Basic equation: $5 - x = A(x + 1) + B(2x - 1)$

When $x = \dfrac{1}{2}$: $4.5 = 1.5A, A = 3$

When $x = -1$: $6 = -3B, B = -2$

$\displaystyle \int \dfrac{5 - x}{2x^2 + x - 1}\, dx = 3 \int \dfrac{1}{2x - 1}\, dx - 2 \int \dfrac{1}{x + 1}\, dx$

$\qquad = \dfrac{3}{2} \ln|2x - 1| - 2 \ln|x + 1| + C$

25. $\dfrac{x^2 + 12x + 12}{x(x + 2)(x - 2)} = \dfrac{A}{x} + \dfrac{B}{x + 2} + \dfrac{C}{x - 2}$

Basic equation: $x^2 + 12x + 12 = A(x + 2)(x - 2) + Bx(x - 2) + Cx(x + 2)$

When $x = 0$: $12 = -4A, A = -3$

When $x = -2$: $-8 = 8B, B = -1$

When $x = 2$: $40 = 8C, C = 5$

$\displaystyle \int \dfrac{x^2 + 12x + 12}{x^3 - 4x}\, dx = 5 \int \dfrac{1}{x - 2}\, dx - \int \dfrac{1}{x + 2}\, dx - 3 \int \dfrac{1}{x}\, dx$

$\qquad = 5 \ln|x - 2| - \ln|x + 2| - 3 \ln|x| + C$

27. $\dfrac{x + 2}{x(x - 4)} = \dfrac{A}{x - 4} + \dfrac{B}{x}$

Basic equation: $x + 2 = Ax + B(x - 4)$

When $x = 4$: $6 = 4A, A = \dfrac{3}{2}$

When $x = 0$: $2 = -4B, B = -\dfrac{1}{2}$

$\displaystyle \int \dfrac{x + 2}{x^2 - 4x}\, dx = \dfrac{1}{2}\left[3 \int \dfrac{1}{x - 4}\, dx - \int \dfrac{1}{x}\, dx\right]$

$\qquad = \dfrac{1}{2}[3 \ln|x - 4| - \ln|x|] + C$

29. $\dfrac{4 - 3x}{(x - 1)^2} = \dfrac{A}{x - 1} + \dfrac{B}{(x - 1)^2}$

Basic equation: $4 - 3x = A(x - 1) + B$

When $x = 1$: $1 = B$

When $x = 0$: $4 = -A + 1 \Longrightarrow A = -3$

$\displaystyle \int \dfrac{4 - 3x}{(x - 1)^2}\, dx = \int \dfrac{-3}{x - 1}\, dx + \int \dfrac{1}{(x - 1)^2}\, dx$

$\qquad = -3 \ln|x - 1| - \dfrac{1}{x - 1} + C$

31. $\dfrac{3x^2 + 3x + 1}{x(x+1)^2} = \dfrac{A}{x} + \dfrac{B}{x+1} + \dfrac{C}{(x+1)^2}$

Basic equation: $A(x+1)^2 + Bx(x+1) + Cx = 3x^2 + 3x + 1$

When $x = 0$: $A = 1$

When $x = -1$: $-C = 3 - 3 + 1 \Longrightarrow C = -1$

When $x = 1$: $4A + 2B + C = 7$

$$4 + 2B - 1 = 7$$

$$B = 2$$

$$\int \frac{3x^2 + 3x + 1}{x(x^2 + 2x + 1)}\,dx = \int \left(\frac{1}{x} + \frac{2}{x+1} + \frac{-1}{(x+1)^2}\right) dx = \ln|x| + 2\ln|x+1| + \frac{1}{x+1} + C$$

33. $\dfrac{1}{9 - x^2} = \dfrac{-1}{(x-3)(x+3)} = \dfrac{A}{x-3} + \dfrac{B}{x+3}$

Basic equation: $-1 = A(x+3) + B(x-3)$

When $x = 3$: $-1 = 6A \Longrightarrow A = -\dfrac{1}{6}$

When $x = -3$: $-1 = -6B \Longrightarrow B = \dfrac{1}{6}$

$$\int_4^5 \frac{1}{9 - x^2}\,dx = \int_4^5 \frac{-1/6}{x-3}\,dx + \int_4^5 \frac{1/6}{x+3}\,dx$$

$$= \frac{1}{6}\left[-\ln(x-3) + \ln(x+3)\right]_4^5$$

$$= \frac{1}{6}[-\ln 2 + \ln 8 - \ln 7]$$

$$= \frac{1}{6}\ln\frac{4}{7}$$

$$\approx -0.093$$

35. $\dfrac{x-1}{x^2(x+1)} = \dfrac{A}{x} + \dfrac{B}{x^2} + \dfrac{C}{x+1}$

Basic equation: $x - 1 = Ax(x+1) + B(x+1) + Cx^2$

When $x = 0$: $B = -1$

When $x = -1$: $C = -2$

When $x = 1$: $0 = 2A + 2B + C$, $0 = 2A - 4$, $A = 2$

$$\int_1^5 \frac{x-1}{x^2(x+1)}\,dx = 2\int_1^5 \frac{1}{x}\,dx - \int_1^5 \frac{1}{x^2}\,dx - 2\int_1^5 \frac{1}{x+1}\,dx$$

$$= \left[2\ln|x| + \frac{1}{x} - 2\ln|x+1|\right]_1^5$$

$$= \left[2\ln\left|\frac{x}{x+1}\right| + \frac{1}{x}\right]_1^5$$

$$= 2\ln\left(\frac{5}{3}\right) - \frac{4}{5}$$

$$\approx 0.222$$

37. $\dfrac{x^3}{x^2 - 2} = x + \dfrac{2x}{x^2 - 2}$

$$\int_0^1 \left(x + \frac{2x}{x^2 - 2}\right) dx = \left[\frac{x^2}{2} + \ln|x^2 - 2|\right]_0^1 = \frac{1}{2} - \ln 2 \approx -0.193$$

39. First divide to obtain $\dfrac{x^3 - 4x^2 - 3x + 3}{x^2 - 3x} = x - 1 - \dfrac{6x - 3}{x^2 - 3x}$.

Use partial fractions for the $\dfrac{6x - 3}{x(x - 3)}$ term.

$$\frac{6x - 3}{x(x - 3)} = \frac{A}{x} + \frac{B}{x - 3}$$

Basic equation: $6x - 3 = A(x - 3) + Bx$

When $x = 3$: $15 = 3B \Longrightarrow B = 5$

When $x = 0$: $-3 = -3A \Longrightarrow A = 1$

$$\int_1^2 \frac{x^3 - 4x^2 - 3x + 3}{x^2 - 3x}\, dx = \int_1^2 \left(x - 1 - \frac{1}{x} - \frac{5}{x - 3} \right) dx$$

$$= \left[\frac{x^2}{2} - x - \ln|x| - 5\ln|x - 3| \right]_1^2$$

$$= (-\ln 2) - \left(-\frac{1}{2} - 5\ln 2 \right)$$

$$= \frac{1}{2} + 4\ln 2$$

$$\approx 3.273$$

41. To find the limits of integration, solve:

$$2 = \frac{14}{16 - x^2}$$

$$32 - 2x^2 = 14$$

$$18 = 2x^2$$

$$x = \pm 3$$

$$A = \int_{-3}^3 \left(2 - \frac{14}{16 - x^2} \right) dx$$

$$= \int_{-3}^3 \left(2 - \frac{7}{4} \cdot \frac{1}{x + 4} + \frac{7}{4} \cdot \frac{1}{x - 4} \right) dx$$

$$= \left[2x - \frac{7}{4}\ln|x + 4| + \frac{7}{4}\ln|x - 4| \right]_{-3}^3$$

$$= \left(6 - \frac{7}{4}\ln 7 \right) - \left(-6 + \frac{7}{4}\ln 7 \right)$$

$$= 12 - \frac{7}{2}\ln 7$$

$$\approx 5.1893$$

43. $A = \displaystyle\int_2^5 \frac{x + 1}{x^2 - x}\, dx$

$$= \int_2^5 \left(-\frac{1}{x} + \frac{2}{x - 1} \right) dx$$

$$= \left[-\ln x + 2\ln(x - 1) \right]_2^5$$

$$= (-\ln 5 + 2\ln 4) - (-\ln 2)$$

$$= 5\ln 2 - \ln 5$$

$$\approx 1.8563$$

45. $\dfrac{1}{a^2 - x^2} = \dfrac{1}{(a + x)(a - x)}$

$$= \frac{A}{a + x} + \frac{B}{a - x}$$

$$= \frac{1}{2a}\left(\frac{1}{a + x} + \frac{1}{a - x} \right)$$

47. $\dfrac{1}{x(a - x)} = \dfrac{A}{x} + \dfrac{B}{a - x}$

Basic equation: $1 = A(a - x) + Bx$

When $x = a$: $1 = Ba \Longrightarrow B = \dfrac{1}{a}$

When $x = 0$: $1 = Aa \Longrightarrow A = \dfrac{1}{a}$

$$\frac{1}{x(a - x)} = \frac{1/a}{x} + \frac{1/a}{a - x}$$

49. $V = \pi \int_{1}^{5} \left[\dfrac{10}{x(x+10)} \right]^2 dx = 100\pi \int_{1}^{5} \dfrac{1}{x^2(x+10)^2} dx$

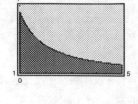

(a) $V \approx 1.773$ cubic units

(b) Let $\dfrac{1}{x^2(x+10)^2} = \dfrac{A}{x} + \dfrac{B}{x^2} + \dfrac{C}{x+10} + \dfrac{D}{(x+10)^2}$.

Basic equation: $1 = Ax(x+10)^2 + B(x+10)^2 + Cx^2(x+10) + Dx^2$

When $x = 0$: $1 = 100B \Rightarrow B = \dfrac{1}{100}$

When $x = -10$: $1 = 100D \Rightarrow D = \dfrac{1}{100}$

When $x = 1$: $1 = 121A + 121B + 11C + D$ $\left.\begin{array}{l} \\ \\ \end{array}\right\}$ $A = -\dfrac{1}{500}$

When $x = -1$: $1 = -81A + 81B + 9C + D$ $\qquad C = \dfrac{1}{500}$

$V = \pi \int_{1}^{5} 100 \left[\dfrac{-1/500}{x} + \dfrac{1/100}{x^2} + \dfrac{1/500}{x+10} + \dfrac{1/100}{(x+10)^2} \right] dx$

$= \pi \int_{1}^{5} \left[-\dfrac{1}{5x} + \dfrac{1}{x^2} + \dfrac{1}{5(x+10)} + \dfrac{1}{(x+10)^2} \right] dx$

$= \pi \left[-\dfrac{1}{5}\ln|x| - \dfrac{1}{x} + \dfrac{1}{5}\ln|x+10| - \dfrac{1}{x+10} \right]_{1}^{5}$

$= \pi \left[\dfrac{1}{5}\ln\left|\dfrac{x+10}{x}\right| - \dfrac{2x+10}{x(x+10)} \right]_{1}^{5}$

$= \pi \left[\left(\dfrac{1}{5}\ln 3 - \dfrac{4}{15} \right) - \left(\dfrac{1}{5}\ln 11 - \dfrac{12}{11} \right) \right] \approx 1.773$ cubic units

51. (a) $V = \pi \int_{-1}^{1} \left(\dfrac{2}{x^2-4} \right)^2 dx \approx 1.9100$

(b) $V = \pi \int_{-1}^{1} \left(\dfrac{2}{x^2-4} \right)^2 dx$

$= \pi \int_{-1}^{2} \left(\dfrac{2}{(x-2)(x+2)} \right)^2 dx$

$\dfrac{4}{(x-2)^2(x+2)^2} = \dfrac{A}{x-2} + \dfrac{B}{(x-2)^2} + \dfrac{C}{x+2} + \dfrac{D}{(x+2)^2}$

Basic equation: $4 = A(x-2)(x+2)^2 + B(x+2)^2 + C(x+2)(x-2)^2 + D(x-2)^2$

When $x = 2$: $4 = 16B \Rightarrow B = \dfrac{1}{4}$

When $x = -2$: $4 = 16D \Rightarrow D = \dfrac{1}{4}$

When $x = 0$: $4 = -8A + 4\left(\dfrac{1}{4}\right) + 8C + 4\left(\dfrac{1}{4}\right) \Rightarrow 1 = -4A + 4C$

When $x = 1$: $4 = -9A + 9\left(\dfrac{1}{4}\right) + 3C + \dfrac{1}{4} \Rightarrow 3 = -18A + 6C$

Solving for A and C, $A = -\dfrac{1}{8}$, $C = \dfrac{1}{8}$. Hence,

$V = \pi \int_{-1}^{1} \left(-\dfrac{1}{8} \cdot \dfrac{1}{x-2} + \dfrac{1}{4} \cdot \dfrac{1}{(x-2)^2} + \dfrac{1}{8} \cdot \dfrac{1}{x+2} + \dfrac{1}{4} \cdot \dfrac{1}{(x+2)^2} \right) dx$

$= \pi \left[-\dfrac{1}{8}\ln|x-2| - \dfrac{1}{4} \cdot \dfrac{1}{x-2} + \dfrac{1}{8}\ln|x+2| - \dfrac{1}{4} \cdot \dfrac{1}{x+2} \right]_{-1}^{1}$

$= \pi \left[\left(\dfrac{1}{4} + \dfrac{1}{8}\ln 3 - \dfrac{1}{12} \right) - \left(-\dfrac{1}{8}\ln 3 + \dfrac{1}{12} - \dfrac{1}{4} \right) \right]$

$= \pi \left[\dfrac{1}{3} + \dfrac{1}{4}\ln 3 \right] \approx 1.9100.$

53. $\dfrac{1}{y(1000 - y)} = \dfrac{A}{y} + \dfrac{B}{1000 - y}$

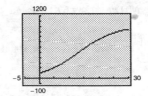

Basic equation: $1 = A(1000 - y) + By$

When $y = 0$: $1 = 1000A, A = \dfrac{1}{1000}$

When $y = 1000$: $1 = 1000B, B = \dfrac{1}{1000}$

$$\int \frac{1}{y(1000 - y)}\, dy = \int k\, dt$$

$$\frac{1}{1000} \int \left(\frac{1}{y} + \frac{1}{1000 - y} \right) dy = \int k\, dt$$

$$\ln|y| - \ln|1000 - y| = 1000(kt + C_1)$$

$$\ln \left| \frac{y}{1000 - y} \right| = 1000(kt + C_1)$$

$$\frac{y}{1000 - y} = e^{1000(kt + C_1)} = e^{1000C_1}e^{1000kt}$$

$$\frac{y}{1000 - y} = Ce^{1000kt}$$

When $t = 0, y = 100$: $\dfrac{1}{9} = C, \dfrac{y}{1000 - y} = \dfrac{1}{9}e^{1000kt}$

When $t = 2, y = 134$: $\dfrac{67}{433} = \dfrac{1}{9}e^{2000k}, \ln \dfrac{603}{433} = 2000k, k = \dfrac{1}{2000} \ln \dfrac{603}{433}$

Thus, $\dfrac{y}{1000 - y} = \dfrac{1}{9}e^{(1/2)\ln(603/433)t}$. Solving for y yields

$$9y \approx e^{0.1656t}(1000 - y)$$

$$y(9 + e^{0.1656t}) = 1000e^{0.1656t}$$

$$y = \frac{1000e^{0.1656t}}{e^{0.1656t} + 9} = \frac{1000}{1 + 9e^{-0.1656t}}.$$

55. $\dfrac{dS}{dt} = \dfrac{2t}{(t + 4)^2}$

$\dfrac{2t}{(t + 4)^2} = \dfrac{A}{t + 4} + \dfrac{B}{(t + 4)^2}$

Basic equation: $2t = A(t + 4) + B$

When $t = -4$: $-8 = B$

When $t = 0$: $0 = 4A - 8 \Longrightarrow A = 2$

$$S = \int \frac{2t}{(t + 4)^2}\, dt = \int \left(\frac{2}{t + 4} - \frac{8}{(t + 4)^2} \right) dt = 2 \ln(t + 4) + \frac{8}{t + 4} + C$$

When $t = 0$: $S = 0 \Longrightarrow 0 - 2 \ln 4 + 2 + C \Longrightarrow C \approx -4.77259$

When $t = 10$: $S \approx 2 \ln(14) + \dfrac{8}{14} - 4.77259 \approx 1.077$ thousand

57. Total number $= \int_5^{13} N(t)\, dt \approx 289.67$ thousand

Average $= \dfrac{1}{13-5}\int_5^{13} N(t)\, dt \approx 36.21$ thousand per year

Section 6.4   Integration Tables and Completing the Square

1. Formula 4: $u = x,\, du = dx,\, a = 2,\, b = 3$

$$\int \frac{x}{(2+3x)^2}\, dx = \frac{1}{9}\left[\frac{2}{2+3x} + \ln|2+3x|\right] + C$$

3. Formula 19: $u = x,\, du = dx,\, a = 2,\, b = 3$

$$\int \frac{x}{\sqrt{2+3x}}\, dx = \frac{-2(4-3x)}{27}\sqrt{2+3x} + C$$

$$= \frac{2(3x-4)}{27}\sqrt{2+3x} + C$$

5. Formula 25: $u = x^2,\, du = 2x\, dx,\, a = 3$

$$\int \frac{2x}{\sqrt{x^4-9}}\, dx = \ln\left|x^2 + \sqrt{x^4-9}\right| + C$$

7. Formula 35: $u = x^2,\, du = 2x\, dx$

$$\int x^3 e^{x^2}\, dx = \frac{1}{2}\int x^2 e^{x^2} 2x\, dx = \frac{1}{2}(x^2-1)e^{x^2} + C$$

9. Formula 10: $u = x,\, du = dx,\, a = b = 1$

$$\int \frac{1}{x(1+x)}\, dx = \ln\left|\frac{x}{1+x}\right| + C$$

11. Formula 26: $u = x,\, du = dx,\, a = 3$

$$\int \frac{1}{x\sqrt{x^2+9}}\, dx = -\frac{1}{3}\ln\left|\frac{1+\sqrt{x^2+1}}{x}\right| + C$$

13. Formula 32: $u = x,\, du = dx,\, a = 2$

$$\int \frac{1}{x\sqrt{4-x^2}}\, dx = -\frac{1}{2}\ln\left|\frac{2+\sqrt{4-x^2}}{x}\right| + C$$

15. Formula 40: $u = x,\, du = dx$

$$\int x \ln x\, dx = \frac{x^2}{4}(-1 + 2\ln x) + C$$

17. Formula 37: $u = 3x^2,\, du = 6x\, dx$

$$\int \frac{6x}{1+e^{3x^2}}\, dx = 3x^2 - \ln(1 + e^{3x^2}) + C$$

19. Formula 21: $u = x^2,\, du = 2x\, dx,\, a = 3$

$$\int x\sqrt{x^4-9}\, dx = \frac{1}{2}\int \sqrt{(x^2)^2 - 3^2}\,(2x)\, dx$$

$$= \frac{1}{4}\left(x^2\sqrt{x^4-9} - 9\ln\left|x^2 + \sqrt{x^4-9}\right|\right) + C$$

21. Formula 8: $u = t,\, du = dt,\, a = 2,\, b = 3$

$$\int \frac{t^2}{(2+3t)^3}\, dt = \frac{1}{27}\left[\frac{4}{2+3t} - \frac{4}{2(2+3t)^2} + \ln|2+3t|\right] + C$$

23. Formula 15: $u = s,\, du = ds,\, a = 3,\, b = 1$

$$\int \frac{s}{s^2\sqrt{3+s}}\, ds = \int \frac{1}{s\sqrt{3+s}}\, ds = \frac{1}{\sqrt{3}}\ln\left|\frac{\sqrt{3+s} - \sqrt{3}}{\sqrt{3+s} + \sqrt{3}}\right| + C$$

25. Formula 9: $u = x,\, du = dx,\, a = 3,\, b = 2,\, n = 5$

$$\int \frac{x^2}{(3+2x)^5}\, dx = \frac{1}{8}\left[\frac{-1}{(3+2x)^2} + \frac{6}{3(3+2x)^3} - \frac{9}{4(3+2x)^4}\right] + C$$

27. Formula 23: $u = x, du = dx, a = 1$

$$\int \frac{1}{x^2\sqrt{1-x^2}} dx = -\frac{\sqrt{1-x^2}}{x} + C$$

29. Formula 41: $u = x, du = dx, n = 2$

$$\int x^2 \ln x \, dx = \frac{x^3}{9}(-1 + 3 \ln x) + C$$

31. Formula 7: $u = x, du = dx, a = -5, b = 3$

$$\int \frac{x^2}{(3x-5)^2} dx = \frac{1}{27}\left[3x - \frac{25}{3x-5} + 10 \ln|3x-5|\right] + C$$

33. Formula 3: $u = \ln x, du = (1/x) \, dx, a = 4, b = 3$

$$\int \frac{\ln x}{x(4 + 3 \ln x)} dx = \int \frac{\ln x}{4 + 3 \ln x}\left(\frac{1}{x} dx\right) = \frac{1}{9}[3 \ln x - 4 \ln|4 + 3 \ln x|] + C$$

35. Formula 19: $u = x, du = dx, a = b = 1$

$$A = \int_0^8 \frac{x}{\sqrt{x+1}} dx = \left[-\frac{2(2-x)}{3}\sqrt{x+1}\right]_0^8 = \frac{40}{3}$$

Approximate area: 13.333

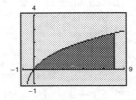

37. Formula 37: $u = x^2, du = 2x \, dx$

$$\int_0^2 \frac{x}{1 + e^{x^2}} dx = \frac{1}{2}\int_0^4 \frac{1}{1 + e^u} du$$

$$= \frac{1}{2}\left[u - \ln(1 + e^u)\right]_0^4$$

$$= \frac{1}{2}[4 - \ln(1 + e^4) + \ln 2]$$

$$= \frac{1}{2}\left[4 + \ln\frac{2}{1 + e^4}\right]$$

Approximate area: 0.3375

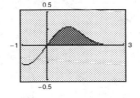

39. Formula 22: $u = x, a = 2, du = dx$

$$\int_0^{\sqrt{5}} x^2\sqrt{x^2 + 4} \, dx = \int_0^{\sqrt{5}} u^2\sqrt{u^2 + 2^2} \, du$$

$$= \frac{1}{8}\left[u(2u^2 + 4)\sqrt{u^2 + 4} - 16 \ln\left|u + \sqrt{u^2 + 4}\right|\right]_0^{\sqrt{5}}$$

$$= \frac{1}{8}\left[\sqrt{5}(14)(3) - 16 \ln\left|\sqrt{5} + 3\right| + 16 \ln 2\right]$$

$$= \frac{1}{4}\left[21\sqrt{5} - 8 \ln(\sqrt{5} + 3) + 8 \ln 2\right]$$

Approximate area: 9.8145

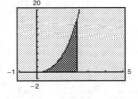

41. Formula 19: $u = x, du = dx, a = 5, b = 2$

$$\int_0^5 \frac{x}{\sqrt{5 + 2x}} dx = \frac{-2(10 - 2x)}{12}\sqrt{5 + 2x}\Big]_0^5 = 0 + \frac{5}{3}\sqrt{5} = \frac{5\sqrt{5}}{3}$$

43. Formula 38: $u = x, du = dx, n = 0.5$

$$6 \int_0^4 \frac{1}{1 + e^{0.5x}} \, dx = 6\left[x - \frac{1}{0.5} \ln(1 + e^{0.5x}) \right]_0^4$$

$$= 6\{[4 - 2\ln(1 + e^2)] - (0 - 2\ln 2)\}$$

$$\approx 6.795$$

45. (a) Formula 36: $u = x, du = dx, n = 2$

$$\int x^2 e^x \, dx = x^2 e^x - 2 \int x e^x \, dx$$

Formula 35: $u = x, du = dx$

$$\int x^2 e^x \, dx = x^2 e^x - 2(x - 1)e^x + C = e^x(x^2 - 2x + 2) + C$$

(b) Let $u = x^2$ and $dv = e^x \, dx$, then $du = 2x \, dx$ and $v = e^x$.

$$\int x^2 e^x \, dx = x^2 e^x - 2 \int x e^x \, dx$$

Let $u = x$ and $dv = e^x \, dx$, then $du = dx$ and $v = e^x$.

$$\int x^2 e^x \, dx = x^2 e^x - 2\left[xe^x - \int e^x \, dx \right] = x^2 e^x - 2xe^x + 2e^x + C = e^x(x^2 - 2x + 2) + C$$

47. Formula 12: $u = x, du = dx, a = 1, b = 1$

(a) $\displaystyle \int \frac{1}{x^2(x + 1)} \, dx = -\left[\frac{1}{x} + \ln\left| \frac{x}{1 + x} \right| \right] + C$

(b) $\displaystyle \frac{1}{x^2(x + 1)} = \frac{A}{x} + \frac{B}{x^2} + \frac{C}{x + 1}$

Basic equation: $1 = Ax(x + 1) + B(x + 1) + Cx^2$

When $x = 0$: $1 = B$

When $x = -1$: $1 = C$

When $x = 1$:
$$1 = 2A + 2B + C = 2A + 3 \Rightarrow A = -1$$

$$\int \frac{1}{x^2(x + 1)} \, dx = \int \left[-\frac{1}{x} + \frac{1}{x^2} + \frac{1}{x + 1} \right] dx$$

$$= -\ln|x| - \frac{1}{x} + \ln|x + 1| + C$$

$$= -\left[\frac{1}{x} + \ln|x| - \ln|x + 1| \right] + C$$

$$= -\left[\frac{1}{x} + \ln\left| \frac{x}{x + 1} \right| \right] + C$$

49. (a) $x^2 + 6x = x^2 + 6x + 9 - 9 = (x + 3)^2 - 9$

(b) $x^2 - 8x + 9 = x^2 - 8x + 16 - 16 + 9$
$$= (x - 4)^2 - 7$$

(c) $x^4 + 2x^2 - 5 = x^4 + 2x^2 + 1 - 1 - 5$
$$= (x^2 + 1)^2 - 6$$

(d) $3 - 2x - x^2 = -(x^2 + 2x - 3)$
$$= -(x^2 + 2x + 1 - 1 - 3)$$
$$= -[(x + 1)^2 - 4]$$
$$= 4 - (x + 1)^2$$

51. Formula 29: $u = x + 3, du = dx, a = \sqrt{17}$

$$\int \frac{1}{x^2 + 6x - 8} \, dx = \int \frac{1}{(x + 3)^2 - 17} \, dx = \frac{1}{2\sqrt{17}} \ln\left| \frac{(x + 3) - \sqrt{17}}{(x + 3) + \sqrt{17}} \right| + C$$

53. Formula 26: $u = x - 1, du = dx, a = 1$

$$\int \frac{1}{(x - 1)\sqrt{x^2 - 2x + 2}} \, dx = \int \frac{1}{(x - 1)\sqrt{(x - 1)^2 + 1}} \, dx = -\ln\left| \frac{1 + \sqrt{x^2 - 2x + 2}}{x - 1} \right| + C$$

55. Formula 29: $u = x - 1, du = dx, a = 2$

$$\int \frac{1}{2x^2 - 4x - 6} \, dx = \frac{1}{2} \int \frac{1}{(x-1)^2 - 4} \, dx = \frac{1}{8} \ln\left|\frac{x-3}{x+1}\right| + C$$

57. Formula 25: $u = x^2 + 1, du = 2x \, dx, a = 1$

$$\int \frac{x}{\sqrt{x^4 + 2x^2 + 2}} \, dx = \int \frac{x}{\sqrt{(x^2+1)^2 + 1}} \, dx$$

$$= \frac{1}{2} \int \frac{2x}{\sqrt{(x^2+1)^2 + 1}} \, dx$$

$$= \frac{1}{2} \ln\left|x^2 + 1 + \sqrt{x^4 + 2x^2 + 2}\right| + C$$

59. Formula 38: $u = 4.8 - 1.9t, du = -1.9 \, dt, n = 1$ (or Formula 37)

$$\text{Average} = \int_3^4 \frac{50}{1 + e^{4.8 - 1.9t}} \, dt$$

$$= -\frac{50}{1.9} \int_3^4 \frac{-1.9}{1 + e^{4.8 - 1.9t}} \, dt$$

$$= -\frac{50}{1.9}\left[4.8 - 1.9t - \ln(1 + e^{4.8 - 1.9t})\right]_3^4$$

$$\approx 42.58$$

$$\approx 43$$

61. $R = \int_0^2 10{,}000\left[1 - \frac{1}{(1 + 0.1t^2)^{1/2}}\right] dt$

$$= 10{,}000t\Big]_0^2 - 10{,}000\int_0^2 \frac{1}{(1 + 0.1t^2)^{1/2}} \, dt$$

Formula 25: $u = \sqrt{0.1}t, du = \sqrt{0.1} \, dt, a = 1$

$$R = 20{,}000 - \frac{10{,}000}{\sqrt{0.1}} \ln\left|\sqrt{0.1}\,t\right| + \sqrt{0.1t^2 + 1}\Big]_0^2$$

$$\approx \$1138.43$$

63. Average net profit $= \dfrac{1}{16 - 8} \displaystyle\int_8^{16} \sqrt{375.67t^2 - 715.86} \, dt \approx \$230.98$ million

Section 6.5   **Numerical Integration**

1. Exact: $\displaystyle\int_0^2 x^2 \, dx = \frac{1}{3}x^3\Big]_0^2 = \frac{8}{3} \approx 2.6667$

Trapezoidal Rule: $\displaystyle\int_0^2 x^2 \, dx \approx \frac{1}{4}\left[0 + 2\left(\frac{1}{2}\right)^2 + 2(1)^2 + 2\left(\frac{3}{2}\right)^2 + (2)^2\right] = \frac{11}{4} = 2.7500$

Simpson's Rule: $\displaystyle\int_0^2 x^2 \, dx \approx \frac{1}{6}\left[0 + 4\left(\frac{1}{2}\right)^2 + 2(1)^2 + 4\left(\frac{3}{2}\right)^2 + (2)^2\right] = \frac{8}{3} \approx 2.6667$

3. Exact: $\displaystyle\int_0^2 (x^4 + 1) \, dx = \frac{x^5}{5} + x\Big]_0^2 = \frac{32}{5} + 2 = \frac{42}{5} = 8.4$

Trapezoidal Rule: $\displaystyle\int_0^2 (x^4 + 1) \, dx \approx \frac{1}{4}\left[1 + 2\left(\frac{1}{16} + 1\right) + 2(1 + 1) + 2\left(\frac{81}{16} + 1\right) + 17\right] = \frac{36.25}{4} = 9.0625$

Simpson's Rule: $\displaystyle\int_0^2 (x^4 + 1) \, dx \approx \frac{1}{6}\left[1 + 4\left(\frac{1}{16} + 1\right) + 2(1 + 1) + 4\left(\frac{81}{16} + 1\right) + 17\right] = \frac{50.5}{6} \approx 8.4167$

5. Exact: $\displaystyle\int_0^2 x^3\,dx = \frac{1}{4}x^4\Big]_0^2 = 4.0000$

Trapezoidal Rule: $\displaystyle\int_0^2 x^3\,dx \approx \frac{1}{8}\left[0 + 2\left(\frac{1}{4}\right)^3 + 2\left(\frac{2}{4}\right)^3 + 2\left(\frac{3}{4}\right)^3 + 2(1)^3 + 2\left(\frac{5}{4}\right)^3 + 2\left(\frac{6}{4}\right)^3 + 2\left(\frac{7}{4}\right)^3 + 8\right] = 4.0625$

Simpson's Rule: $\displaystyle\int_0^2 x^3\,dx \approx \frac{1}{12}\left[0 + 4\left(\frac{1}{4}\right)^3 + 2\left(\frac{2}{4}\right)^3 + 4\left(\frac{3}{4}\right)^3 + 2(1)^3 + 4\left(\frac{5}{4}\right)^3 + 2\left(\frac{6}{4}\right)^3 + 4\left(\frac{7}{4}\right)^3 + 8\right] = 4.0000$

7. Exact: $\displaystyle\int_1^2 \frac{1}{x^2}\,dx = \frac{-1}{x}\Big]_1^2 = 0.5000$

Trapezoidal Rule: $\displaystyle\int_1^2 \frac{1}{x^2}\,dx \approx \frac{1}{8}\left[1 + 2\left(\frac{4}{5}\right)^2 + 2\left(\frac{4}{6}\right)^2 + 2\left(\frac{4}{7}\right)^2 + \frac{1}{4}\right] \approx 0.5090$

Simpson's Rule: $\displaystyle\int_1^2 \frac{1}{x^2}\,dx \approx \frac{1}{12}\left[1 + 4\left(\frac{4}{5}\right)^2 + 2\left(\frac{4}{6}\right)^2 + 4\left(\frac{4}{7}\right)^2 + \frac{1}{4}\right] \approx 0.5004$

9. Exact: $\displaystyle\int_0^1 \frac{1}{1+x}\,dx = \ln|1+x|\,\Big]_0^1 \approx 0.6931$

Trapezoidal Rule: $\displaystyle\int_0^1 \frac{1}{1+x}\,dx \approx \frac{1}{8}\left[1 + 2\left(\frac{4}{5}\right) + 2\left(\frac{2}{3}\right) + 2\left(\frac{4}{7}\right) + \frac{1}{2}\right] \approx 0.6970$

Simpson's Rule: $\displaystyle\int_0^1 \frac{1}{1+x}\,dx \approx \frac{1}{12}\left[1 + 4\left(\frac{4}{5}\right) + 2\left(\frac{2}{3}\right) + 4\left(\frac{4}{7}\right) + \frac{1}{2}\right] \approx 0.6933$

11. (a) Trapezoidal Rule: $\displaystyle\frac{1}{8}\left[1 + 2\left(\frac{1}{1+\frac{1}{16}}\right) + 2\left(\frac{1}{1+\frac{1}{4}}\right) + 2\left(\frac{1}{1+\frac{9}{16}}\right) + \frac{1}{2}\right] \approx 0.783$

   (b) Simpson's Rule: $\displaystyle\frac{1}{12}\left[1 + 4\left(\frac{1}{1+\frac{1}{16}}\right) + 2\left(\frac{1}{1+\frac{1}{4}}\right) + 4\left(\frac{1}{1+\frac{9}{16}}\right) + \frac{1}{2}\right] \approx 0.785$

13. (a) Trapezoidal Rule: $\displaystyle\frac{1}{8}\left[1 + 2\sqrt{\frac{15}{16}} + 2\sqrt{\frac{3}{4}} + 2\sqrt{\frac{7}{16}} + 0\right] \approx 0.749$

   (b) Simpson's Rule: $\displaystyle\frac{1}{12}\left[1 + 4\sqrt{\frac{15}{16}} + 2\sqrt{\frac{3}{4}} + 4\sqrt{\frac{7}{16}} + 0\right] \approx 0.771$

15. (a) Trapezoidal Rule: $\frac{2}{4}\left[e^0 + 2e^{-1} + e^{-4}\right] \approx 0.877$

   (b) Simpson's Rule: $\frac{2}{6}\left[e^0 + 4e^{-1} + e^{-4}\right] \approx 0.830$

17. (a) Trapezoidal Rule: $\frac{1}{4}\left[\frac{1}{2} + 2(0.8) + 2(1) + 2(0.8) + 2(0.5) + 2(0.30769) + 0.2\right] \approx 1.88$

   (b) Simpson's Rule: $\frac{1}{6}\left[\frac{1}{2} + 4(0.8) + 2(1) + 4(0.8) + 2(0.5) + 4(0.30769) + 0.2\right] \approx 1.89$

19. $\displaystyle V = \int_0^{t_1} c(t)e^{-rt}\,dt$

$\displaystyle = \int_0^4 \left(6000 + 200\sqrt{t}\right)e^{-0.07t}\,dt$

$\displaystyle \approx \frac{4-0}{24}\left[\left(6000 + 200\sqrt{0}\right)e^0 + 4\left(6000 + 200\sqrt{\frac{1}{2}}\right)e^{-0.07(1/2)} + 2\left(6000 + 200\sqrt{1}\right)e^{-0.07} + 4\left(6000 + 200\sqrt{\frac{3}{2}}\right)e^{-0.07(3/2)}\right.$

$\displaystyle \qquad + 2\left(6000 + 200\sqrt{2}\right)e^{-0.14} + 4\left(6000 + 200\sqrt{\frac{5}{2}}\right)e^{-0.07(5/2)} + 2\left(6000 + 200\sqrt{3}\right)e^{-0.21}$

$\displaystyle \qquad \left. + 4\left(6000 + 200\sqrt{\frac{7}{2}}\right)e^{-0.07(7/2)} + \left(6000 + 200\sqrt{4}\right)e^{-0.28}\right] \approx \$21,831.20$

21. $\Delta R = \int_{14}^{16} 5\sqrt{8000 - x^3}\, dx$

$\approx \dfrac{16 - 14}{12}\left[5\sqrt{8000 - 14^3} + 4(5)\sqrt{8000 - \left(\dfrac{29}{2}\right)^3} + 2(5)\sqrt{8000 - 15^3} + 4(5)\sqrt{8000 - \left(\dfrac{31}{2}\right)^3} + 5\sqrt{8000 - 16^3}\right]$

$\approx \$678.36$

23. $P(a \leq x \leq b) = \int_a^b \dfrac{1}{\sqrt{2\pi}}e^{-x^2/2}\, dx$

$P(0 \leq x \leq 1) = \dfrac{1}{\sqrt{2\pi}}\int_0^1 e^{-x^2/2}\, dx$

$\approx \dfrac{1}{\sqrt{2\pi}}\left(\dfrac{1}{18}\right)\left[e^0 + 4e^{-(1/6)^2/2} + 2e^{-(1/3)^2/2} + 4e^{-(1/2)^2/2} + 2e^{-(2/3)^2/2} + 4e^{-(5/6)^2/2} + e^{-1/2}\right]$

$= \dfrac{1}{18\sqrt{2\pi}}\left[1 + 4e^{-1/72} + 2e^{-1/18} + 4e^{-1/8} + 2e^{-2/9} + 4e^{-25/72} + e^{-1/2}\right]$

≈ 0.3413

$= 34.13\%$

25. $A \approx \dfrac{1000}{3(10)}\left[125 + 4(125) + 2(120) + 4(112) + 2(90) + 4(90) + 2(95) + 4(88) + 2(75) + 4(35) + 0\right]$

$= 89{,}500$ square feet

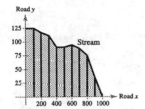

27. $f(x) = x^4$

$f'(x) = 4x^3$

$f''(x) = 12x^2$

$f'''(x) = 24x$

$f^{(4)}(x) = 24$

(a) Trapezoidal Rule: Since $f''(x)$ is maximum in $[0, 2]$ when $x = 2$, we have

$$|\text{Error}| \leq \dfrac{(2 - 0)^3}{48(4^2)}(48) = 0.2.$$

(b) Simpson's Rule: Since $f^{(4)}(x) = 24$, we have

$$|\text{Error}| \leq \dfrac{(2 - 0)^5}{180(4^4)}(24) = 0.017.$$

29. $f(x) = e^{x^3}$

$f'(x) = 3x^2 e^{x^3}$

$f''(x) = 3(3x^4 + 2x)e^{x^3}$

$f'''(x) = 3(9x^6 + 18x^3 + 2)e^{x^3}$

$f^{(4)}(x) = 3(27x^8 + 108x^5 + 60x^2)e^{x^3}$

(a) Trapezoidal Rule: Since $|f''(x)|$ is maximum in $[0, 1]$ when $x = 1$, we have

$$|\text{Error}| \leq \dfrac{(1 - 0)^3}{12(4^2)}(15e) = \dfrac{5e}{64} \approx 0.212.$$

(b) Simpson's Rule: Since $|f^{(4)}(x)|$ is maximum in $[0, 1]$ when $x = 1$, we have

$$|\text{Error}| \leq \dfrac{(1 - 0)^5}{180(4^4)}(585e) = \dfrac{13e}{1024} \approx 0.035.$$

31. $f(x) = x^4$

$f'(x) = 4x^3$

$f''(x) = 12x^2$

$f'''(x) = 24x$

$f^{(4)}(x) = 24$

(a) Trapezoidal Rule: Since $|f''(x)|$ is maximum in $[0, 1]$ when $x = 1$, and $|f''(1)| = 12$, we have:

$$|\text{Error}| \leq \frac{(1 - 0)^3}{12n^2}(12) \leq 0.0001$$

$$\frac{1}{n^2} < 0.0001$$

$$n^2 > 10,000$$

$$n > 100$$

Let $n = 101$.

(b) Simpson's Rule: Since $|f^{(4)}(x)| = 24$ in $[0, 1]$, we have:

$$|\text{Error}| \leq \frac{1}{180n^4}(24) < 0.0001$$

$$n^4 > 1333.33$$

$$n > 6.04$$

Let $n = 8$ (n must be even).

33. $f(x) = e^{2x}$

$f'(x) = 2e^{2x}$

$f''(x) = 4e^{2x}$

$f'''(x) = 8e^{2x}$

$f^{(4)}(x) = 16e^{2x}$

(a) Trapezoidal Rule: Since $|f''(x)|$ is maximum in $[1, 3]$ when $x = 3$, and $|f''(3)| = 4e^6 \approx 1613.715$, we have:

$$|\text{Error}| \leq \frac{(3 - 1)^3}{12n^2}(1613.715) < 0.0001$$

$$n^2 > 10,758,100$$

$$n > 3279.95$$

Take $n = 3280$.

(b) Simpson's Rule: Since $|f^{(4)}(x)|$ is maximum in $[1, 3]$ when $x = 3$, and $|f^{(4)}(3)| = 16e^6 \approx 6454.861$, we have:

$$|\text{Error}| \leq \frac{(3 - 1)^5}{180n^4}(6454.861) < 0.0001$$

$$n^4 > 11,475,308.44$$

$$n > 58.2$$

Take $n = 60$ (n must be even).

35. $\displaystyle\int_1^4 x\sqrt{x + 4}\, dx \approx 19.5215$ $(n = 100)$

37. $\displaystyle\int_2^5 10xe^{-x}\, dx \approx 3.6558$ $(n = 100)$

39. $P_3(x) = ax^3 + bx^2 + cx + d$

$P_3'(x) = 3ax^2 + 2bx + c$

$P_3''(x) = 6ax + 2b$

$P_3'''(x) = 6a$

$P_3^{(4)}(x) = 0$

$$|\text{Error}| \leq \frac{(b - a)^5}{180n^4}[\max|P_3^{(4)}(x)|] = \frac{(b - a)^5}{180n^4}(0) = 0$$

Therefore, Simpson's Rule is exact when used to approximate a cubic polynomial.

$$\int_0^1 x^3\, dx = \frac{1 - 0}{3(2)}\left[0^3 + 4\left(\frac{1}{2}\right)^3 + (1)^3\right] = \frac{1}{6}\left[0 + \frac{1}{2} + 1\right] = \frac{1}{4}$$

The exact value of this integral is

$$\int_0^1 x^3\, dx = \frac{x^4}{4}\bigg]_0^1 = \frac{1}{4}$$

which is the same as the Simpson approximation.

41. $y = \dfrac{x^2}{800}$, $y' = \dfrac{x}{400}$, $1 + (y')^2 = 1 + \dfrac{x^2}{160,000}$

$$S = \int_{-200}^{200} \sqrt{\dfrac{160,000 + x^2}{160,000}}\, dx$$

$$= \int_{-200}^{200} \dfrac{\sqrt{160,000 + x^2}}{400}\, dx$$

$$= \dfrac{1}{200} \int_{0}^{200} \sqrt{160,000 + x^2}\, dx$$

$$\approx \dfrac{200 - 0}{36(200)}\left[400 + 4\sqrt{160,000 + \left(\dfrac{50}{3}\right)^2} + 2\sqrt{160,000 + \left(\dfrac{100}{3}\right)^2} + 4\sqrt{160,000 + (50)^2} + 2\sqrt{160,000 + \left(\dfrac{200}{3}\right)^2} \right.$$

$$+ 4\sqrt{160,000 + \left(\dfrac{250}{3}\right)^2} + 2\sqrt{160,000 + (100)^2} + 4\sqrt{160,000 + \left(\dfrac{350}{3}\right)^2} + 2\sqrt{160,000 + \left(\dfrac{400}{3}\right)^2}$$

$$\left. + 4\sqrt{160,000 + (150)^2} + 2\sqrt{160,000 + \left(\dfrac{500}{3}\right)^2} + 4\sqrt{160,000 + \left(\dfrac{550}{3}\right)^2} + \sqrt{160,000 + (200)^2} \right]$$

$$\approx 416.1 \text{ feet}$$

Using Formula 21: $u = x$, $du = dx$, $a = 40$

$$\dfrac{1}{200} \int_{0}^{200} \sqrt{160,000 + x^2}\, dx = \dfrac{1}{400}\left[x\sqrt{160,000 + x^2} + 160,000 \ln\left|x + \sqrt{160,000 + x^2}\right| \right]_{0}^{200}$$

$$= \dfrac{1}{400}\left[200\sqrt{200,000} + 160,000 \ln\left|200 + \sqrt{200,000}\right| - 160,000 \ln\sqrt{160,000} \right]$$

$$\approx 416.1$$

43. $C = \displaystyle\int_{0}^{12} [8 - \ln(t^2 - 2t + 4)]\, dt \approx 58.876$ grams

(Simpson's Rule with $n = 100$)

45. $\displaystyle\int_{0}^{6} 1000t^2 e^{-t}\, dt \approx 1876$ subscribers

(Simpson's Rule with $n = 100$)

Section 6.6   Improper Integrals

1. This integral converges since

$$\int_{0}^{\infty} e^{-x}\, dx = \lim_{b \to \infty} \left[-e^{-x} \right]_{0}^{b} = 0 + 1 = 1.$$

3. This integral converges since

$$\int_{1}^{\infty} \dfrac{1}{x^2}\, dx = \lim_{b \to \infty} \left[-\dfrac{1}{x} \right]_{1}^{b} = 0 + 1 = 1.$$

5. This integral diverges since

$$\int_{0}^{\infty} e^{x/3}\, dx = \lim_{b \to \infty} \left[3e^{x/3} \right]_{0}^{b} = \infty.$$

7. This integral diverges since

$$\int_{5}^{\infty} \dfrac{x}{\sqrt{x^2 - 16}}\, dx = \lim_{b \to \infty} \left[\sqrt{x^2 - 16} \right]_{5}^{b} = \infty.$$

9. This integral diverges since

$$\int_{-\infty}^{0} e^{-x}\, dx = \lim_{a \to -\infty} \left[-e^{-x} \right]_{a}^{0} = -1 + \infty = \infty.$$

11. This integral diverges since

$$\int_{-\infty}^{\infty} e^{|x|}\, dx = \int_{-\infty}^{0} e^{-x}\, dx + \int_{0}^{\infty} e^{x}\, dx$$

$$= \lim_{a \to -\infty} \left[-e^{-x} \right]_{a}^{0} + \lim_{b \to \infty} \left[e^{x} \right]_{0}^{b}$$

$$= (-1 + \infty) + (\infty - 1)$$

$$= \infty.$$

13. This integral converges since

$$\int_{-\infty}^{\infty} 2xe^{-3x^2}\, dx = \int_{-\infty}^{0} 2xe^{-3x^2}\, dx + \int_{0}^{\infty} 2xe^{-3x^2}\, dx$$

$$= \lim_{a \to -\infty} \left[-\frac{1}{3} e^{-3x^2} \right]_{a}^{0} + \lim_{b \to \infty} \left[-\frac{1}{3} e^{-3x^2} \right]_{0}^{b}$$

$$= \left(-\frac{1}{3} + 0 \right) + \left(0 + \frac{1}{3} \right)$$

$$= 0.$$

15. This integral converges since $\displaystyle\int_{0}^{4} \frac{1}{\sqrt{x}}\, dx = \lim_{b \to 0^{+}} \left[2\sqrt{x} \right]_{b}^{4} = 4.$

17. This integral converges since

$$\int_{0}^{2} \frac{1}{(x-1)^{2/3}}\, dx = \int_{0}^{1} \frac{1}{(x-1)^{2/3}}\, dx + \int_{1}^{2} \frac{1}{(x-1)^{2/3}}\, dx$$

$$= \lim_{b \to 1^{-}} \left[3(x-1)^{1/3} \right]_{0}^{b} + \lim_{a \to 1^{+}} \left[3(x-1)^{1/3} \right]_{a}^{2}$$

$$= 3 + 3$$

$$= 6.$$

19. This integral diverges since

$$\int_{0}^{1} \frac{1}{1-x}\, dx = \lim_{b \to 1^{-}} \left[-\ln|1-x| \right]_{0}^{b} = \infty.$$

21. This integral converges since

$$\int_{0}^{9} \frac{1}{\sqrt{9-x}}\, dx = \lim_{b \to 9^{-}} \left[-2\sqrt{9-x} \right]_{0}^{b}$$

$$= 0 - \left(-2\sqrt{9} \right)$$

$$= 6.$$

23. This integral diverges since $\displaystyle\int_{0}^{1} \frac{1}{x^2}\, dx = \lim_{b \to 0^{+}} \left[-\frac{1}{x} \right]_{b}^{1} = \infty.$

25. This integral converges since

$$\int_{0}^{2} \frac{1}{\sqrt[3]{x-1}}\, dx = \int_{0}^{1} \frac{1}{\sqrt[3]{x-1}}\, dx + \int_{1}^{2} \frac{1}{\sqrt[3]{x-1}}\, dx$$

$$= \lim_{b \to 1^{-}} \left[\frac{3}{2}(x-1)^{2/3} \right]_{0}^{b} + \lim_{a \to 1^{+}} \left[\frac{3}{2}(x-1)^{2/3} \right]_{a}^{2}$$

$$= -\frac{3}{2} + \frac{3}{2}$$

$$= 0.$$

27. This integral converges since

$$\int_{3}^{4} \frac{1}{\sqrt{x^2-9}}\, dx = \lim_{a \to 3^{+}} \left[\ln\left(x + \sqrt{x^2-9} \right) \right]_{a}^{4}$$

$$= \ln\left(4 + \sqrt{7} \right) - \ln 3$$

$$\approx 0.7954.$$

29. $a = 1, n = 1$: $\displaystyle\lim_{x \to \infty} xe^{-x}$

x	1	10	25	50
xe^{-x}	0.3679	0.0005	0.0000	0.0000

31. $a = \frac{1}{2}, n = 2$: $\lim\limits_{x \to \infty} x^2 e^{-(1/2)x}$

x	1	10	25	50
$x^2 e^{-(1/2)x}$	0.6065	0.6738	0.0023	0.0000

33. $\displaystyle\int_0^\infty x^2 e^{-x}\, dx = \lim\limits_{b \to \infty}\left[-x^2 e^{-x} - 2xe^{-x} - 2e^{-x}\right]_0^b = 2$

35. $\displaystyle\int_0^\infty xe^{-2x}\, dx = \lim\limits_{b \to \infty}\left[-\frac{1}{2}xe^{-2x} - \frac{1}{4}e^{-2x}\right]_0^b = (0 - 0) - \left(0 - \frac{1}{4}\right) = \frac{1}{4}$

37. (a) Present value $= \displaystyle\int_0^{20} 500{,}000e^{-0.09t}\, dt = \frac{500{,}000}{-0.09}e^{-0.09t}\Big]_0^{20} \approx \$4{,}637{,}228.40$

(b) Present value $= \displaystyle\int_0^\infty 500{,}000e^{-0.09t}\, dt = \lim\limits_{b \to \infty}\frac{500{,}000}{-0.09}e^{-0.09t}\Big]_0^b \approx \$5{,}555{,}555.56$

39. $C = 650{,}000 + \displaystyle\int_0^n 25{,}000e^{-0.10t}\, dt = 650{,}000 - \left[250{,}000e^{-0.10t}\right]_0^n$

(a) For $n = 5$, we have $C = 650{,}000 - [250{,}000(e^{-0.50} - 1)] \approx \$748{,}367.34$.

(b) For $n = 10$, we have $C = 650{,}000 - [250{,}000(e^{-1} - 1)] \approx \$808{,}030.14$.

(c) For $n = \infty$, we have

$$C = 650{,}000 - \lim\limits_{n \to \infty}\left[250{,}000e^{-0.10t}\right]_0^n = 650{,}000 - 250{,}000(0 - 1) = \$900{,}000.$$

41. (a) $A = \displaystyle\int_1^\infty \frac{1}{x^2}\, dx = \lim\limits_{b \to \infty}\left[-\frac{1}{x}\right]_1^b = 1$

(b) $V = \pi \displaystyle\int_1^\infty \left(\frac{1}{x^2}\right)^2 dx = \pi \lim\limits_{b \to \infty}\left[-\frac{1}{3x^3}\right]_1^b = \frac{\pi}{3}$

43. $\mu = 64.5$, $\sigma = 2.4$

$$f(x) = \frac{1}{\sigma\sqrt{2\pi}}e^{-(x-\mu)^2/2\sigma^2} = 0.166226e^{-(x-64.5)^2/11.52}$$

(a) $\displaystyle\int_{60}^{72} f(x)\, dx \approx 0.9687$

(b) $\displaystyle\int_{68}^\infty f(x)\, dx \approx 0.0724$

(c) $\displaystyle\int_{72}^\infty f(x)\, dx \approx 0.0009$

Review Exercises for Chapter 6

1. $\displaystyle\int dt = t + C$

3. $\displaystyle\int (x + 5)^3\, dx = \frac{(x + 5)^4}{4} + C$

5. $\displaystyle\int e^{10x}\, dx = \frac{1}{10}e^{10x} + C$

7. $\displaystyle\int \frac{1}{5x}\, dx = \frac{1}{5}\ln|x| + C$

9. $\displaystyle\int x\sqrt{x^2 + 4}\, dx = \frac{1}{2}(x^2 + 4)^{3/2}\left(\frac{2}{3}\right) + C$

$\qquad\qquad\qquad\qquad = \frac{1}{3}(x^2 + 4)^{2/3} + C$

11. $\displaystyle\int \frac{2e^x}{3 + e^x}\, dx = 2\ln(3 + e^x) + C$

13. $u = x - 2, x = u + 2, du = dx$

$$\int x(x - 2)^3 \, dx = \int (u + 2)u^3 \, du$$

$$= \int (u^4 + 2u^3) \, du$$

$$= \frac{u^5}{5} + \frac{u^4}{2} + C$$

$$= \frac{(x - 2)^5}{5} + \frac{(x - 2)^4}{2} + C$$

15. $u = x + 1, x = u - 1, du = dx$

$$\int x\sqrt{x + 1} \, dx = \int (u - 1)u^{1/2} \, du$$

$$= \int (u^{3/2} - u^{1/2}) \, du$$

$$= \frac{2}{5}u^{5/2}\left(\frac{-2}{3}\right)u^{3/2} + C$$

$$= \frac{2}{5}(x + 1)^{5/2} - \frac{2}{3}(x + 1)^{3/2} + C$$

$$= \frac{2}{15}(x + 1)^{3/2}[3(x + 1) - 5] + C$$

$$= \frac{2}{15}(x + 1)^{3/2}(3x - 2) + C$$

17. $\displaystyle\int 2x\sqrt{x - 3} \, dx = 2\int (u + 3)\sqrt{u} \, du$

$$= 2\int (u^{3/2} + 3u^{1/2}) \, du$$

$$= 2\left(\frac{2}{5}u^{5/2} + 2u^{3/2}\right) + C$$

$$= \frac{4}{5}u^{3/2}(u + 5) + C$$

$$= \frac{4}{5}(x - 3)^{3/2}(x + 2) + C$$

19. Let $u = 1 - x$, then $x = 1 - u$ and $dx = -du$.

$$\int (x + 1)\sqrt{1 - x} \, dx = -\int (2 - u)\sqrt{u} \, du$$

$$= -\int (2u^{1/2} - u^{3/2}) \, du$$

$$= -\left(\frac{4}{3}u^{3/2} - \frac{2}{5}u^{5/2}\right) + C$$

$$= -\frac{2}{15}u^{3/2}(10 - 3u) + C$$

$$= -\frac{2}{15}(1 - x)^{3/2}(3x + 7) + C$$

21. $u = x - 2, x = u + 2, du = dx$

$$\int_2^3 x\sqrt{x - 2} \, dx = \int_0^1 (u + 2)u^{1/2} \, du$$

$$= \int_0^1 (u^{3/2} + 2u^{1/2}) \, du$$

$$= \left[\frac{2}{5}u^{5/2} + \frac{4}{3}u^{3/2}\right]_0^1$$

$$= \frac{2}{5} + \frac{4}{3}$$

$$= \frac{26}{15}$$

23. $u = x - 1, x = u + 1, du = dx$

$$\int_1^3 x^2(x - 1)^3 \, dx = \int_0^2 (u + 1)^2 u^3 \, du$$

$$= \int_0^2 (u^5 + 2u^4 + u^3) \, du$$

$$= \left[\frac{u^6}{6} + \frac{2u^5}{5} + \frac{u^4}{4}\right]_0^2$$

$$= \frac{32}{3} + \frac{64}{4} + 4$$

$$= \frac{412}{15}$$

$$\approx 27.467$$

25. (a) $P(0 \le x \le 0.80) = \displaystyle\int_0^{0.8} \frac{105}{16}x^2\sqrt{1 - x} \, dx, \quad u = 1 - x, x = 1 - u, du = -dx$

$$\int_1^{0.2} \frac{105}{16}(1 - u)^2 u^{1/2}(-du) = -\frac{105}{16}\int_1^{0.2} (u^{1/2} - 2u^{3/2} + u^{5/2}) \, du$$

$$= -\frac{105}{16}\left[\frac{2}{3}u^{3/2} - \frac{4}{5}u^{5/2} + \frac{2}{7}u^{7/2}\right]_1^{0.2}$$

$$\approx 0.696$$

—CONTINUED—

25. —CONTINUED—

(b) Solve the equation $\int_0^b \frac{105}{16} x^2 \sqrt{1 - x} \, dx = 0.5$ for b. From part (a):

$$\frac{-105}{16} \left[\frac{2}{3} u^{3/2} - \frac{4}{5} u^{5/2} + \frac{2}{7} u^{7/2} \right]_1^{1-b} = 0.5$$

$$\left[\frac{2}{3}(1 - b)^{3/2} - \frac{4}{5}(1 - b)^{5/2} + \frac{2}{7}(1 - b)^{7/2} \right] - \left[\frac{2}{3} - \frac{4}{5} + \frac{2}{7} \right] = \frac{-8}{105}$$

$$\left[\frac{2}{3}(1 - b)^{3/2} - \frac{4}{5}(1 - b)^{5/2} + \frac{2}{7}(1 - b)^{7/2} \right] = \frac{8}{105}$$

Using a graphing utility to solve for b, we obtain $b \approx 0.693$.

27. Using integration by parts with $u = \ln t$, $du = (1/t) \, dt$, $dv = t^2 \, dt$, and $v = t^3/3$,

$$\int t^2 \ln t \, dt = \frac{t^3}{3} \ln t - \int \frac{t^3}{3} \frac{1}{t} \, dt$$

$$= \frac{t^3}{3} \ln t - \frac{t^3}{9} + C.$$

(a) Average $= \frac{1}{16 - 7} \int_7^{16} P \, dt$

$$= \frac{1}{9} \left[449.63t + 4.15 \left(\frac{t^3}{3} \ln t - \frac{t^3}{9} \right) \right]_7^{16}$$

$$\approx \$1900.30 \text{ million}$$

(b) Total $= \int_7^{16} P \, dt \approx \$17,102.68$ million

29. Let $u = \ln x$, $dv = \left(1/\sqrt{x} \right) dx$, $du = (1/x) \, dx$, $v = 2\sqrt{x}$.

$$\int \frac{\ln x}{\sqrt{x}} \, dx = 2\sqrt{x} \ln x - \int 2\sqrt{x} \left(\frac{1}{x} \right) dx$$

$$= 2\sqrt{x} \ln x - \int 2x^{-1/2} \, dx$$

$$= 2\sqrt{x} \ln x - 4\sqrt{x} + C$$

31. Use integration by parts and let $u = x - 1$ and $dv = e^x \, dx$, then $du = dx$ and $v - e^x$.

$$\int (x - 1)e^x \, dx = (x - 1)e^x - \int e^x \, dx$$

$$= (x - 1)e^x - e^x + C$$

$$= (x - 2)e^x + C$$

33. Let $u = 2x^2$, $dv = e^{2x}$, $du = 4x \, dx$, $v = \frac{1}{2}e^{2x}$.

$$\int 2x^2 e^{2x} \, dx = (2x^2) \left(\frac{1}{2}e^{2x} \right) - \int \frac{1}{2}e^{2x}(4x \, dx)$$

$$= x^2 e^{2x} - \int 2xe^{2x} \, dx$$

Use integration by parts again for the integral on the right: $u = 2x$, $dv = e^{2x}$, $du = 2 \, dx$, $v = \frac{1}{2}e^{2x}$.

$$\int 2x^2 e^{2x} \, dx = x^2 e^{2x} - \left[(2x) \left(\frac{1}{2}e^{2x} \right) - \int \frac{1}{2}e^{2x}(2) \, dx \right]$$

$$= x^2 e^{2x} - xe^{2x} + \int e^{2x} \, dx$$

$$= x^2 e^{2x} - xe^{2x} + \frac{1}{2}e^{2x} + C$$

35. Present value $= \int_0^{t_1} c(t)e^{-rt} \, dt$

$$= \int_0^5 10,000e^{-0.04t} \, dt$$

$$= -250,000e^{-0.04t} \Big]_0^5$$

$$= \$45,317.31$$

37. Present value $= \int_0^{t_1} c(t)e^{-rt} \, dt$

$$= \int_0^{10} 12,000e^{-0.05t} \, dt$$

$$= (-240,000t - 4,800,000)e^{-0.05t} \Big]_0^{10}$$

$$= \$432,979.25$$

39. Present value = $\int_0^{20} 100,000e^{-0.08t}\, dt \approx \$997,629.35$

41. Present value = $\int_0^{10} 6000e^{-0.06t}\, dt = \$45,118.84$

43. Use partial fractions.

$$\frac{1}{x(x+5)} = \frac{A}{x} + \frac{B}{x+5}$$

Basic equation: $1 = A(x+5) + Bx$

When $x = 0$: $1 = 5A, A = \dfrac{1}{5}$

When $x = -5$: $1 = -5B, B = -\dfrac{1}{5}$

$$\int \frac{1}{x(x+5)}\, dx = \frac{1}{5} \int \left[\frac{1}{x} - \frac{1}{x+5}\right] dx$$

$$= \frac{1}{5}\Big[\ln|x| - \ln|x+5|\Big] + C$$

$$= \frac{1}{5} \ln\left|\frac{x}{x+5}\right| + C$$

45. Partial fractions: $\dfrac{4x - 13}{x^2 - 3x - 10} - \dfrac{A}{x-5} + \dfrac{B}{x+2}$

Basic equation: $4x - 13 = A(x+2) + B(x-5)$

When $x = -2$: $-21 = -7B, B = 3$

When $x = 5$: $7 = 7A, A = 1$

$$\int \frac{4x-13}{x^2-3x-10}\, dx = \int \left(\frac{1}{x-5} + \frac{3}{x+2}\right) dx$$

$$= \ln|x-5| + 3\ln|x+2| + C$$

47. $\dfrac{x^2}{x^2 + 2x - 15} = 1 - \dfrac{2x - 15}{(x+5)(x-3)}$

Use partial fractions.

$$\frac{2x-15}{(x+5)(x-3)} = \frac{A}{x+5} + \frac{B}{x-3}$$

Basic equation: $2x - 15 = A(x-3) + B(x+5)$

When $x = -5$: $-25 = -8A, A = \dfrac{25}{8}$

When $x = 3$: $-9 = 8B, B = -\dfrac{9}{8}$

$$\int \frac{x^2}{x^2+2x-15}\, dx = \int \left[1 - \frac{25}{8}\left(\frac{1}{x+5}\right) + \frac{9}{8}\left(\frac{1}{x-3}\right)\right] dx$$

$$= x - \frac{25}{8}\ln|x+5| + \frac{9}{8}\ln|x-3| + C$$

49. (a) $y = \dfrac{L}{1 + be^{-kt}} = \dfrac{10,000}{1 + be^{-kt}}$

Since $y = 1250$ when $t = 0$,

$$\frac{10,000}{1+b} = 1250 \Rightarrow 1 + b = 8 \Rightarrow b = 7.$$

When $t = 26$ (6 months), $y = 6500 = \dfrac{10,000}{1 + 7e^{-k(26)}}$.

Solving for k: $1 + 7e^{-26k} = \dfrac{100}{65}$

$$ye^{-26k} = \frac{35}{65}$$

$$e^{26k} = \frac{5}{65} = \frac{1}{13}$$

$$k = \frac{-1}{26}\ln\left(\frac{1}{13}\right) \approx 0.098652$$

Thus, $y = \dfrac{10,000}{1 + 7e^{-0.098652t}}$.

(b)

Time, t	0	3	6	12	24
Sales, y	1250	1611	2052	3182	6039

(c) The sales will be 7500 when $t \approx 31$ weeks.

51. Use Formula 23 from the tables and let $u = x$, $a = 5$, and $du = dx$.

$$\int \frac{\sqrt{x^2 + 25}}{x} dx = \sqrt{x^2 + 25} - 5 \ln\left|\frac{5 + \sqrt{x^2 + 25}}{x}\right| + C$$

53. Use Formula 29 and let $u = x$, $a = 2$.

$$\int \frac{1}{x^2 - 4} dx = \frac{1}{4} \ln\left|\frac{x - 2}{x + 2}\right| + C$$

55. $\displaystyle\int_0^3 \frac{x}{\sqrt{1 + x}} dx = \left[-\frac{2(2 - x)}{3}\sqrt{1 + x}\right]_0^3 = \frac{8}{3}$

57. Use Formula 17 and let $u = x$, $a = 1$, $b = 1$.

$$\int \frac{\sqrt{x + 1}}{x} dx = 2\sqrt{1 + x} + \int \frac{1}{x\sqrt{1 + x}} dx$$

$$= 2\sqrt{1 + x} + \ln\left|\frac{\sqrt{1 + x} - 1}{\sqrt{1 + x} + 1}\right| + C \qquad \text{(Formula 15)}$$

59. Use Formula 36 and $u = x - 5$, $n = 3$, $du = dx$.

$$\int (x - 5)^3 e^{x-5} dx = (x - 5)^3 e^{x-5} - 3\int (x - 5)^2 e^{x-5} dx$$

$$= (x - 5)^3 e^{x-5} - 3\left[\int (x - 5)^2 e^{x-5} - 2\int (x - 5)e^{x-5} dx\right]$$

$$= (x - 5)^3 e^{x-5} - 3(x - 5)^2 e^{x-5} + 6(x - 6)e^{x-5} + C \qquad \text{(Formula 35)}$$

61.
$$\frac{1}{x^2 + 4x - 21} = \frac{1}{x^2 + 4x + 4 - 25} = \frac{1}{(x + 2)^2 - 25}$$

$$\int \frac{1}{(x + 1)^2 - 25} dx = \frac{1}{10} \ln\left|\frac{(x + 2) - 5}{(x + 2) + 5}\right| + C \qquad \text{(Formula 29)}$$

$$= \frac{1}{10} \ln\left|\frac{x - 3}{x + 7}\right| + C$$

63. $\qquad x^2 - 10x = x^2 - 10x + 25 - 25 = (x - 5)^2 - 25$

$$\int \sqrt{x^2 - 10x}\, dx = \int \sqrt{(x - 5)^2 - 5^2}\, dx$$

$$= \frac{1}{2}\left[(x - 5)\sqrt{(x - 5)^2 - 5^2} + 25 \ln\left|(x - 5) + \sqrt{(x - 5)^2 - 5^2}\right|\right] + C \qquad \text{(Formula 21)}$$

65. $\displaystyle\int_1^3 \frac{1}{x^2} dx \approx \frac{3 - 1}{2(4)}\left[1 + 2\left(\frac{4}{9}\right) + 2\left(\frac{1}{4}\right) + 2\left(\frac{4}{25}\right) + \frac{1}{9}\right] = 0.705$

67. $\displaystyle\int_1^2 \frac{1}{1 + \ln x} dx = \frac{1}{8}\left(1 + 2\left[\frac{1}{1 + \ln(5/4)}\right] + 2\left[\frac{1}{1 + \ln(3/2)}\right] + 2\left[\frac{1}{1 + \ln(7/4)}\right] + \frac{1}{1 + \ln 2}\right) \approx 0.741$

69. $\displaystyle\int_1^2 \frac{1}{x^3} dx \approx \frac{2 - 1}{3(4)}\left[1 + 4\left(\frac{4}{5}\right)^3 + 2\left(\frac{2}{3}\right)^3 + 4\left(\frac{4}{7}\right)^3 + \frac{1}{8}\right] \approx 0.376$

71. $\displaystyle\int_0^1 \frac{x^{3/2}}{2 - x^2} dx \approx \frac{1}{12}\left[0 + \frac{8}{31} + \frac{2\sqrt{2}}{7} + \frac{24\sqrt{3}}{23} + 1\right] \approx 0.289$

73. $f(x) = e^{2x}$

$f'(x) = 2e^{2x}$

$f''(x) = 4e^{2x}$

$|f''(x)| \leq 4e^{2(2)} \approx 219$ on $[0, 2]$.

$|E| \leq \dfrac{(2-0)^3}{12(4)^2}(219) \leq 9.125$

75. $f(x) = \dfrac{1}{x-1} = (x-1)^{-1}$

$f'(x) = -(x-1)^{-2}$

$f''(x) = 2(x-1)^{-3}$

$f'''(x) = -6(x-1)^{-4}$

$f^{(4)}(x) = 24(x-1)^{-5} = \dfrac{24}{(x-1)^5}$

$|f^{(4)}(x)| \leq 24$ on $[2, 4]$.

$|E| \leq \dfrac{(4-2)^5}{180(4)^4}(24) \approx 0.017$

77. $\displaystyle\int_0^\infty 4xe^{-2x^2}\,dx = \lim_{b\to\infty} -\int_0^b e^{-2x^2}(-4x\,dx)$

$\qquad = \lim_{b\to\infty}\left[-e^{2x^2}\right]_0^b$

$\qquad = 1$

79. $\displaystyle\int_{-\infty}^0 \dfrac{1}{3x^2}\,dx = \lim_{b\to-\infty}\int_b^{-1}\dfrac{1}{3}x^{-2}\,dx + \lim_{a\to0^-}\int_{-1}^a \dfrac{1}{3}x^{-2}\,dx$

$\qquad = \lim_{b\to-\infty}\left[\dfrac{-1}{3x}\right]_b^{-1} + \lim_{x\to0^-}\left[\dfrac{-1}{3x}\right]_{-1}^0$

(diverges)

81. $\displaystyle\int_0^4 \dfrac{1}{\sqrt{4x}}\,dx = \lim_{a\to0^+}\int_a^4 \dfrac{1}{2}x^{-1/2}\,dx$

$\qquad = \lim_{a\to0^+}\left[x^{1/2}\right]_a^4$

$\qquad = 2$

83. $\displaystyle\int_2^3 \dfrac{1}{\sqrt{x-2}}\,dx = \lim_{b\to2^+}\left[2(x-2)^{1/2}\right]_b^3 = 2$

85. (a) Present value $= \displaystyle\int_0^{15} 50{,}000e^{-0.06t}\,dt$

$\qquad = \left[\dfrac{50{,}000}{-0.06}e^{-0.06t}\right]_0^{15}$

$\qquad \approx \$494{,}525.28$

(b) Present value $= \displaystyle\int_0^\infty 50{,}000e^{-0.06t}\,dt$

$\qquad = \lim_{b\to\infty}\left[\dfrac{50{,}000}{-0.06}e^{-0.06t}\right]_0^b$

$\qquad \approx \$833{,}333.33$

87. $P(a \leq x \leq b) = \displaystyle\int_a^b \dfrac{1}{2988.40\sqrt{2\pi}}e^{-(x-21{,}875.30)^2/2(2988.40)^2}\,dx$

Using numerical integration or a graphing utility:

(a) $P(25{,}000 \leq x) \approx 0.148$

(b) $P(30{,}000 \leq x) \approx 0.003$

Practice Test for Chapter 6

1. Evaluate $\int x\sqrt{x+3}\,dx$.

2. Evaluate $\int \dfrac{x}{(x-2)^3}\,dx$.

3. Evaluate $\int \dfrac{1}{3x+\sqrt{x}}\,dx$.

4. Evaluate $\int \dfrac{\ln 7x}{x}\,dx$.

5. Evaluate $\int xe^{2x}\,dx$.

6. Evaluate $\int x^3 \ln x\,dx$.

7. Evaluate $\int x^2\sqrt{x-6}\,dx$.

8. Evaluate $\int x^2 e^{4x}\,dx$.

9. Evaluate $\int \dfrac{-5}{x^2+x-6}\,dx$.

10. Evaluate $\int \dfrac{x+12}{x^2+4x}\,dx$.

11. Evaluate $\int \dfrac{5x+3}{(x+2)^2}\,dx$.

12. Evaluate $\int \dfrac{3x^3+9x^2-x+3}{x(x+3)}\,dx$.

13. Evaluate $\int \dfrac{1}{x^2\sqrt{16-x^2}}\,dx$. (Use tables.)

14. Evaluate $\int (\ln x)^3\,dx$. (Use tables.)

15. Evaluate $\int \dfrac{1200}{1+e^{0.06x}}\,dx$. (Use tables.)

16. Approximate the integral using (a) the Trapezoidal Rule and (b) Simpson's Rule.

$$\int_0^4 \sqrt{3+x^3}\,dx, \quad n=8$$

17. Approximate the integral using (a) the Trapezoidal Rule and (b) Simpson's Rule.

$$\int_0^2 e^{-x^2/2}\,dx, \quad n=4$$

18. Determine the divergence or convergence of the integral. Evaluate the integral if it converges.

$$\int_0^9 \dfrac{1}{\sqrt{x}}\,dx$$

19. Determine the divergence or convergence of the integral. Evaluate the integral if it converges.

$$\int_1^\infty \dfrac{1}{x-3}\,dx$$

20. Determine the divergence or convergence of the integral. Evaluate the integral if it converges.

$$\int_{-\infty}^0 e^{-3x}\,dx$$

Graphing Calculator Required

21. Use a program similar to that on page 426 of the textbook to approximate

$$\int_0^3 \frac{1}{\sqrt{x^3 + 1}}\, dx$$

when $n = 50$ and $n = 100$.

22. Complete the following chart using Simpson's Rule to determine the convergence or divergence of

$$\int_0^1 \frac{1}{\sqrt{1 - x^2}}\, dx.$$

n	100	1000	10,000
$\displaystyle\int_1^{0.9999999} \frac{1}{\sqrt{1 - x^2}}\, dx$			

Note: The upper limit cannot equal 1 to avoid division by zero. Let $b = 0.9999999$.

C H A P T E R 7
Functions of Several Variables

CHAPTER 7
Functions of Several Variables

Section 7.1 The Three-Dimensional Coordinate System

Solutions to Odd-Numbered Exercises

1. (a) and (b)

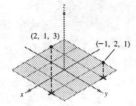

3. (a) and (b)

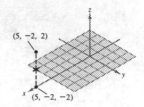

5. $d = \sqrt{(8-4)^2 + (2-1)^2 + (6-5)^2}$

$ = \sqrt{18}$

$ = 3\sqrt{2}$

7. $d = \sqrt{(-3+1)^2 + (4+5)^2 + (-4-7)^2}$

$ = \sqrt{206}$

9. Midpoint $= \left(\dfrac{6 + (-2)}{2}, \dfrac{-9 + (-1)}{2}, \dfrac{1+5}{2} \right)$

$\phantom{\text{Midpoint}} = (2, -5, 3)$

11. Midpoint $= \left(\dfrac{-5+6}{2}, \dfrac{-2+3}{2}, \dfrac{5 + (-7)}{2} \right)$

$\phantom{\text{Midpoint}} = \left(\dfrac{1}{2}, \dfrac{1}{2}, -1 \right)$

13. $(2, -1, 3) = \left(\dfrac{x + (-2)}{2}, \dfrac{y+1}{2}, \dfrac{z+1}{2} \right)$

$2 = \dfrac{x-2}{2} \qquad -1 = \dfrac{y+1}{2} \qquad 3 = \dfrac{z+1}{2}$

$4 = x - 2 \qquad -2 = y + 1 \qquad 6 = z + 1$

$x = 6 \qquad\quad y = -3 \qquad\quad z = 5(x, y, z)$

$ = (6, -3, 5)$

15. $\left(\dfrac{3}{2}, 1, 2 \right) = \left(\dfrac{x+2}{2}, \dfrac{y+0}{2}, \dfrac{z+3}{2} \right)$

$\dfrac{3}{2} = \dfrac{x+2}{2} \qquad 1 = \dfrac{y}{2} \qquad 2 = \dfrac{z+3}{2}$

$x = 1 \qquad\quad y = 2 \qquad z = 1(x, y, z) = (1, 2, 1)$

17. Let $A = (0, 0, 0)$, $B = (2, 2, 1)$, and $C = (2, -4, 4)$. Then we have the following.

$d(AB) = \sqrt{(2-0)^2 + (2-0)^2 + (1-0)^2} = 3$

$d(AC) = \sqrt{(2-0)^2 + (-4-0)^2 + (4-0)^2} = 6$

$d(BC) = \sqrt{(2-2)^2 + (-4-2)^2 + (4-1)^2}$

$ = 3\sqrt{5}$

The triangle is a right triangle since

$d^2(AB) + d^2(AC) = (3)^2 + (6)^2$

$ = 9 + 36$

$ = 45$

$ = d^2(BC).$

19. Let $A = (-2, 2, 4)$, $B = (-2, 2, 6)$, and $C = (-2, 4, 8)$. Then we have the following.

$d(AB) = \sqrt{[-2-(-2)]^2 + (2-2)^2 + (6-4)^2}$

$ = 2$

$d(AC) = \sqrt{[-2-(-2)]^2 + (4-2)^2 + (8-4)^2}$

$ = 2\sqrt{5}$

$d(BC) = \sqrt{[-2-(-2)]^2 + (4-2)^2 + (8-6)^2}$

$ = 2\sqrt{2}$

The triangle is neither right nor isosceles.

21. $x^2 + (y - 2)^2 + (z - 2)^2 = 4$

23. The midpoint of the diameter is the center.

$$\text{Center} = \left(\frac{2 + 1}{2}, \frac{1 + 3}{2}, \frac{3 - 1}{2}\right) = \left(\frac{3}{2}, 2, 1\right)$$

The radius is the distance between the center and either endpoint.

$$\text{Radius} = \sqrt{\left(2 - \frac{3}{2}\right)^2 + (1 - 2)^2 + (3 - 1)^2}$$

$$= \sqrt{\frac{1}{4} + 1 + 4}$$

$$= \frac{\sqrt{21}}{2}\left(x - \frac{3}{2}\right)^2 + (y - 2)^2 + (z - 1)^2$$

$$= \frac{21}{4}$$

25. $(x - 1)^2 + (y - 1)^2 + (z - 5)^2 = 9$

27. The midpoint of the diameter is the center.

$$\text{Center} = \left(\frac{2 + 0}{2}, \frac{0 + 6}{2}, \frac{0 + 0}{2}\right) = (1, 3, 0)$$

The radius is the distance between the center and either endpoint.

$$\text{Radius} = \sqrt{(1 - 2)^2 + (3 - 0)^2 + (0 - 0)^2} = \sqrt{10}$$

$$(x - 1)^2 + (y - 3)^2 + (z - 0)^2 = \left(\sqrt{10}\right)^2$$

$$(x - 1)^2 + (y - 3)^2 + z^2 = 10$$

29. $(x - 1)^2 + (y + 3)^2 + (z + 4)^2 = 25$

Center: $(1, -3, -4)$

Radius: 5

31. $(x - 0)^2 + (y - 4)^2 + (z - 0)^2 = 16$

Center: $(0, 4, 0)$

Radius: 4

33. $(x - 1)^2 + (y - 3)^2 + (z - 2)^2 = -\frac{3}{2} + 1 + 9 + 4 = \frac{25}{2}$

Center: $(1, 3, 2)$

Radius: $\dfrac{5}{\sqrt{2}}$

35. To find the xy-trace, we let $z = 0$.

$$(x - 1)^2 + (y - 3)^2 + (0 - 2)^2 = 25$$

$$(x - 1)^2 + (y - 3)^2 = 21$$

The xy-trace is a circle centered at $(1, 3)$ with radius of $\sqrt{21}$ in the xy-plane.

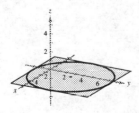

37. $(x - 2)^2 + (y - 2)^2 + (z - 3)^2 = 29$

To find the yz-trace, we let $x = 0$.

$$(0 - 2)^2 + (y - 2)^2 + (z - 3)^2 = 29$$

$$4 + (y - 2)^2 + (z - 3)^2 = 29$$

$$(y - 2)^2 + (z - 3)^2 = 25$$

The yz-trace is a circle centered at $(2, 3)$ with a radius of 5 in the yz-plane.

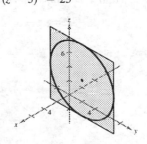

39. (a) To find the trace with $z = 3$, replace z with 3 in the equation of the sphere

$$x^2 + y^2 + 3^2 = 25$$

$$x^2 + y^2 = 16.$$

The trace is a circle centered at $(0, 0, 3)$ with a radius of 4.

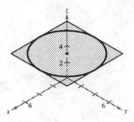

(b) To find the trace with $x = 4$, replace x with 4 in the equation of the sphere

$$4^2 + y^2 + z^2 = 25$$

$$y^2 + z^2 = 9.$$

The trace is a circle centered at $(4, 0, 0)$ with a radius of 3.

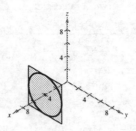

41. (a) Let $x = 2$.

$$4 + y^2 + z^2 - 8 - 6y + 9 = 0$$

$$(y - 3)^2 + z^2 = 2^2$$

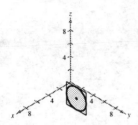

(b) Let $y = 3$.

$$x^2 + 9 + z^2 - 4x - 18y + 9 = 0$$

$$(x - 2)^2 + z^2 = 2^2$$

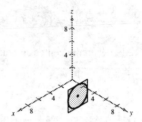

43. Since the crystal is a cube, $A = (3, 3, 0)$. Thus, $(x, y, z) = (3, 3, 3)$.

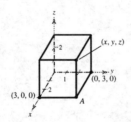

Section 7.2   Surfaces in Space

1. To find the x-intercept, let $y = 0$ and $z = 0$.

$$4x = 12 \Longrightarrow x = 3$$

To find the y-intercept, let $x = 0$ and $z = 0$.

$$2y = 12 \Longrightarrow y = 6$$

To find the z-intercept, let $x = 0$ and $y = 0$.

$$6z = 12 \Longrightarrow z = 2$$

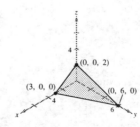

3. To find the *x*-intercept, let $y = 0$ and $z = 0$.

$$3x = 15 \Rightarrow x = 5$$

To find the *y*-intercept, let $x = 0$ and $z = 0$.

$$3y = 15 \Rightarrow y = 5$$

To find the *z*-intercept, let $x = 0$ and $y = 0$.

$$5z = 15 \Rightarrow z = 3$$

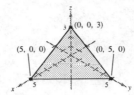

5. To find the *x*-intercept, let $y = 0$ and $z = 0$.

$$2x = 8 \Rightarrow x = 4$$

To find the *y*-intercept, let $x = 0$ and $z = 0$.

$$-y = 8 \Rightarrow y = -8$$

To find the *z*-intercept, let $x = 0$ and $y = 0$.

$$3z = 8 \Rightarrow z = \frac{8}{3}$$

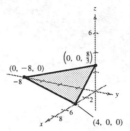

7. Since the coefficients of *x* and *y* are zero, the only intercept is the *z*-intercept of 3. The plane is parallel to the *xy*-plane.

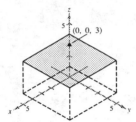

9. The *y*- and *z*-intercepts are both 5, and the plane is parallel to the *x*-axis.

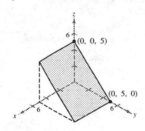

11. The only intercept is the origin. The *xy*-trace is the line $x + y = 0$. The *xz*-trace is the line $x - z = 0$. The *yz*-trace is the line $y - z = 0$.

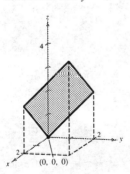

13. For the first plane, $a_1 = 5$, $b_1 = -3$, and $c_1 = 1$, and for the second plane, $a_2 = 1$, $b_2 = 4$, and $c_2 = 7$. Therefore, we have

$$a_1 a_2 + b_1 b_2 + c_1 c_2 = (5)(1) + (-3)(4) + (1)(7)$$

$$= 5 - 12 + 7$$

$$= 0.$$

The planes are perpendicular.

15. For the first plane, $a_1 = 1$, $b_1 = -5$, and $c_1 = -1$, and for the second plane, $a_2 = 5$, $b_2 = -25$, and $c_2 = -5$. Therefore, we have

$$a_2 = 5a_1, \quad b_2 = 5b_1, \quad c_2 = 5c_1.$$

The planes are parallel.

17. For the first plane, $a_1 = 1$, $b_1 = 2$, and $c_1 = 0$, and for the second plane, $a_2 = 4$, $b_2 = 8$, and $c_2 = 0$. Therefore, we have

$$a_2 = 4a_1, \quad b_2 = 4b_1, \quad c_2 = 4c_1.$$

The planes are parallel.

19. For the first plane, $a_1 = 2$, $b_1 = 1$, and $c_1 = 0$, and for the second plane, $a_2 = 3$, $b_2 = 0$, and $c_2 = -5$. The planes are not parallel since

$$3a_1 = 2a_2 \quad \text{and} \quad 3b_1 \neq 2b_2$$

and the planes are not perpendicular since

$$a_1a_2 + b_1b_2 + c_1c_2 = (2)(3) + (1)(0) + (0)(-5)$$
$$= 6$$
$$\neq 0.$$

21. $D = \dfrac{|ax_0 + by_0 + cz_0 + d|}{\sqrt{a^2 + b^2 + c^2}}$

$= \dfrac{|2(0) + 3(0) + 1(0) - 12|}{\sqrt{(2)^2 + (3)^2 + (1)^2}}$

$= \dfrac{12}{\sqrt{4 + 9 + 1}}$

$= \dfrac{12}{\sqrt{14}}$

$= \dfrac{6\sqrt{14}}{7}$

23. $D = \dfrac{|ax_0 + by_0 + cz_0 + d|}{\sqrt{a^2 + b^2 + c^2}}$

$= \dfrac{|2(1) - 1(2) + 1(3) - 4|}{\sqrt{(2)^2 + (-1)^2 + (1)^2}}$

$= \dfrac{1}{\sqrt{6}}$

$= \dfrac{\sqrt{6}}{6}$

25. $D = \dfrac{|ax_0 + by_0 + cz_0 + d|}{\sqrt{a^2 + b^2 + c^2}}$

$= \dfrac{|2(1) - 4(0) + 3(-1) + (-12)|}{\sqrt{4 + 16 + 9}}$

$= \dfrac{13}{\sqrt{29}}$

$= \dfrac{13\sqrt{29}}{29}$

27. The graph is an ellipsoid that matches graph (c).

29. The graph of

$$x^2 + y^2 - \frac{z^2}{4} = 1$$

is a hyperboloid of one sheet that matches graph (f).

31. The graph of

$$y = \frac{x^2}{1} + \frac{z^2}{4}$$

is an elliptic paraboloid that matches graph (d).

33. The graph of

$$z = \frac{y^2}{4} - \frac{x^2}{1}$$

is a hyperbolic paraboloid that matches graph (a).

35. $x^2 - y - z^2 = 0$

Trace in xy-plane ($z = 0$): $y = x^2$ Parabola

Trace in plane $y = 1$: $x^2 - z^2 = 1$ Hyperbola

Trace in yz-plane ($x = 0$): $y = -z^2$ Parabola

37. $\dfrac{x^2}{4} + y^2 + z^2 = 1$

Trace in xy-plane ($z = 0$): $\dfrac{x^2}{4} + y^2 = 1$ Ellipse

Trace in xz-plane ($y = 0$): $\dfrac{x^2}{4} + z^2 = 1$ Ellipse

Trace in yz-plane ($x = 0$): $y^2 + z^2 = 1$ Circle

39. The graph is an ellipsoid.

41. $\dfrac{x^2}{1/5} + \dfrac{y^2}{1/5} - \dfrac{z^2}{5} = 1$

The graph is a hyperboloid of one sheet.

43. The graph of $y = x^2 + z^2$ is an elliptic paraboloid.

45. The graph of $z = y^2 - x^2$ is a hyperbolic paraboloid.

47. $\dfrac{y^2}{16} - \dfrac{x^2}{4} - \dfrac{z^2}{4} = 1$

The graph is a hyperboloid of two sheets.

49. The graph of $-z^2 + 9x^2 + y^2 = 0$ is an elliptic cone.

51. The graph of $z = \dfrac{x^2}{3} - \dfrac{y^2}{3}$ is a hyperbolic paraboloid.

53.

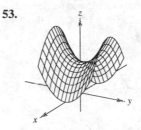

55.

57. $\dfrac{x^2}{3969^2} + \dfrac{y^2}{3969^2} + \dfrac{z^2}{3942^2} = 1$

Section 7.3   Functions of Several Variables

1. (a) $f(3, 2) = \dfrac{3}{2}$ (b) $f(-1, 4) = -\dfrac{1}{4}$

   (c) $f(30, 5) = 6$ (d) $f(5, y) = \dfrac{5}{y}$

   (e) $f(x, 2) = \dfrac{x}{2}$ (f) $f(5, t) = \dfrac{5}{t}$

3. (a) $f(5, 0) = 5$ (b) $f(3, 2) = 3e^2$

   (c) $f(2, -1) = \dfrac{2}{e}$ (d) $f(5, y) = 5e^y$

   (e) $f(x, 2) = xe^2$ (f) $f(t, t) = te^t$

5. (a) $h(2, 3, 9) = \frac{2}{3}$

   (b) $h(1, 0, 1) = 0$

7. (a) $V(3, 10) = \pi(3)^2(10) = 90\pi$

   (b) $V(5, 2) = \pi(5)^2(2) = 50\pi$

9. (a) $A(100, 0.10, 10) = 100\left[\left(1 + \dfrac{0.10}{12}\right)^{120} - 1\right]\left(1 + \dfrac{12}{0.10}\right) \approx \$20{,}655.20$

   (b) $A(275, 0.0925, 40) = 275\left[\left(1 + \dfrac{0.0925}{12}\right)^{480} - 1\right]\left(1 + \dfrac{12}{0.0925}\right) \approx \$1{,}397{,}672.67$

11. (a) $f(1, 2) = \displaystyle\int_1^2 (2t - 3)\, dt = \left[(t^2 - 3t)\right]_1^2 = (-2) - (-2) = 0$

   (b) $f(1, 4) = \displaystyle\int_1^4 (2t - 3)\, dt = \left[(t^2 - 3t)\right]_1^4 = 6$

13. (a) $f(x + \Delta x, y) = (x + \Delta x)^2 - 2y = x^2 + 2x\Delta x + (\Delta x)^2 - 2y$

   (b) $\dfrac{f(x, y + \Delta y) - f(x, y)}{\Delta y} = \dfrac{[x^2 - 2(y + \Delta y)] \quad (x^2 - 2y)}{\Delta y}$

   $= \dfrac{x^2 - 2y - 2\Delta y - x^2 + 2y}{\Delta y}$

   $= -\dfrac{2\Delta y}{\Delta y}$

   $= -2, \quad \Delta y \neq 0$

15. The domain is the set of all points inside and on the circle $x^2 + y^2 = 16$ since $16 - x^2 - y^2 \geq 0$ and the range is $[0, 4]$.

17. The domain is the set of all points above or below the x-axis since $y \neq 0$. The range is $(0, \infty)$.

19. The domain is the set of all points inside and on the ellipse $9x^2 + y^2 = 9$ since $9 - 9x^2 - y^2 \leq 0$.

21. Since $y \neq 0$, the domain is the set of all points above or below the x-axis.

23. The domain is the set of all points in the xy-plane except those on the x- and y-axes.

25. The domain is the set of all points in the xy-plane such that $y \geq 0$.

27. The domain is the half plane below the line $y = -x + 4$ since $4 - x - y > 0$.

29. The contour map consists of ellipses

$$x^2 + \frac{y^2}{4} = C.$$

Matches (b).

31. The contour map consists of curves $e^{1-x^2-y^2} = C$, or

$$1 - x^2 - y^2 = \ln C \Longrightarrow x^2 + y^2 = 1 - \ln C, \text{ circles.}$$

Matches (a).

33. $c = -1, \qquad -1 = x + y, \qquad y = -x - 1$

$ c = 0, \qquad\quad 0 = x + y, \qquad\quad y = -x$

$ c = 2, \qquad\quad 2 = x + y, \qquad\quad y = -x + 2$

$ c = 4, \qquad\quad 4 = x + y, \qquad\quad y = -x + 4$

The level curves are parallel lines.

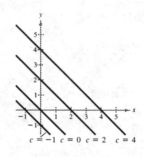

35. $c = 0, \qquad 0 = \sqrt{16 - x^2 - y^2}, \qquad x^2 + y^2 = 16$

$ c = 1, \qquad 1 = \sqrt{16 - x^2 - y^2}, \qquad x^2 + y^2 = 15$

$ c = 2, \qquad 2 = \sqrt{16 - x^2 - y^2}, \qquad x^2 + y^2 = 12$

$ c = 3, \qquad 3 = \sqrt{16 - x^2 - y^2}, \qquad x^2 + y^2 = 7$

$ c = 4, \qquad 4 = \sqrt{16 - x^2 - y^2}, \qquad x^2 + y^2 = 0$

The level curves are circles.

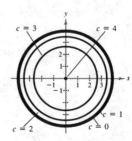

37. $c = 1, \qquad\quad xy = 1$

$ c = -1, \qquad xy = -1$

$ c = 2, \qquad\quad xy = 2$

$ c = -2, \qquad xy = -2$

$ c = \pm 3, \qquad xy = \pm 3$

$ c = \pm 4, \qquad xy = \pm 4$

$ c = \pm 5, \qquad xy = \pm 5$

$ c = \pm 6, \qquad xy = \pm 6$

The level curves are hyperbolas.

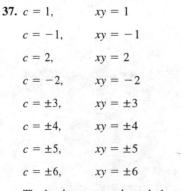

39. $c = \frac{1}{2}$, $\frac{1}{2} = \frac{x}{x^2 + y^2}$, $(x - 1)^2 + y^2 = 1$

$c = -\frac{1}{2}$, $-\frac{1}{2} = \frac{x}{x^2 + y^2}$, $(x + 1)^2 + y^2 = 1$

$c = 1$, $1 = \frac{x}{x^2 + y^2}$, $\left(x - \frac{1}{2}\right)^2 + y^2 = \frac{1}{4}$

$c = -1$, $-1 = \frac{x}{x^2 + y^2}$, $\left(x + \frac{1}{2}\right)^2 + y^2 = \frac{1}{4}$

$c = \frac{3}{2}$, $\frac{3}{2} = \frac{x}{x^2 + y^2}$, $\left(x - \frac{1}{3}\right)^2 + y^2 = \frac{1}{9}$

$c = -\frac{3}{2}$, $-\frac{3}{2} = \frac{x}{x^2 + y^2}$, $\left(x + \frac{1}{3}\right)^2 + y^2 = \frac{1}{9}$

$c = 2$, $2 = \frac{x}{x^2 + y^2}$, $\left(x - \frac{1}{4}\right)^2 + y^2 = \frac{1}{16}$

$c = -2$, $-2 = \frac{x}{x^2 + y^2}$, $\left(x + \frac{1}{4}\right)^2 + y^2 = \frac{1}{16}$

The level curves are circles.

41. $f(1500, 1000) = 100(1500)^{0.75}(1000)^{0.25}$

$\approx 135{,}540$ units

43. $C(80, 20) = 27\sqrt{(80)(20)} + 195(80) + 215(20) + 980$

$\approx \$21{,}960.00$

45. $P(x_1, x_2) = 15(x_1 + x_2) - (0.02x_1^2 + 4x_1 + 500) - (0.05x_2^2 + 4x_2 + 275)$

$= -0.02x_1^2 - 0.05x_2^2 + 11x_1 + 11x_2 - 775$

(a) $P(250, 150) = \$1250.00$

(b) $P(300, 200) = \$925.00$

47.

	Inflation Rate, I		
	0.00	0.03	0.05
Tax Rate, R 0.00	\$2593.74	\$1929.99	\$1592.33
0.28	\$2004.23	\$1491.34	\$1230.42
0.35	\$1877.14	\$1396.77	\$1152.40

49. (a) $z = 0.186(8) + 0.071(6) - 0.396 \approx \$1.52$ per share

(b) The variable x has more influence because its coefficient is larger.

Section 7.4 Partial Derivatives

1. $f_x(x, y) = 2$

$f_y(x, y) = -3$

3. $f_x(x, y) = \frac{5}{2\sqrt{x}}$

$f_y(x, y) = -12y$

5. $f_x(x, y) = \frac{1}{y}$

$f_y(x, y) = -xy^{-2} = \frac{-x}{y^2}$

7. $f_x(x, y) = \frac{1}{2}(x^2 + y^2)^{-1/2}(2x) = \frac{x}{\sqrt{x^2 + y^2}}$

$f_y(x, y) = \frac{1}{2}(x^2 + y^2)^{-1/2}(2y) = \frac{y}{\sqrt{x^2 + y^2}}$

9. $\dfrac{\partial z}{\partial x} = 2xe^{2y}$

$\dfrac{\partial z}{\partial y} = 2x^2e^{2y}$

11. $h_x(x, y) = -2xe^{-(x^2+y^2)}$

$h_y(x, y) = -2ye^{-(x^2+y^2)}$

13. $z = \ln\dfrac{x - y}{(x + y)^2} = \ln(x - y) - 2\ln(x + y)$

$\dfrac{\partial z}{\partial x} = \dfrac{1}{x - y} - \dfrac{2}{x + y} = \dfrac{3y - x}{x^2 - y^2}$

$\dfrac{\partial x}{\partial y} = \dfrac{-1}{x - y} - \dfrac{2}{x + y} = \dfrac{y - 3x}{x^2 - y^2}$

15. $f_x(x, y) = 6xye^{x-y} + 3x^2ye^{x-y}$

$= 3xye^{x-y}(2 + x)$

17. $g_x(x, y) = 3y^2e^{y-x} - 3xy^2e^{y-x}$

$= 3y^2e^{y-x}(1 - x)$

19. Using the solution from Exercise 15, $f_x(1, 1) = 9$.

21. $f_x(x, y) = 6x + y,$ $\quad f_x(2, 1) = 13$

$f_y(x, y) = x - 2y,$ $\quad f_y(2, 1) = 0$

23. $f_x(x, y) = 3ye^{3xy},$ $\quad f_x(0, 4) = 3(4)e^0 = 12$

$f_y(x, y) = 3xe^{3xy},$ $\quad f_y(0, 4) = 0$

25. $f_x(x, y) = \dfrac{(x - y)y - xy}{(x - y)^2} = -\dfrac{y^2}{(x - y)^2},$ $\quad f_x(2, -2) = -\dfrac{4}{16} = -\dfrac{1}{4}$

$f_y(x, y) = \dfrac{(x - y)x - xy(-1)}{(x - y)^2} = \dfrac{x^2}{(x - y)^2},$ $\quad f_y(2, -2) = \dfrac{4}{16} = \dfrac{1}{4}$

27. $f_x(x, y) = \dfrac{1}{x^2 + y^2}(2x) = \dfrac{2x}{x^2 + y^2},$ $\quad f_x(1, 0) = \dfrac{2}{1 + 0} = 2$

$f_y(x, y) = \dfrac{1}{x^2 + y^2}(2y) = \dfrac{2y}{x^2 + y^2},$ $\quad f_y(1, 0) = 0$

29. $w_x = \dfrac{x}{\sqrt{x^2 + y^2 + z^2}},$ $\quad$ at $(2, -1, 2),$ $\quad w_x = \dfrac{2}{3}$

$w_y = \dfrac{y}{\sqrt{x^2 + y^2 + z^2}},$ $\quad$ at $(2, -1, 2),$ $\quad w_y = -\dfrac{1}{3}$

$w_z = \dfrac{z}{\sqrt{x^2 + y^2 + z^2}},$ $\quad$ at $(2, -1, 2),$ $\quad w_z = \dfrac{2}{3}$

31. $w = \ln\sqrt{x^2 + y^2 + z^2} = \dfrac{1}{2}\ln(x^2 + y^2 + z^2)$

$w_x = \dfrac{x}{x^2 + y^2 + z^2},$ $\quad w_x(3, 0, 4) = \dfrac{3}{25}$

$w_y = \dfrac{y}{x^2 + y^2 + z^2},$ $\quad w_y(3, 0, 4) = 0$

$w_z(x, y, z) = \dfrac{z}{x^2 + y^2 + z^2},$ $\quad w_z(3, 0, 4) = \dfrac{4}{25}$

33. $w_x = 2z^2 + 3yz,$ $\quad w_x(1, -1, 2) = 8 - 6 = 2$

$w_y = 3xz - 12yz,$ $\quad w_y(1, -1, 2) = 6 + 24 = 30$

$w_z = 4xz + 3xy - 6y^2,$ $\quad w_z(1, -1, 2) = 8 - 3 - 6 = -1$

35. $\left.\begin{array}{l} f_x(x, y) = 2x + 4y - 4 = 0 \\ f_y(x, y) = 4x + 2y + 16 = 0 \end{array}\right\}$

$\begin{array}{r} -4x - 8y = -8 \\ 4x + 2y = -16 \\ \hline -6y = -24 \\ y = 4 \\ x = -6 \end{array}$

Solution: $(-6, 4)$

37. $\left.\begin{array}{l} f_x(x, y) = -\dfrac{1}{x^2} + y = 0 \Longrightarrow x^2y = 1 \\ f_y(x, y) = -\dfrac{1}{y^2} + x = 0 \Longrightarrow y^2x = 1 \end{array}\right\}$ $x = y = 1$

Solution: $(1, 1)$

39. (a) $\dfrac{\partial z}{\partial x} = 2$; at $(2, 1, 6)$, $\dfrac{\partial z}{\partial x} = 2$

(b) $\dfrac{\partial z}{\partial y} = -3$; at $(2, 1, 6)$, $\dfrac{\partial z}{\partial y} = -3$

41. (a) $\dfrac{\partial z}{\partial x} = 2x$; at $(3, 1, 0)$, $\dfrac{\partial z}{\partial x} = 6$

(b) $\dfrac{\partial z}{\partial y} = -18y$; at $(3, 1, 0)$, $\dfrac{\partial z}{\partial y} = -18$

43. (a) $\dfrac{\partial z}{\partial x} = -\dfrac{x}{\sqrt{25 - x^2 - y^2}}$; at $(3, 0, 4)$, $\dfrac{\partial z}{\partial x} = -\dfrac{3}{4}$

(b) $\dfrac{\partial z}{\partial y} = -\dfrac{y}{\sqrt{25 - x^2 - y^2}}$; at $(3, 0, 4)$, $\dfrac{\partial z}{\partial y} = 0$

45. (a) $\dfrac{\partial z}{\partial x} = -2x$; at $(1, 1, 2)$, $\dfrac{\partial z}{\partial x} = -2$

(b) $\dfrac{\partial z}{\partial y} = -2y$; at $(1, 1, 2)$, $\dfrac{\partial z}{\partial y} = -2$

47. $\dfrac{\partial z}{\partial x} = 2x - 2y$, $\dfrac{\partial^2 z}{\partial y \partial x} = -2$

$\dfrac{\partial z}{\partial y} = -2x + 6y$, $\dfrac{\partial^2 z}{\partial x \partial y} = -2$

49. $\dfrac{\partial z}{\partial x} = \dfrac{e^{2xy}(2xy - 1)}{4x^2}$, $\dfrac{\partial^2 z}{\partial y \partial x} = ye^{2xy}$

$\dfrac{\partial z}{\partial y} = \dfrac{1}{2}e^{2xy}$, $\dfrac{\partial^2 z}{\partial x \partial y} = ye^{2xy}$

51. The first partial derivatives are

$$\frac{\partial z}{\partial x} = 3x^2 \quad \text{and} \quad \frac{\partial z}{\partial y} = -8y$$

and the second partial derivatives are

$$\frac{\partial^2 z}{\partial x^2} = 6x, \quad \frac{\partial^2 z}{\partial y \partial x} = 0 = \frac{\partial^2 z}{\partial x \partial y}, \quad \text{and} \quad \frac{\partial^2 z}{\partial y^2} = -8.$$

53. The first partial derivatives are

$$\frac{\partial z}{\partial x} = 12x^2 + 3y^2 \quad \text{and} \quad \frac{\partial z}{\partial y} = 6xy - 12y^2$$

and the second partial derivatives are

$$\frac{\partial^2 z}{\partial x^2} = 24x, \quad \frac{\partial^2 z}{\partial y \partial x} = 6y = \frac{\partial^2 z}{\partial x \partial y}, \quad \text{and} \quad \frac{\partial^2 z}{\partial y^2} = 6x - 24y.$$

55. From Exercise 25, the first partial derivatives are

$$\frac{\partial z}{\partial x} = \frac{-y^2}{(x - y)^2} \quad \text{and} \quad \frac{\partial z}{\partial y} = \frac{x^2}{(x - y)^2}$$

and the second partial derivatives are

$$\frac{\partial^2 z}{\partial x^2} = \frac{2y^2}{(x - y)^3}, \quad \frac{\partial^2 z}{\partial y \partial x} = \frac{-2xy}{(x - y)^3} = \frac{\partial^2 z}{\partial x \partial y},$$

and

$$\frac{\partial^2 z}{\partial y^2} = \frac{2x^2}{(x - y)^3}.$$

57. The first partial derivatives are

$$\frac{\partial z}{\partial x} = e^{-y^2} \quad \text{and} \quad \frac{\partial z}{\partial y} = -2xye^{-y^2}$$

and the second partial derivatives are

$$\frac{\partial^2 z}{\partial x^2} = 0, \quad \frac{\partial^2 z}{\partial y \partial x} = -2ye^{-y^2} = \frac{\partial^2 z}{\partial x \partial y},$$

and

$$\frac{\partial^2 z}{\partial y^2} = -2xe^{-y^2} + 4xy^2 e^{-y^2} = 2xe^{-y^2}(2y^2 - 1).$$

59. $f_x(x, y) = 4x^3 - 6xy^2$ $f_y(x, y) = -6x^2y + 2y$

$f_{xx}(x, y) = 12x^2 - 6y^2,$ $f_{xx}(1, 0) = 12$

$f_{xy}(x, y) = -12xy,$ $f_{xy}(1, 0) = 0$

$f_{yx}(x, y) = -12xy,$ $f_{yx}(1, 0) = 0$

$f_{yy}(x, y) = -6x^2 + 2,$ $f_{yy}(1, 0) = -4$

61. $f_x(x, y) = \dfrac{1}{x - y}$ $f_y(x, y) = \dfrac{-1}{x - y}$

$f_{xx}(x, y) = \dfrac{-1}{(x - y)^2},$ $f_{xx}(2, 1) = -1$

$f_{xy}(x, y) = \dfrac{1}{(x - y)^2},$ $f_{xy}(2, 1) = 1$

$f_{yx}(x, y) = \dfrac{1}{(x - y)^2},$ $f_{yx}(2, 1) = 1$

$f_{yy}(x, y) = \dfrac{-1}{(x - y)^2},$ $f_{yy}(2, 1) = -1$

63. $w_x = 6xy - 5yz$

$w_y = 3x^2 - 5xz + 10z^2$

$w_z = -5xy + 20yz$

65. $w_x = \dfrac{y(y + z)}{(x + y + z)^2}$

$w_y = \dfrac{x(x + z)}{(x + y + z)^2}$

$w_z = \dfrac{-xy}{(x + y + z)^2}$

67. $\dfrac{\partial C}{\partial x} = \dfrac{5y}{\sqrt{xy}} + 149;$ at $(120, 160),$ $\dfrac{\partial C}{\partial x} \approx 154.77$

$\dfrac{\partial C}{\partial y} = \dfrac{5x}{\sqrt{xy}} + 189;$ at $(120, 160),$ $\dfrac{\partial C}{\partial y} \approx 193.33$

69. (a) $\dfrac{\partial f}{\partial x} = 60x^{-0.4}y^{0.4} = 60\left(\dfrac{y}{x}\right)^{0.4};$ at $(1000, 500),$ $\dfrac{\partial f}{\partial x} \approx 45.47$

   (b) $\dfrac{\partial f}{\partial y} = 40x^{0.6}y^{-0.6} = 40\left(\dfrac{x}{y}\right)^{0.6};$ at $(1000, 500),$ $\dfrac{\partial f}{\partial y} \approx 60.63$

71. (a) Complementary since $\dfrac{\partial x_1}{\partial p_2} = -\dfrac{5}{2} < 0$ and $\dfrac{\partial x_2}{\partial p_1} = -\dfrac{3}{2} < 0.$

   (b) Substitute since $\dfrac{\partial x_1}{\partial p_2} = 1.8 > 0$ and $\dfrac{\partial x_2}{\partial p_1} = 0.75 > 0.$

   (c) Complementary since $\dfrac{\partial x_1}{\partial p_2} = \dfrac{-500}{p_1^2 p_2 \sqrt{p_1 p_2}} < 0$ and $\dfrac{\partial x_2}{\partial p_1} = \dfrac{-375}{p_1 p_2 \sqrt{p_1}} < 0.$

73. Since both first partials are negative, an increase in the charge for food and housing or tuition will cause a decrease in the number of applicants.

75. (a) $U_x = -10x + y$

   (b) $U_y = x - 6y$

   (c) When $x = 2$ and $y = 3$, $U_x = -17$ and $U_y = -16$.
   The person should consume one more unit of good y, since the rate of decrease of satisfaction is less for y.

(d)

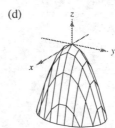

Section 7.5   Extrema of Functions of Two Variables

1. The first partial derivatives of f, $f_x(x, y) = 2x + 4$ and $f_y(x, y) = -2y - 8$, are zero at the point $(-2, -4)$. Moreover, since

$$f_{xx}(x, y) = 2, \ f_{yy}(x, y) = -2, \text{ and } f_{xy}(x, y) = 0,$$

it follows that

$$f_{xx}(2, 4)f_{yy}(2, 4) - [f_{xy}(2, 4)]^2 = -4 < 0.$$

Thus, $(-2, -4, 1)$ is a saddle point. There are no relative extrema.

3. The first partial derivatives of f,

$$f_x(x, y) = \frac{x}{\sqrt{x^2 + y^2 + 1}} \text{ and } f_y(x, y) = \frac{y}{\sqrt{x^2 + y^2 + 1}},$$

are zero at the point $(0, 0)$. Moreover, since

$$f_{xx}(x, y) = \frac{y^2 + 1}{(x^2 + y^2 + 1)^{3/2}}, \ f_{yy}(x, y) = \frac{x^2 + 1}{(x^2 + y^2 + 1)^{3/2}}, \text{ and } f_{xy}(x, y) = \frac{-xy}{(x^2 + y^2 + 1)^{3/2}},$$

it follows that

$$f_{xx}(0, 0) = 1 > 0 \text{ and } f_{xx}(0, 0)f_{yy}(0, 0) - [f_{xy}(0, 0)]^2 = 1 > 0.$$

Thus, $(0, 0, 1)$ is a relative minimum.

5. The first partial derivatives of f, $f_x(x, y) = 2(x - 1)$ and $f_y(x, y) = 2(y - 3)$, are zero at the point $(1, 3)$. Moreover, since

$$f_{xx}(x, y) = 2, \; f_{yy}(x, y) = 2, \text{ and } f_{xy}(x, y) = 0,$$

it follows that

$$f_{xx}(1, 3) > 0 \text{ and } f_{xx}(1, 3)f_{yy}(1, 3) - [f_{xy}(1, 3)]^2 = 4 > 0.$$

Thus, $(1, 3, 0)$ is a relative minimum.

7. The first partial derivatives of f, $f_x(x, y) = 4x + 2y + 2$ and $f_y(x, y) = 2x + 2y$, are zero at the point $(-1, 1)$. Moreover, since

$$f_{xx}(x, y) = 4, \; f_{yy}(x, y) = 2, \text{ and } f_{xy}(x, y) = 2,$$

it follows that

$$f_{xx}(-1, 1) > 0 \text{ and } f_{xx}(-1, 1)f_{yy}(-1, 1) - [f_{xy}(-1, 0)]^2 = 4 > 0.$$

Thus, $(-1, 1, -4)$ is a relative minimum.

9. The first partial derivatives of f, $f_x(x, y) = -10x + 4y + 16$ and $f_y(x, y) = 4x - 2y$, are zero at the point $(8, 16)$. Moreover, since

$$f_{xx}(x, y) = -10, \; f_{yy}(x, y) = -2, \text{ and } f_{xy}(x, y) = 4,$$

it follows that

$$f_{xx}(8, 16) < 0 \text{ and } f_{xx}(8, 16)f_{yy}(8, 16) - [f_{xy}(8, 16)]^2 = 4 > 0.$$

Thus, $(8, 16, 74)$ is a relative maximum.

11. The first partial derivatives of f, $f_x(x, y) = 6x - 12$ and $f_y(x, y) = 4y - 4$, are zero at the point $(2, 1)$. Moreover, since

$$f_{xx}(x, y) = 6, \; f_{yy}(x, y) = 4, \text{ and } f_{xy}(x, y) = 0,$$

it follows that

$$f_{xx}(2, 1) > 0 \text{ and } f_{xx}(2, 1)f_{yy}(2, 1) - [f_{xy}(2, 1)]^2 = 24 > 0.$$

Thus, $(2, 1, -7)$ is a relative minimum.

13. The first partial derivatives of f, $f_x(x, y) = 2x + 4$ and $f_y(x, y) = -2y - 4 = -2(y + 2)$, are zero at the point $(-2, -2)$. Moreover, since

$$f_{xx}(x, y) = 2, \; f_{yy}(x, y) = -2, \text{ and } f_{xy}(x, y) = 0,$$

it follows that

$$f_{xx}(1, -2)f_{yy}(1, -2) - [f_{xy}(1, -2)]^2 = -4 < 0.$$

Thus, $(-2, -2, -8)$ is a saddle point.

15. The first partial derivatives of f, $f_x(x, y) = y$ and $f_y(x, y) = x$, are zero at the point $(0, 0)$. Moreover, since

$$f_{xx}(x, y) = 0, \; f_{yy}(x, y) = 0, \text{ and } f_{xy}(x, y) = 1,$$

it follows that

$$f_{xx}(0, 0)f_{yy}(0, 0) - [f_{xy}(0, 0)]^2 = -1 < 0.$$

Thus, $(0, 0, 0)$ is a saddle point.

17. The first partial derivatives of f,

$$f_x(x, y) = 2xe^{1-x^2-y^2}(1 - x^2 - 4y^2) \text{ and } f_y(x, y) = 2ye^{1-x^2-y^2}(4 - x^2 - 4y^2)$$

are zero at $(0, 0)$, $(0, \pm 1)$, and $(\pm 1, 0)$. Moreover, since

$$f_{xx}(x, y) = 2e^{1-x^2-y^2}(1 - 5x^2 + 2x^4 - 4y^2 + 8x^2y^2),$$

$$f_{yy}(x, y) = 2e^{1-x^2-y^2}(4 - x^2 - 20y^2 + 8y^4 + 2x^2y^2), \text{ and}$$

$$f_{xy}(x, y) = -4xye^{1-x^2-y^2}(5 - x^2 - 4y^2),$$

we can determine that $(0, 0, 0)$ is a relative minimum, $(0, \pm 1, 4)$ are relative maxima, and $(\pm 1, 0, 1)$ are saddle points.

19. The first partial derivatives of f, $f_x(x, y) = ye^{xy}$ and $f_y(x, y) = xe^{xy}$, are zero at the point $(0, 0)$. Moreover, since

$$f_{xx}(x, y) = y^2 e^{xy}, \quad f_{yy}(x, y) = x^2 e^{xy}, \quad \text{and} \quad f_{xy}(x, y) = e^{xy}(1 + xy),$$

it follows that

$$f_{xx}(0, 0) = 0 \quad \text{and} \quad f_{xx}(0, 0)f_{yy}(0, 0) - [f_{xy}(0, 0)]^2 = -1 < 0.$$

Thus, $(0, 0, 1)$ is a saddle point.

21. Since

$$d = f_{xx}(x_0, y_0)f_{yy}(x_0, y_0) - [f_{xy}(x_0, y_0)]^2$$
$$= (16)(4) - (8)^2$$
$$= 0,$$

there is insufficient information.

23. Since

$$d = f_{xx}(x_0, y_0)f_{yy}(x_0, y_0) - [f_{xy}(x_0, y_0)]^2$$
$$= (-7)(4) - (9)^2$$
$$< 0,$$

$f(x_0, y_0)$ is a saddle point.

25. The first partial derivatives of f, $f_x(x, y) = 2xy^2$ and $f_y(x, y) = 2x^2y$, are zero at the points $(a, 0)$ and $(0, b)$ where a and b are any real numbers. Since

$$f_{xx}(x, y) = 2y^2, \quad f_{yy}(x, y) = 2x^2, \quad \text{and} \quad f_{xy}(x, y) = 4xy,$$

it follows that

$$f_{xx}(a, 0)f_{yy}(a, 0) - [f_{xy}(a, 0)]^2 = 0 \quad \text{and} \quad f_{xx}(0, b)f_{yy}(0, b) - [f_{xy}(0, b)]^2 = 0$$

and the Second-Derivative Test fails. We note that $f(x, y) = (xy)^2$ is nonnegative for all $(a, 0, 0)$ and $(0, b, 0)$ where a and b are real numbers. Therefore, $(a, 0, 0)$ and $(0, b, 0)$ are relative minima.

27. The first partial derivatives of f, $f_x(x, y) = 3x^2$ and $f_y(x, y) = 3y^2$, are zero at $(0, 0)$. Moreover, since

$$f_{xx}(x, y) = 6x, \quad f_{yy}(x, y) = 6y, \quad f_{xy}(x, y) = 0, \quad \text{and} \quad f_{xx}(0, 0)f_{yy}(0, 0) - [f_{xy}(0, 0)]^2 = 0,$$

the Second-Partials Test fails. By testing "nearby" points, we conclude that $(0, 0, 0)$ is a saddle point.

29. The first partial derivatives of f,

$$f_x(x, y) = \frac{2}{3\sqrt[3]{x}} \quad \text{and} \quad f_y(x, y) = \frac{2}{3\sqrt[3]{y}},$$

are undefined at the point $(0, 0)$. Since

$$f_{xx}(x, y) = -\frac{2}{9x^{4/3}}, \quad f_{yy}(x, y) = -\frac{2}{9y^{4/3}}, \quad f_{xy}(x, y) = 0$$

and $f_{xx}(0, 0)$ is undefined, the Second-Derivative Test fails. Since $f(x, y) \geq 0$ for all points in the xy-coordinate plane, $(0, 0, 0)$ is a relative minimum.

31. Critical point: $(x, y, z) = (1, -3, 0)$

Relative minimum

33. False. (a, b) must be a critical point.

35. False. The function need not have partial derivatives.

37. Let x, y, and z be the numbers. The sum is given by $x + y + z = 30$ and $z = 30 - x - y$, and the product is given by $P = xyz = 30xy - x^2y - xy^2$. The first partial derivatives of P are

$$P_x = 30y - 2xy - y^2 = y(30 - 2x - y)$$

$$P_y = 30x - x^2 - 2xy = x(30 - x - 2y).$$

Setting these equal to zero produces the system

$$2x + y = 30$$

$$x + 2y = 30.$$

Solving the system, we have $x = 10$, $y = 10$, and $z = 10$.

39. The sum is given by

$$x + y + z = 30$$
$$z = 30 - x - y$$

and the sum of the squares is given by $S = x^2 + y^2 + z^2 = x^2 + y^2 + (30 - x - y)^2$. The first partial derivatives of S are

$$S_x = 2x - 2(30 - x - y) = 4x + 2y - 60 \text{ and}$$
$$S_y = 2y - 2(30 - x - y) = 2x + 4y - 60.$$

Setting these equal to zero produces the system

$$2x + y = 30$$
$$x + 2y = 30.$$

Solving this system yields $x = 10$ and $y = 10$. Thus, the sum of squares is a minimum when $x = y = z = 10$.

41. The first partial derivatives of R are

$$R_{x_1} = -10x_1 - 2x_2 + 42 \text{ and}$$
$$R_{x_2} = -16x_2 - 2x_1 + 102.$$

Setting these equal to zero produces the system

$$5x_1 + x_2 = 21$$
$$x_1 + 8x_2 = 51$$

which yields $x_1 = 3$ and $x_2 = 6$. By the Second-Partials Test, it follows that the revenue is maximized when $x_1 = 3$ and $x_2 = 6$.

43. The revenue function is given by

$$R = x_1 p_1 + x_2 p_2 = 1000p_1 + 1500p_2 + 3p_1 p_2 - 2p_1^2 - 1.5p_2^2$$

and the first partials of R are $R_{p_1} = 1000 + 3p_2 - 4p_1$ and $R_{p_2} = 1500 + 3p_1 - 3p_2$. Setting these equal to zero produces the system

$$4p_1 - 3p_2 = 1000$$
$$-3p_1 + 3p_2 = 1500.$$

Solving this system yields $p_1 = 2500$ and $p_2 = 3000$, and, by the Second-Partials Test, we conclude that the revenue is maximized when $p_1 = 2500$ and $p_2 = 3000$.

45. The profit is given by

$$P = R - C_1 - C_2$$
$$= [225 - 0.4(x_1 + x_2)](x_1 + x_2) - (0.05x_1^2 + 15x_1 + 5400) - (0.03x_2^2 + 15x_2 + 6100)$$
$$= -0.45x_1^2 - 0.43x_2^2 - 0.8x_1 x_2 + 210x_1 + 210x_2 - 11,500$$

and the first partial derivatives of P are $P_{x_1} = -0.9x_1 - 0.8x_2 + 210$ and $P_{x_2} = -0.86x_2 - 0.8x_1 + 210$. By setting these equal to zero, we obtain the system

$$0.9x_1 + 0.8x_2 = 210$$
$$0.8x_1 + 0.86x_2 = 210.$$

Solving this system yields $x_1 \approx 94$ and $x_2 \approx 157$, and, by the Second-Partials Test, we conclude that the profit is maximum when $x_1 \approx 94$ and $x_2 \approx 157$.

47. Let x = length, y = width, and z = height. The sum of length and girth is given by

$$x + (2y + 2z) = 108$$
$$x = 108 - 2y - 2z$$

and the volume of the package is given by $V = xyz = 108yz - 2zy^2 - 2yz^2$. The first partial derivatives of V are

$$V_y = 108z - 4yz - 2z^2 = (108 - 4y - 2z) \text{ and } V_z = 108y - 2y^2 - 4yz = y(108 - 2y - 4z).$$

Setting these equal to zero produces the system

$$4y + 2z = 108$$
$$2y + 4z = 108$$

which yields the solution $x = 36$ inches, $y = 18$ inches, and $z = 18$ inches.

49. $P(p, q, r) = 2pq + 2pr + 2qr$

Since $p + q + r = 1$, $r = 1 - p - q$, and

$$P = 2pq + 2p(1 - p - q) + 2q(1 - p - q)$$
$$= -2p^2 + 2p - 2q^2 + 2q - 2pq$$

$$P_p = -4p + 2 - 2q$$
$$P_q = -4q + 2 - 2p.$$

Solving $P_p = P_q = 0$, we obtain $p = q = \frac{1}{3}$, and hence $r = \frac{1}{3}$. Finally, the maximum proportion is

$$P = 2\left(\frac{1}{3}\right)\left(\frac{1}{3}\right) + 2\left(\frac{1}{3}\right)\left(\frac{1}{3}\right) + 2\left(\frac{1}{3}\right)\left(\frac{1}{3}\right) = \frac{6}{9} = \frac{2}{3}.$$

51. $D_x(x, y) = 2x - 18 + 2y = 0$

$D_y(x, y) = 4y - 24 + 2x = 0$

To solve $x + y = 9$ and $x + 2y = 12$, subtract the equations $-y = -3$ or $y = 3$. Hence, $x = 6$.

Section 7.6   Lagrange Multipliers

1. $F(x, y, \lambda) = xy - \lambda(x + y - 10)$

$\quad F_x = y - \lambda = 0, \qquad\qquad y = \lambda$

$\quad F_y = x - \lambda = 0, \qquad\qquad x = \lambda$

$\quad F_\lambda = -(x + y - 10) = 0, \qquad 2\lambda = 10$

Thus, $\lambda = 5$, $x = 5$, $y = 5$, and $f(x, y)$ is maximum at $(5, 5)$. The maximum is $f(5, 5) = 25$.

3. $F(x, y, \lambda) = x^2 + y^2 - \lambda(x + y - 4)$

$\quad F_x = 2x - \lambda = 0, \qquad\qquad x = \frac{1}{2}\lambda$

$\quad F_y = 2y - \lambda = 0, \qquad\qquad y = \frac{1}{2}\lambda$

$\quad F_\lambda = -(x + y - 4) = 0, \qquad \lambda = 4$

Thus, $\lambda = 4$, $x = 2$, $y = 2$, and $f(x, y)$ is minimum at $(2, 2)$. The minimum is $f(2, 2) = 8$.

5. $F(x, y, \lambda) = x^2 - y^2 - \lambda(y - x^2)$

$\quad F_x = 2x + 2x\lambda = 0, \qquad 2x(1 + \lambda) = 0$

$\quad F_y = -2y - \lambda = 0, \qquad\qquad y = -\frac{1}{2}\lambda$

$\quad F_\lambda = -(y - x^2) = 0, \qquad\qquad x = \sqrt{y}$

Thus, $\lambda = -1$, $x = \sqrt{2}/2$, $y = 1/2$, and $f(x, y)$ is maximum at $\left(\sqrt{2}/2, 1/2\right)$. The maximum is $f\left(\sqrt{2}/2, 1/2\right) = 1/4$.

7. $F(x, y, \lambda) = 3x + xy + 3y - \lambda(x + y - 25)$

$\quad F_x = 3 + y - \lambda = 0, \qquad\qquad y = \lambda - 3$

$\quad F_y = 3 + x - \lambda = 0, \qquad\qquad x = \lambda - 3$

$\quad F_\lambda = -(x + y - 25) = 0, \qquad 2\lambda - 6 = 25$

Thus, $\lambda = \frac{31}{2}$, $x = \frac{25}{2}$, $y = \frac{25}{2}$, and $f(x, y)$ is maximum at $\left(\frac{25}{2}, \frac{25}{2}\right)$. The maximum is $f\left(\frac{25}{2}, \frac{25}{2}\right) = 231.25$.

9. *Note:* $f(x, y)$ has a maximum value when $g(x, y) = 6 - x^2 - y^2$ is maximum.

$F(x, y, \lambda) = 6 - x^2 - y^2 - \lambda(x + y - 2)$

$\quad F_x = -2x - \lambda = 0, \qquad -2x = \lambda$

$\quad F_y = -2y - \lambda = 0, \qquad -2y = \lambda$ $\left.\right\}$ $x = y$

$\quad F_\lambda = -(x + y - 2) = 0, \qquad 2x = 2$

Thus, $x = y = 1$ and $f(x, y)$ is maximum at $(1, 1)$. The maximum is $f(1, 1) = 2$.

11. $F(x, y, \lambda) = e^{xy} - \lambda(x^2 + y^2 - 8)$

$$\left. \begin{array}{ll} F_x = ye^{xy} - 2x\lambda = 0, & e^{xy} = \dfrac{2x\lambda}{y} \\[3mm] F_y = xe^{xy} - 2y\lambda = 0, & e^{xy} = \dfrac{2y\lambda}{x} \end{array} \right\} x - y$$

$$F_\lambda = -(x^2 + y^2 - 8) = 0, \quad 2x^2 = 8$$

Thus, $x = y = 2$ and $f(x, y)$ is maximum at $(2, 2)$. The maximum is $f(2, 2) = e^4$.

13. $F(x, y, z, \lambda) = 2x^2 + 3y^2 + 2z^2 - \lambda(x + y + z - 24)$

$$\begin{array}{ll} F_x = 4x - \lambda = 0, & \lambda = 4x \\[2mm] F_y = 6y - \lambda = 0, & \lambda = 6y \\[2mm] F_z = 4z - \lambda = 0, & \lambda = 4z \end{array}$$

$$F_\lambda = -(x + y + z - 24) = 0$$

$$\frac{\lambda}{4} + \frac{\lambda}{6} + \frac{\lambda}{4} = 24$$

$$8\lambda = 24 \cdot 12$$

$$\lambda = 36$$

Thus, $x = 9$, $y = 6$, $z = 9$, and $f(x, y, z)$ is minimum at $(9, 6, 9)$. The minimum is $f(9, 6, 9) = 432$.

15. $F(x, y, z, \lambda) = x^2 + y^2 + z^2 - \lambda(x + y + z - 1)$

$$\left. \begin{array}{l} F_x = 2x - \lambda = 0 \\ F_y = 2y - \lambda = 0 \\ F_z = 2z - \lambda = 0 \end{array} \right\} x = y = z$$

$$F_\lambda = -(x + y + z - 1) = 0, \quad 3x = 1$$

Thus, $x = y = z = \frac{1}{3}$ and $f(x, y, z)$ is minimum at $f\left(\frac{1}{3}, \frac{1}{3}, \frac{1}{3}\right) = \frac{1}{3}$.

17. $F(x, y, z, \lambda) = x + y + z - \lambda(x^2 + y^2 + z^2 - 1)$

$$\left. \begin{array}{l} F_x = 1 - 2x\lambda = 0 \\ F_y = 1 - 2y\lambda = 0 \\ F_z = 1 - 2z\lambda = 0 \end{array} \right\} x = y = z$$

$$F_\lambda = -(x^2 + y^2 + z^2 - 1) = 0$$

$$3x^2 = 1 \Longrightarrow x = \frac{1}{\sqrt{3}} = y = z$$

$f(x, y, z)$ is maximum at

$$f\left(\frac{1}{\sqrt{3}}, \frac{1}{\sqrt{3}}, \frac{1}{\sqrt{3}}\right) = \frac{3}{\sqrt{3}} = \sqrt{3}$$

19. $F(x, y, z, w, \lambda) = 2x^2 + y^2 + z^2 + 2x^2 - \lambda(2x + 2y + z + w - z)$

$$\begin{array}{ll} F_x = 4x - 2\lambda = 0, & x = \frac{1}{2}\lambda \\[2mm] F_y = 2y - 2\lambda = 0, & y = \lambda = 2x \\[2mm] F_z = 2z - \lambda = 0, & z = \frac{1}{2}\lambda = x \\[2mm] F_w = 4w - \lambda = 0, & w = \frac{1}{4}\lambda = \frac{1}{2}x \end{array}$$

$$F_\lambda = -(2x + 2y + z + w) = 0$$

$$2x + 2(2x) + x + \frac{1}{2}x = 2 \Longrightarrow x = \frac{4}{15}$$

The maximum is $f\left(\frac{4}{15}, \frac{8}{15}, \frac{4}{15}, \frac{2}{15}\right) = \frac{8}{15}$.

21. $F(x, y, z, \lambda, \eta) = xyz - \lambda(x + y + z - 24) - \eta(x - y + z - 12)$

$$\left. \begin{array}{l} F_x = yz - \lambda - \eta = 0 \\ F_y = xz - \lambda + \eta = 0 \\ F_z = xy - \lambda - \eta = 0 \end{array} \right\} x = z$$

$$F_\lambda = -(x + y + z - 24) = 0, \quad x + (2x - 12) + x = 24$$

$$F_\eta = -(x - y + z - 12) = 0, \quad y = 2x - 12$$

Thus, $x = 9$, $y = 6$, and $z = 9$. The maximum is $f(9, 6, 9) = 486$.

23. $F(x, y, z, \lambda, \eta) = xyz - \lambda(x^2 + z^2 - 5) - \eta(x - 2y)$

$\quad F_x = yz - 2x\lambda - \eta = 0$

$\quad F_y = xz + 2\eta = 0, \qquad \eta = -\dfrac{xz}{2}$

$\quad F_z = xy - 2z\lambda = 0, \qquad \lambda = \dfrac{xy}{2z}$

$\quad F_\lambda = -(x^2 + z^2 - 5) = 0, \qquad z = \sqrt{5 - x^2}$

$\quad F_\eta = -(x - 2y) = 0, \qquad y = \dfrac{x}{2}$

From F_x, we can write:

$$\frac{x\sqrt{5 - x^2}}{2} - \frac{x^3}{2\sqrt{5 - x^2}} + \frac{x\sqrt{5 - x^2}}{2} = 0$$

$$x\sqrt{5 - x^2} = \frac{x^3}{2\sqrt{5 - x^2}}$$

$$2x(5 - x^2) = x^3$$

$$3x^3 - 10x = 0$$

$$x(3x^2 - 10) = 0$$

Since x, y, and z are positive, we have

$$x = \sqrt{\frac{10}{3}}, y = \frac{1}{2}\sqrt{\frac{10}{3}}, \text{ and } z = \sqrt{\frac{5}{3}}.$$

$$f\left(\sqrt{\frac{10}{3}}, \frac{1}{2}\sqrt{\frac{10}{3}}, \sqrt{\frac{5}{3}}\right) = \frac{5\sqrt{15}}{9}$$

25. Maximize $f(x, y, z) = xyz$ subject to the constraint $x + y + z = 120$.

$\quad F(x, y, z, \lambda) = xyz - \lambda(x + y + z - 120)$

$\quad \left.\begin{array}{l} F_x = yz - \lambda = 0 \\ F_y = xz - \lambda = 0 \\ F_z = xy - \lambda = 0 \end{array}\right\} yz = xz = xy \Rightarrow x = y = z$

$\quad F_\lambda = -(x + y + z - 120) = 0, \quad 3x = 120$

Thus, $x = y = z = 40$.

27. Maximize $f(x, y, z) = xyz$ subject to the constraint $x + y + z = S$.

$\quad F(x, y, z, \lambda) = xyz - \lambda(x + y + z - S)$

$\quad \left.\begin{array}{l} F_x = yz - \lambda = 0 \\ F_y = xz - \lambda = 0 \\ F_z = xy - \lambda = 0 \end{array}\right\} yz = xz = xy \Rightarrow x = y = z$

$\quad F_\lambda = -(x + y + z - S) = 0, \quad 3x = S$

Thus, $x = y = z = S/3$.

29. $F(x, y, \lambda) = x^2 + y^2 - \lambda(x + 2y - 5)$

$\quad F_x = 2x - \lambda = 0, \qquad x = \dfrac{\lambda}{2}$

$\quad F_y = 2y - 2\lambda = 0, \qquad y = \lambda = 2x$

$\quad F_\lambda = -(x + 2y - 5) = 0$

$\quad x + 2(2x) = 5 \Rightarrow x = 1, y = 2$

The minimum distance is $\sqrt{x^2 + y^2} = \sqrt{1 + 4} = \sqrt{5}$.

31. $F(x, y, z, \lambda) = (x - 2)^2 + (y - 1)^2 + (z - 1)^2 - \lambda(x + y + z - 1)$

$\quad \left.\begin{array}{l} F_x = 2(x - 2) - \lambda = 0 \\ F_y = 2(y - 1) - \lambda = 0 \\ F_z = 2(z - 1) - \lambda = 0 \end{array}\right\} \begin{array}{l} x - 2 = y - 1 = z - 1 \\ x - 1 = y = z \end{array}$

$\quad F_\lambda = -(x + y + z - 1) = 0$

Thus, $x = 1$, $y = z = x - 1 = 0$, and $d = \sqrt{(1 - 2)^2 + (0 - 1)^2 + (0 - 1)^2} = \sqrt{3}$.

33. $F(x, y, z, \lambda) = xyz - \lambda(x + 2y + 2z - 108)$

$\quad \left.\begin{array}{l} F_x = yz - \lambda = 0 \\ F_y = xz - 2\lambda = 0 \end{array}\right\} x = 2y$

$\quad F_z = xy - 2\lambda = 0, \qquad y = z$

$\quad F_\lambda = -(x + 2y + 2z - 108) = 0, \qquad 6y = 108$

Thus, $x = 36$, $y = 18$, and $z = 18$. The volume is maximum when the dimensions are $36 \times 18 \times 18$ inches.

35. Minimize $C(x, y, z) = 5xy + 3(xy + 2xz + 2yz) = 8xy + 6xz + 6yz$ subject to the constraint $xyz = 480$.

$F(x, y, z, \lambda) = 8xy + 6xz + 6yz - \lambda(xyz - 480)$

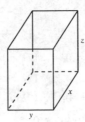

$$\left.\begin{array}{l} F_x = 8y + 6z - \lambda yz = 0 \\ F_y = 8x + 6y - \lambda xy = 0 \\ F_z = 6x + 6y - \lambda xy = 0 \end{array}\right\} \begin{array}{l} x = y \\ 4y = 3z \end{array}$$

$F_\lambda = -(xyz - 480) = 0,\ y(y)\left(\tfrac{4}{3}y\right) = 480,\ y = 2\sqrt[3]{45}$

Thus, the dimensions are $2\sqrt[3]{45}$ feet $\times\ 2\sqrt[3]{45}$ feet $\times \tfrac{8}{3}\sqrt[3]{45}$ feet.

37. $F(x_1, x_2, \lambda) = 0.25x_1{}^2 + 25x_1 + 0.05\lambda_2{}^2 + 12x_2 - \lambda(x_1 + x_2 - 2000)$

$F_{x_1} = 0.5x_1 + 10 - \lambda = 0, \qquad x_1 = 2\lambda - 20$

$F_{x_2} = 0.3x_2 + 12 - \lambda = 0, \qquad x_2 = \tfrac{10}{3}\lambda - 40$

$F_\lambda = -(x_1 + x_2 - 2000)$

$(2\lambda - 20) + \left(\tfrac{10}{3}\lambda - 40\right) = 2000$

$$\tfrac{16}{3}\lambda = 2060$$

$$\lambda = 386.25$$

Hence, $x_1 = 725.5$ and $x_2 = 1287.5$. To minimize cost, let $x_1 = 753$ units and $x_2 = 1287$ units.

39. (a) $F(x, y, \lambda) = 48x + 36y - \lambda(x^{0.25}y^{0.75} - 200)$

$\quad F_x = 48 - 0.25\lambda x^{-0.75}y^{0.75} = 0$

$\quad F_y = 36 - 0.75\lambda x^{0.25}y^{-0.25} = 0$

$\quad F_\lambda = -(x^{0.25}y^{0.75} - 200) = 0$

This produces $\left(\dfrac{y}{x}\right)^{0.75} = \dfrac{48}{0.25\lambda}$ and $\left(\dfrac{y}{x}\right)^{0.25} = \dfrac{0.75\lambda}{36}$.

Thus,

$$\frac{y}{x} = \left(\frac{48}{0.25\lambda}\right)\left(\frac{0.75\lambda}{36}\right) = 4$$

$$x = \frac{200}{4^{0.75}} = \frac{200}{2\sqrt{2}} = 50\sqrt{2} \approx 71$$

$$y = 4x = 200\sqrt{2} \approx 283.$$

(b) $\dfrac{f_x(x, y)}{f_y(x, y)} = \dfrac{48}{36}$

$\dfrac{25x^{-0.75}y^{0.75}}{75x^{0.25}y^{-0.25}} = \dfrac{48}{36}$

$\dfrac{y}{x} = 4$

Thus, $y = 4x$ and the conditions of part (a) are met.

41. (a) From Exercise 39, we have $y = 4x$.

$\quad F(x, y, \lambda) = 100x^{0.25}y^{0.75} - \lambda(48x + 36y - 100{,}000)$

$\quad F_\lambda = -(48x + 36y - 100{,}000) = 0$

Thus, $x = \tfrac{3125}{6}$, $y = \tfrac{6250}{3}$, and $f\left(\tfrac{3125}{6}, \tfrac{6250}{3}\right) \approx 147{,}313.91 \approx 147{,}314$.

(b) $F_x = 25x^{-0.75}y^{0.75} - 48\lambda = 0$

$\quad \lambda = \dfrac{25x^{-0.75}y^{0.75}}{48} = \dfrac{25}{48}\left(\dfrac{3125}{6}\right)^{-0.75}\left(\dfrac{6250}{3}\right)^{0.75} \approx 1.4731$

(c) $147{,}314 + 25{,}000\lambda \approx 147{,}314 + 25{,}000(1.4731) \approx 184{,}141.5 \approx 184{,}142$ units

43. Minimize cost $= x + 2y + 3z$.

Constraint: $12xyz = 0.13$

$F(x, y, z, \lambda) = x + 2y + 3z - \lambda(12xyz - 0.13)$

$$\left.\begin{array}{ll} F_x = 1 - 12\lambda yz = 0, & 12\lambda yz = 1 \\ F_y = 2 - 12\lambda xz = 0, & 12\lambda xz = 2 \\ F_z = 3 - 12\lambda xy = 0, & 12\lambda xy = 3 \end{array}\right\} \begin{array}{l} x = 2y \\ \\ x = 3z \end{array}$$

$F_\lambda = -(12xyz - 0.13) = 0$

$12x\left(\dfrac{x}{2}\right)\left(\dfrac{x}{3}\right) = 0.13, \qquad x = \sqrt[3]{0.065} \approx 0.402$

$2x^3 = 0.13, \qquad y = \dfrac{1}{2}\sqrt[3]{0.065} \approx 0.201$

$z = \dfrac{1}{3}\sqrt[3]{0.065} \approx 0.134$

45. Maximize $F(D, C, P) = 0.039D + 0.045C + 0.040P$.

Constraints: $D + C + P = 300{,}000$

$3000D - 3000C + P^2 = 0$

$0.039 = \lambda + 3000\mu$

$0.045 = \lambda - 3000\mu$

$0.040 = \lambda + 2P\mu$

From Equations 1 and 2:

$\lambda = 0.042$

$0.039 = 0.042 + 3000\mu \Longrightarrow \mu = \dfrac{-0.003}{3000}$

From Equation 3:

$0.040 = 0.042 + 2P\left(\dfrac{-0.003}{3000}\right)$

$0.002 = 2P\left(\dfrac{0.003}{3000}\right)$

$P = \$3000$

Finally, $C = \$150{,}000$ and $D = \$147{,}000$.

47. Let $f(x, y) = xy$ be the area function.

Constraint: $g(x, y) = 2x + 2y - P$

$F(x, y, \lambda) = xy - \lambda(2x + 2y - P)$

$F_x = y - 2\lambda = 0$

$F_y = x - 2\lambda = 0$

$F_\lambda = -2x - 2y + P = 0$

Solving this system, $y = 2\lambda$, $x = 2\lambda$, and

$-2x - 2y + P = 0$

$-2(2\lambda) - 2(2\lambda) + P = 0$

$P = 8\lambda$

$\lambda = \dfrac{P}{8}.$

Thus,

$x = y = \dfrac{P}{4}$ (square!)

and

$\text{Area} = f\left(\dfrac{P}{4}, \dfrac{P}{4}\right) = \dfrac{P^2}{16}.$

49. $f(x, y) = \text{Cost} = 10(2x + 2y) + 4x = 24x + 20y$

Constraint: $g(x, y) = 2xy - 6000 = 0$

$F(x, y, \lambda) = 24x + 20y - \lambda(2xy - 6000)$

$\left.\begin{array}{l} F_x = 24 - 2\lambda y = 0 \\ F_y = 20 - 2\lambda x = 0 \end{array}\right\} \begin{array}{l} y = 12/\lambda \\ x = 10/\lambda \end{array}$

$F_\lambda = -2xy + 6000 = 0$

$2xy = 6000$

$2\left(\dfrac{10}{\lambda}\right)\left(\dfrac{12}{\lambda}\right) = 6000$

$\dfrac{1}{\lambda^2} = 25$

$\lambda = \dfrac{1}{5} \Longrightarrow x = 50$ and $y = 60$

Dimensions: 50 feet by 120 feet

Cost $= 24(50) + 20(60) = \$2400$

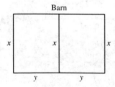

Section 7.7   Least Squares Regression Analysis

1. (a) $\sum x_i = 0$

$\sum y_i = 4$

$\sum x_i y_i = 6$

$\sum x_i^2 = 8$

$a = \dfrac{3(6) - 0(4)}{3(8) - 0^2} = \dfrac{3}{4}$

$b = \dfrac{1}{3}\left[4 - \dfrac{3}{4}(0)\right] = \dfrac{4}{3}$

The regression line is $y = \frac{3}{4}x + \frac{4}{3}$.

(b) $\left(-\dfrac{3}{2} + \dfrac{4}{3} - 0\right)^2 + \left(\dfrac{4}{3} - 1\right)^2 + \left(\dfrac{3}{2} + \dfrac{4}{3} - 3\right)^2 = \dfrac{1}{6}$

3. (a) $\sum x_i = 4$

$\sum y_i = 8$

$\sum x_i y_i = 4$

$\sum x_i^2 = 6$

$a = \dfrac{4(4) - 4(8)}{4(6) - 4^2} = -2$

$b = \dfrac{1}{4}[8 + 2(4)] = 4$

The regression line is $y = -2x + 4$.

(b) $(4 - 4)^2 + (2 - 3)^2 + (2 - 1)^2 + (0 - 0)^2 = 2$

5. $\sum x_i = 0$

$\sum y_i = 7$

$\sum x_i y_i = 7$

$\sum x_i^2 = 10$

Since $a = \frac{7}{10}$ and $b = \frac{7}{5}$, the regression line is $y = \frac{7}{10}x + \frac{7}{5}$.

7. $\sum x_i = 3$

$\sum y_i = 15$

$\sum x_i y_i = 29$

$\sum x_i^2 = 17$

Since $a = 1$ and $b = 4$, the regression line is $y = x + 4$.

9. $\sum x_i = 0$

$\sum y_i = 7$

$\sum x_i y_i = -13$

$\sum x_i^2 = 20$

Since $a = -\frac{13}{20}$ and $b = \frac{7}{4}$, the regression line is $y = -\frac{13}{20}x + \frac{7}{4}$.

11. $\sum x_i = 13$

$\sum y_i = 12$

$\sum x_i y_i = 46$

$\sum x_i^2 = 51$

Since $a = \frac{37}{43}$ and $b = \frac{7}{43}$, the regression line is $y = \frac{37}{43}x + \frac{7}{43}$.

13. $\sum x_i = 27$

$\sum y_i = 0$

$\sum x_i y_i = -70$

$\sum x_i^2 = 205$

Since $a = -\frac{175}{148}$ and $b = \frac{945}{148}$, the regression line is $y = -\frac{175}{148}x + \frac{945}{148}$.

15. The sum of the squared errors is as follows.

$$S = (-2a + b + 1)^2 + (0a + b)^2 + (2a + b - 3)^2$$

$$\frac{\partial S}{\partial a} = 2(-2a + b + 1)(-2) + 2(2a + b - 3)(2) = 16a - 16$$

$$\frac{\partial S}{\partial b} = 2(-2a + b + 1) + 2b + 2(2a + b - 3) = 6b - 4$$

Setting these partial derivatives equal to zero produces $a = 1$ and $b = \frac{2}{3}$. Thus, $y = x + \frac{2}{3}$.

17. The sum of the squared errors is as follows.

$$S = (-2a + b - 4)^2 + (-a + b - 1)^2 + (b + 1)^2 + (a + b + 3)^2$$

$$\frac{\partial S}{\partial a} = -4(-2a + b - 4) - 2(-a + b - 1) + 2(a + b + 3) = 12a - 4b + 24$$

$$\frac{\partial S}{\partial b} = 2(-2a + b - 4) + 2(-a + b - 1) + 2(b + 1) + 2(a + b + 3) = -4a + 8b - 2$$

Setting these partial derivatives equal to zero produces:

$$12a - 4b = -24$$

$$-4a + 8b = 2$$

Thus, $a = -2.3$ and $b = -0.9$, and $y = -2.3x - 0.9$.

19. $\sum x_i = 0$

$\sum y_i = 8$

$\sum x_i^2 = 10$

$\sum x_i^3 = 0$

$\sum x_i^4 = 34$

$\sum x_i y_i = 12$

$\sum x_i^2 y_i = 22$

This produces the system

$$34a + 10c = 22$$
$$10b = 12$$
$$10a + 5c = 8$$

which yields $a = \frac{3}{7}$, $b = \frac{6}{5}$, and $c = \frac{26}{35}$, and we have $y = \frac{3}{7}x^2 + \frac{6}{5}x + \frac{26}{35}$.

21. $\sum x_i = 10$

$\sum y_i = 21$

$\sum x_i^2 = 30$

$\sum x_i^3 = 100$

$\sum x_i^4 = 354$

$\sum x_i y_i = 75$

$\sum x_i^2 y_i = 275$

This produces the system

$$354a + 100b + 30c = 275$$
$$100a + 30b + 10c = 75$$
$$30a + 10b + 4c = 21$$

which yields $a = 1.25$, $b = -1.75$, and $c = 0.25$, and we have $y = 1.25x^2 - 1.75x + 0.25$.

23. Linear: $y = 1.4286x + 6$

Quadratic: $y = 0.1190x^2 + 1.6667x + 5.6429$

25. Linear: $y = -68.9143x + 753.9524$

Quadratic: $y = 2.8214x^2 - 83.0214x + 763.3571$

27. $(1, 450), (1.25, 375), (1.5, 330)$

(a) $\sum x_i = 3.75$

$\sum y_i = 1155$

$\sum x_i y_i = 1413.75$

$\sum x_i^2 = 4.8125$

Thus, $a = -240$, $b = 685$, and $y = -240x + 685$.

(b) When $x = 1.4$, $y = 349$.

(c) $y = 500$ when $x \approx 0.77$.

29. (a) Using a graphing utility, the least squares regression line is $y = 13.8x + 22.1$.

(b) If $x = 1.6$, $y \approx 44.18$ bushels per acre.

31. (a) Using a graphing utility, the least squares regression line is $y = -0.5156t + 19.5599$ ($t = 0$ is 1970). In 2000, $t = 30$ and $y \approx 4.1$ deaths.

(b) The least squares regression quadratic is $y = 0.0004095t^2 - 0.5174t + 19.4624$. In 2000, $t = 30$ and $y \approx 4.3$ deaths.

33. (a) $(0, 0), (2, 15), (4, 30), (6, 50), (8, 65), (10, 70)$

$$\sum x_i = 30$$
$$\sum y_i = 230$$
$$\sum x_i^2 = 220$$
$$\sum x_i^3 = 1800$$
$$\sum x_i^4 = 15,664$$
$$\sum x_i y_i = 1670$$
$$\sum x_i^2 y_i = 13,500$$

$$15,664a + 1800b + 220c = 13,500$$
$$1800a + 220b + 30c = 1670$$
$$220a + 30b + 6c = 230$$

Thus, $a = -\frac{25}{112}$, $b = \frac{541}{56}$, and $c = -\frac{25}{14}$, and we have $y = -\frac{25}{112}x^2 + \frac{541}{56}x - \frac{25}{14}$.

(b) When $x = 5$, $y \approx 40.9$ mph.

35. Linear: $y = 3.7569x + 9.0347$

Quadratic: $y = 0.006316x^2 + 3.6252x + 9.4282$

37. Linear: $y = 0.9374x + 6.2582$

Quadratic: $y = -0.08715x^2 + 2.8159x + 0.3975$

39. Positive correlation

$r = 0.9981$

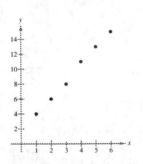

41. No correlation

$r = 0$

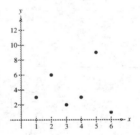

43. No, the slope is positive.

45. Yes, $|r| \approx 1$.

47. Using a graphing utility, you obtain $y = -43.4286x^2 + 3920.2857x - 42,120$. For $x = 28$, $y \approx \$33,600$.

Section 7.8   Double Integrals and Area in the Plane

1. $\displaystyle\int_0^x (2x - y)\, dy = \left(2xy - \frac{y^2}{2}\right)\Big]_0^x = \frac{3x^2}{2}$

3. $\displaystyle\int_1^{2y} \frac{y}{x}\, dx = y \ln|x| \Big]_1^{2y} = y \ln|2y|$

5. $\displaystyle\int_0^{\sqrt{9-x^2}} x^2 y\, dy = \frac{x^2 y^2}{2}\Big]_0^{\sqrt{9-x^2}} = \frac{x^2(9 - x^2)}{2} = \frac{9x^2 - x^4}{2}$

7. $\displaystyle\int_{e^y}^{y} \frac{y \ln x}{x}\, dx = \frac{y(\ln x)^2}{2}\Big]_{e^y}^{y} = \frac{y}{2}[(\ln y)^2 - y^2]$

9. $\displaystyle\int_0^{x^3} ye^{-y/x}\, dy = -xye^{-y/x}\Big]_0^{x^3} + x\int_0^{x^3} e^{-y/x}\, dy$

$$= -x^4 e^{-x^2} - \left[x^2 e^{-y/x}\right]_0^{x^3}$$

$$= -x^4 e^{-x^2} - x^2 e^{-x^2} + x^2$$

$$= x^2(1 - e^{-x^2} - x^2 e^{-x^2})$$

11. $\displaystyle\int_0^2 \int_0^1 (x - y)\, dy\, dx = \int_0^2 \left[xy - \frac{y^2}{2}\right]_0^1 dx$

$$= \int_0^2 \left(x - \frac{1}{2}\right) dx$$

$$= \left[\frac{x^2}{2} - \frac{1}{2}x\right]_0^2$$

$$= 2 - 1 = 1$$

13. $\displaystyle\int_0^4 \int_0^3 xy\, dy\, dx = \int_0^4 \left[\frac{xy^2}{2}\right]_0^3 dx = \frac{9}{2}\int_0^4 x\, dx = \frac{9}{2}\left[\frac{x^2}{2}\right]_0^4 = 36$

15. $\displaystyle\int_0^1 \int_0^{\sqrt{1-y^2}} (x+y)\, dx\, dy = \int_0^1 \left[\frac{x^2}{2} + xy\right]_0^{\sqrt{1-y^2}} dy$

$$= \int_0^1 \left[\frac{1}{2}(1-y^2) + y\sqrt{1-y^2}\right] dy$$

$$= \left[\frac{1}{2}\left(y - \frac{y^3}{3}\right) - \frac{1}{2}\left(\frac{2}{3}\right)(1-y^2)^{3/2}\right]_0^1$$

$$= \frac{1}{2}\left[y - \frac{y^3}{3} - \frac{2}{3}(1-y^2)^{3/2}\right]_0^1$$

$$= \frac{2}{3}$$

17. $\displaystyle\int_1^2 \int_0^4 (x^2 - 2y^2 + 1)\, dx\, dy = \int_1^2 \left[\frac{x^3}{3} - 2xy^2 + x\right]_0^4 dy$

$$= \int_1^2 \left(\frac{64}{3} - 8y^2 + 4\right) dy$$

$$= \left[\frac{76}{3}y - \frac{8y^3}{3}\right]_1^2$$

$$= \left[\frac{4}{3}(19y - 2y^3)\right]_1^2$$

$$= \frac{20}{3}$$

19. $\displaystyle\int_0^2 \int_0^{\sqrt{1-y^2}} -5xy\, dx\, dy = -5\int_0^2 \left[\frac{x^2}{2}y\right]_0^{\sqrt{1-y^2}} dy$

$$= -5\int_0^2 \frac{(1-y^2)y}{2}\, dy$$

$$= -\frac{5}{2}\left[\frac{y^2}{2} - \frac{y^4}{4}\right]_0^2$$

$$= -\frac{5}{2}(2-4)$$

$$= 5$$

21. $\displaystyle\int_0^2 \int_0^{4-x^2} x^3\, dy\, dx = \int_0^2 \left[x^3 y\right]_0^{4-x^2} dx$

$$= \int_0^2 (4x^3 - x^5)\, dx$$

$$= \left[\left(x^4 - \frac{x^6}{6}\right)\right]_0^2$$

$$= \frac{16}{3}$$

23. Since (for fixed x)

$$\lim_{b\to\infty}\left[-2e^{-(x+y)/2}\right]_0^b = 2e^{-x/2},$$

we have the following.

$$\int_0^\infty \int_0^\infty e^{-(x+y)/2}\, dy\, dx = \int_0^\infty 2e^{-x/2}\, dx$$

$$= \lim_{b\to\infty}\left[-4e^{-x/2}\right]_0^b$$

$$= 4$$

25. $\displaystyle\int_0^1 \int_0^2 dy\, dx = \int_0^1 2\, dx = 2$

$\displaystyle\int_0^2 \int_0^1 dx\, dy = \int_0^2 dy = 2$

27. $\displaystyle\int_0^1 \int_{2y}^2 dx\, dy = \int_0^1 (2-2y)\, dy$

$$= (2y - y^2)\Big]_0^1 = 1$$

$\displaystyle\int_0^2 \int_0^{x/2} dy\, dx = \int_0^2 \frac{x}{2}\, dx$

$$= \frac{x^2}{4}\Big]_0^2 = 1$$

29. $\int_0^2 \int_{x/2}^1 dy\, dx = \int_0^2 \left(1 - \frac{x}{2}\right) dx = \left(x - \frac{x^2}{4}\right)\Big]_0^2 = 1$

$\int_0^1 \int_0^{2y} dx\, dy = \int_0^1 2y\, dy = y^2 \Big]_0^1 = 1$

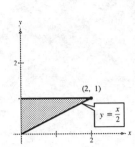

31. $\int_0^1 \int_{y^2}^{\sqrt[3]{y}} dx\, dy = \int_0^1 \left(\sqrt[3]{y} - y^2\right) dy$

$\qquad = \left[\left(\frac{3}{4}y^{4/3} - \frac{y^3}{3}\right)\right]_0^1$

$\qquad = \frac{5}{12}$

$\int_0^1 \int_{x^3}^{\sqrt{x}} dy\, dx = \int_0^1 \left(\sqrt{x} - x^3\right) dx$

$\qquad = \left[\left(\frac{2}{3}x^{3/2} - \frac{x^4}{4}\right)\right]_0^1$

$\qquad = \frac{5}{12}$

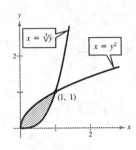

33. $\int_0^3 \int_y^3 e^{x^2} dx\, dy = \int_0^3 \int_0^x e^{x^2} dy\, dx$

$\qquad = \int_0^3 \left[ye^{x^2}\right]_0^x dx$

$\qquad = \int_0^3 xe^{x^2} dx$

$\qquad = \frac{1}{2}e^{x^2} \Big]_0^3$

$\qquad = \frac{1}{2}(e^9 - 1)$

$\qquad \approx 4051.042$

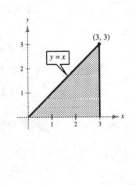

35. $A = \int_0^8 \int_0^3 dy\, dx$

$\qquad = \int_0^8 3\, dx$

$\qquad = 3x \Big]_0^8$

$\qquad = 24$

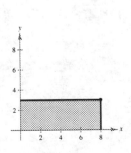

37. $A = \int_0^2 \int_0^{4-x^2} dy\, dx$

$\qquad = \int_0^2 (4 - x^2)\, dx$

$\qquad = \left[4x - \frac{x^3}{3}\right]_0^2$

$\qquad = \frac{16}{3}$

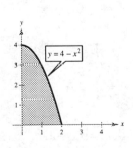

39. $A = \int_0^4 \int_0^{(2-\sqrt{x})^2} dy\, dx$

$= \int_0^4 \left(4 - 4\sqrt{x} + x\right) dx$

$= \left[4x - \dfrac{8}{3}x^{3/2} + \dfrac{x^2}{2}\right]_0^4$

$= \dfrac{8}{3}$

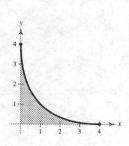

41. $A = \int_{-5}^5 \int_0^{25-x^2} dy\, dx$

$= \int_{-5}^5 (25 - x^2)\, dx$

$= \left[25x - \dfrac{x^3}{3}\right]_{-5}^5$

$= \left(125 - \dfrac{125}{3}\right) - \left(-125 + \dfrac{125}{3}\right)$

$= \dfrac{500}{3}$

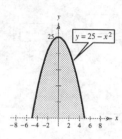

43. The point of intersection of the two graphs is found by equating $y = \dfrac{5}{2}x$ and $y = 3 - x$, which yields $x = \dfrac{6}{7}$ and $y = \dfrac{15}{7}$.

$A = \int_0^{15/7} \int_{2y/5}^{3-y} dx\, dy$

$= \int_0^{15/7} \left(3 - y - \dfrac{2y}{5}\right) dy$

$= \int_0^{15/7} \left(3 - \dfrac{7y}{5}\right) dy$

$= \left[3y - \dfrac{7y^2}{10}\right]_0^{15/7}$

$= 3\left(\dfrac{15}{7}\right) - \dfrac{7}{10}\left(\dfrac{15}{7}\right)^2 = \dfrac{45}{14}$

Note: Area of triangle is $\dfrac{1}{2}(3)\left(\dfrac{15}{7}\right) = \dfrac{45}{14}$.

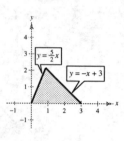

45. $A = \int_0^2 \int_x^{2x} dy\, dx$

$= \int_0^2 (2x - x)\, dx$

$= \int_0^2 x\, dx$

$= \dfrac{x^2}{2}\Big]_0^2$

$= 2$

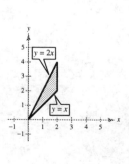

47. $\int_0^1 \int_0^2 e^{-x^2 - y^2}\, dx\, dy \approx 0.65876$

49. $\int_1^2 \int_0^x e^{xy}\, dy\, dx \approx 8.1747$

51. $\int_0^1 \int_x^1 \sqrt{1 - x^2}\, dy\, dx = \dfrac{\pi}{4} - \dfrac{1}{3} \approx 0.4521$

53. False

Section 7.9   Applications of Double Integrals

1. $\displaystyle\int_0^2\int_0^1 (3x + 4y)\, dy\, dx = \int_0^2 (3xy + 2y^2)\Big]_0^1 dx$

$\displaystyle\qquad = \int_0^2 (3x + 2)\, dx$

$\displaystyle\qquad = \left[\left(\frac{2}{3}x^2 + 2x\right)\right]_0^2$

$\displaystyle\qquad = 10$

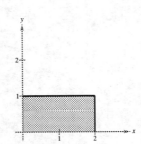

3. $\displaystyle\int_0^1\int_y^{\sqrt{y}} x^2 y^2\, dx\, dy = \int_0^1 \frac{x^3 y^2}{3}\Big]_y^{\sqrt{y}} dy$

$\displaystyle\qquad = \frac{1}{3}\int_0^1 (y^{7/2} - y^5)\, dy$

$\displaystyle\qquad = \frac{1}{3}\left[\frac{2}{9}y^{9/2} - \frac{1}{6}y^6\right]_0^1$

$\displaystyle\qquad = \frac{1}{54}$

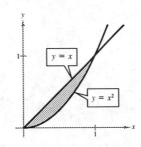

$y = x$

$y = x^2$

5. $\displaystyle\int_0^1\int_0^{\sqrt{1-x^2}} y\, dy\, dx = \int_0^1 \frac{y^2}{2}\Big]_0^{\sqrt{1-x^2}} dx$

$\displaystyle\qquad = \frac{1}{2}\int_0^1 (1 - x^2)\, dx$

$\displaystyle\qquad = \frac{1}{2}\left(x - \frac{x^3}{3}\right)\Big]_0^1$

$\displaystyle\qquad = \frac{1}{3}$

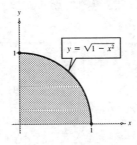

$y = \sqrt{1 - x^2}$

7. $\displaystyle\int_0^3\int_0^5 xy\, dy\, dx = \int_0^5\int_0^3 xy\, dx\, dy$

$\displaystyle\int_0^3\int_0^5 xy\, dy\, dx = \int_0^3 \frac{xy^2}{2}\Big]_0^5 dx$

$\displaystyle\qquad = \int_0^3 \frac{25}{2}x\, dx$

$\displaystyle\qquad = \frac{25}{4}x^2\Big]_0^3$

$\displaystyle\qquad = \frac{225}{4}$

9. $\displaystyle\int_0^4\int_0^{\sqrt{x}} \frac{y}{1 + x^2}\, dy\, dx = \int_0^2\int_{y^2}^4 \frac{y}{1 + x^2}\, dx\, dy$

$\displaystyle\int_0^4\int_0^{\sqrt{x}} \frac{y}{1 + x^2}\, dy\, dx - \int_0^4\left[\frac{y^2}{2(1 + x^2)}\right]_0^{\sqrt{x}} dx$

$\displaystyle\qquad = \int_0^4 \frac{x}{2(1 + x^2)}\, dx$

$\displaystyle\qquad = \frac{1}{4}\ln(1 + x^2)\Big]_0^4$

$\displaystyle\qquad = \frac{1}{4}\ln 17$

$\displaystyle\qquad \approx 0.708$

11. $\int_0^1 \int_{y/2}^{1/2} e^{-x^2}\, dx\, dy = \int_0^{1/2} \int_0^{2x} e^{-x^2}\, dy\, dx$

$\qquad\qquad\qquad = \int_0^{1/2} 2x e^{-x^2}\, dx$

$\qquad\qquad\qquad = -e^{-x^2}\Big]_0^{1/2}$

$\qquad\qquad\qquad = 1 - e^{-1/4}$

$\qquad\qquad\qquad \approx 0.2212$

13. $V = \int_0^2 \int_0^4 \dfrac{y}{2}\, dx\, dy$

$\qquad = \int_0^2 \dfrac{xy}{2}\Big]_0^4\, dy$

$\qquad = \int_0^2 2y\, dy$

$\qquad = y^2\Big]_0^2$

$\qquad = 4$

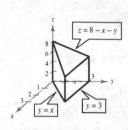

$0 \le x \le 4$

$0 \le y \le 2$

15. $V = \int_0^3 \int_x^3 (8 - x - y)\, dy\, dx$

$\qquad = \int_0^3 \left(8y - xy - \dfrac{y^2}{2}\right)\Big]_x^3\, dx$

$\qquad = \int_0^3 \left[\left(24 - 3x - \dfrac{9}{2}\right) - \left(8x - x^2 - \dfrac{x^2}{2}\right)\right] dx$

$\qquad = \int_0^3 \left(\dfrac{39}{2} - 11x + \dfrac{3}{2}x^2\right) dx$

$\qquad = \left[\dfrac{39}{2}x - \dfrac{11x^2}{2} + \dfrac{x^3}{2}\right]_0^3$

$\qquad = 22.5$

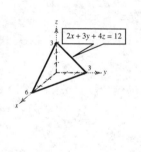

17. $V = \int_0^6 \int_0^{4-(2x/3)} \left(3 - \dfrac{x}{2} - \dfrac{3y}{4}\right) dy\, dx$

$\qquad = \int_0^6 \left[3y - \dfrac{xy}{2} - \dfrac{3y^2}{8}\right]_0^{4-(2x/3)}\, dx$

$\qquad = \int_0^6 \left(6 - 2x + \dfrac{x^2}{6}\right) dx$

$\qquad = \left[6x - x^2 + \dfrac{x^3}{18}\right]_0^6$

$\qquad = 12$

19. $V = \int_0^1 \int_0^y (1 - xy)\, dx\, dy$

$\qquad = \int_0^1 \left[x - \dfrac{x^2 y}{2}\right]_0^y\, dy$

$\qquad = \int_0^1 \left(y - \dfrac{y^3}{2}\right) dy$

$\qquad = \left[\dfrac{y^2}{2} - \dfrac{y^4}{8}\right]_0^1$

$\qquad = \dfrac{3}{8}$

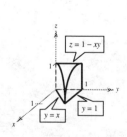

21. $V = 4 \int_0^1 \int_0^1 (4 - x^2 - y^2) \, dy \, dx$

$= 4 \int_0^1 \left[4y - x^2 y - \frac{y^3}{3} \right]_0^1 dx$

$= 4 \int_0^1 \left[4 - x^2 - \frac{1}{3} \right] dx$

$= 4 \int_0^1 \left(\frac{11}{3} - x^2 \right) dx$

$= 4 \left[\frac{11x}{3} - \frac{x^3}{3} \right]_0^1$

$= 4 \left(\frac{11}{3} - \frac{1}{3} \right)$

$= 4 \left(\frac{10}{3} \right)$

$= \frac{40}{3}$

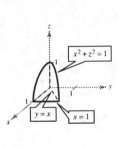

$-1 \leq x \leq 1$

$-1 \leq y \leq 1$

23. $V = \int_0^1 \int_0^x \sqrt{1 - x^2} \, dy \, dx$

$= \int_0^1 x \sqrt{1 - x^2} \, dx$

$= -\frac{1}{3} (1 - x^2)^{3/2} \Big]_0^1$

$= \frac{1}{3}$

25. $V = \int_0^4 \int_0^1 xy \, dx \, dy$

$= \int_0^4 \frac{x^2 y}{2} \Big]_0^1 dy$

$= \frac{1}{2} \int_0^4 y \, dy$

$= \frac{y^2}{4} \Big]_0^4$

$= 4$

27. $V = \int_0^2 \int_0^4 x^2 \, dy \, dx$

$= \int_0^2 x^2 y \Big]_0^4 dx$

$= \int_0^2 4x^2 \, dx$

$= \frac{4x^3}{3} \Big]_0^2$

$= \frac{32}{3}$

29. Average $= \frac{1}{8} \int_0^4 \int_0^2 x \, dy \, dx$

$= \frac{1}{8} \int_0^4 2x \, dx$

$= \frac{x^2}{8} \Big]_0^4$

$= 2$

31. Average $= \frac{1}{4} \int_0^2 \int_0^2 (x^2 + y^2) \, dx \, dy$

$= \frac{1}{4} \int_0^2 \left[\frac{x^3}{3} + xy^2 \right]_0^2 dy$

$= \frac{1}{4} \int_0^2 \left(\frac{8}{3} + 2y^2 \right) dy$

$= \left[\frac{1}{4} \left(\frac{8}{3} y + \frac{2}{3} y^3 \right) \right]_0^2$

$= \frac{8}{3}$

33. Average $= \dfrac{1}{1250} \displaystyle\int_{100}^{150}\int_{50}^{75} [(500-3p_1)p_1 + (750-2.4p_2)p_2]\, dp_1\, dp_2$

$\quad\quad\quad\quad = \dfrac{1}{1250} \displaystyle\int_{100}^{150}\int_{50}^{75} [-3p_1{}^2 + 500p_1 - 2.4p_2{}^2 + 750p_2]\, dp_1\, dp_2$

$\quad\quad\quad\quad = \dfrac{1}{1250} \displaystyle\int_{100}^{150} \left[-p_1{}^3 + 250p_1{}^2 - 2.4p_1 p_2{}^2 + 750p_1 p_2 \right]_{50}^{75} dp_2$

$\quad\quad\quad\quad = \dfrac{1}{1250} \displaystyle\int_{100}^{150} [484{,}375 - 60p_2{}^2 + 18{,}750p_2]\, dp_2$

$\quad\quad\quad\quad = \dfrac{1}{1250} \left[484{,}375p_2 - 20p_2{}^3 + 9375p_2{}^2 \right]_{100}^{150}$

$\quad\quad\quad\quad = \$75{,}125$

Review Exercises for Chapter 7

1.

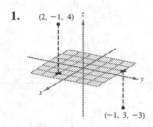

(2, −1, 4)

(−1, 3, −3)

3. $d = \sqrt{(2-0)^2 + (5-0)^2 + (9-0)^2}$

$\quad\quad = \sqrt{4 + 25 + 81}$

$\quad\quad = \sqrt{110}$

5. Midpoint $= \left(\dfrac{2-4}{2}, \dfrac{6+2}{2}, \dfrac{4+8}{2} \right) = (-1, 4, 6)$

7. $(x-0)^2 + (y-1)^2 + (z-0)^2 = 5^2$

$\quad\quad x^2 + (y-1)^2 + z^2 = 25$

9. Center $= \left(\dfrac{3+5}{2}, \dfrac{4+8}{2}, \dfrac{0+2}{2} \right) = (4, 6, 1)$

Radius $= \sqrt{(4-3)^2 + (6-4)^2 + (1-0)^2} = \sqrt{6}$

Circle: $(x-4)^2 + (y-6)^2 + (z-1)^2 = 6$

11. $x^2 + 4x + 4 + y^2 - 2y + 1 + z^2 - 8z + 16 = -5 + 4 + 1 + 16$

$\quad\quad\quad (x+2)^2 + (y-1)^2 + (z-4)^2 = 16 = 4^2$

Center: $(-2, 1, 4)$

Radius: 4

13. Let $z = 0$.

$\quad\quad (x+2)^2 + (y-1)^2 + (0-3)^2 = 25$

$\quad\quad\quad\quad (x+2)^2 + (y-1)^2 = 16 = 4^2$

Circle of radius 4

15. x-intercept: $(6, 0, 0)$

y-intercept: $(0, 3, 0)$

z-intercept: $(0, 0, 2)$

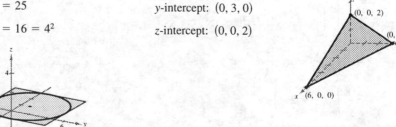

17. *x*-intercept: $(2, 0, 0)$

  y-intercept: $(0, 4, 0)$

  z-intercept: $(0, 0, -2)$

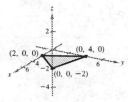

19. The graph is a sphere whose standard equation is

$$(x - 1)^2 + (y + 2)^2 + (z - 3)^2 = 9.$$

21. The graph is an ellipsoid.

23. The graph is an elliptic paraboloid.

25. The graph is the top half of a circular cone whose standard equation is

$$x^2 + y^2 - z^2 = 0.$$

27. $f(x, y) = xy^2$

  (a) $f(2, 3) = 2(3)^2 = 18$

  (b) $f(0, 1) = 0(1)^2 = 0$

  (c) $f(-5, 7) = -5(7)^2 = -245$

  (d) $f(-2, -4) = -2(-4)^2 = -32$

29. The domain is the set of all points inside or on the circle $x^2 + y^2 = 1$ and the range is $[0, 1]$.

31. The level curves are lines of slope $-\frac{2}{5}$.

  $c = 0$: $10 - 2x - 5y = 0$

  $\qquad\qquad 2x + 5y = 10$

  $c = 2$: $10 - 2x - 5y = 2$

  $\qquad\qquad 2x + 5y = 8$

  $c = 4$: $10 - 2x - 5y = 4$

  $\qquad\qquad 2x + 5y = 6$

  $c = 5$: $10 - 2x - 5y = 5$

  $\qquad\qquad 2x + 5y = 5$

  $c = 10$: $10 - 2x - 5y = 10$

  $\qquad\qquad 2x + 5y = 0$

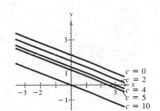

33. $z = (xy)^2$

  $c = 1$: $(xy)^2 = 1$,

  $c = 4$: $(xy)^2 = 4$,

  $c = 9$: $(xy)^2 = 9$,

  $c = 12$: $(xy)^2 = 12$,

  $c = 16$: $(xy)^2 = 16$,

  $y^2 = \dfrac{1}{x^2}, \ y = \perp \dfrac{1}{x}$

  $y = \pm\dfrac{3}{x}$

  $y = \pm\dfrac{2\sqrt{3}}{x}$

  $y = \pm\dfrac{4}{x}$

  The level curves are hyperbolas.

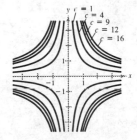

35. (a) The level curves represent lines of equal rainfall, and separate the four colors.

  (b) The small eastern portion containing Davenport

  (c) The northwestern portion containing Sioux City

37. Southwest

39. $P = \dfrac{MV}{T} = \dfrac{(2500)6}{6000} = \$2.50$

41. $f(x, y) = x^2y + 3xy + 2x - 5y$

$f_x = 2xy + 3y + 2$

$f_y = x^2 + 3x - 5$

43. $z = 6x^2\sqrt{y} + 3\sqrt{xy} - 7xy$

$z_x = 12x\sqrt{y} + \dfrac{3}{2}\sqrt{\dfrac{y}{x}} - 7y$

$z_y = 3\dfrac{x^2}{\sqrt{y}} + \dfrac{3}{2}\sqrt{\dfrac{x}{y}} - 7x$

45. $f(x, y) = \ln(2x + 3y)$

$f_x = \dfrac{2}{2x + 3y}$

$f_y = \dfrac{3}{2x + 3y}$

47. $f_x = 2xe^y - y^2e^x$

$f_y = x^2e^y - 2ye^x$

49. $\dfrac{\partial w}{\partial x} = yz^2$

$\dfrac{\partial w}{\partial y} = xz^2$

$\dfrac{\partial w}{\partial z} = 2xyz$

51. (a) $z_x = 3$

(b) $z_y = -4$

53. (a) $z_x = -2x;$ At $(1, 2, 3)$, $z_x = -2$.

(b) $z_y = -2y;$ At $(1, 2, 3)$, $z_y = -4$.

55. $f_x = 3x^2 - 4y^2$

$f_y = -8xy + 3y^2$

$f_{xx} = 6x$

$f_{yy} = -8x + 6y$

$f_{xy} = f_{yx} = -8y$

57. $f_x = \dfrac{-x}{\sqrt{64 - x^2 - y^2}}$

$f_y = \dfrac{-y}{\sqrt{64 - x^2 - y^2}}$

$f_{xx} = \dfrac{y^2 - 64}{(64 - x^2 - y^2)^{3/2}}$

$f_{yy} = \dfrac{x^2 - 64}{(64 - x^2 - y^2)^{3/2}}$

$f_{xy} = \dfrac{-xy}{(64 - x^2 - y^2)^{3/2}} = f_{yx}$

59. $C_x = 5x^{-2/3}y^{1/3} + 99,$ $C_x(250, 175) \approx 99.70$

$C_y = 5x^{1/3}y^{-2/3} + 139,$ $C_y(250, 175) \approx 140.04$

61. $\dfrac{\partial A}{\partial w} = 101.4(0.425)w^{0.425-1}h^{0.725} = 43.095w^{-0.575}h^{0.725}$

$\dfrac{\partial A}{\partial h} = 101.4(0.725)w^{0.425}h^{0.725-1} = 73.515w^{0.425}h^{-0.275}$

$\dfrac{\partial A}{\partial w}(180, 70) = 43.095(180)^{-0.575}(70)^{0.725} \approx 47.35$

The surface area increases approximately 47 cm^2 per pound for a human weighing 180 pounds and 70 inches tall.

63. The first partial derivatives of f, $f_x(x, y) = 2x + 2y$ and $f_y(x, y) = 2x + 2y$, are zero when $y = -x$. The points $(x, -x, 0)$ are relative minima.

Note: The Second-Partials Test fails since $d = 0$.

65. The first partial derivatives of f, $f_x = 2x + 6y + 6$ and $f_y = 6x + 6y$, are zero at $\left(\frac{3}{2}, -\frac{3}{2}\right)$. The second partials are $f_{xx} = 2, f_{xy} = 6$, and $f_{yy} = 6$. Since $f_{xx} > 0$ and $d = f_{xx}f_{yy} - (f_{xy})^2 < 0, f\left(\frac{3}{2}, -\frac{3}{2}\right)$ is a saddle point. No relative extrema.

67. The first partial derivatives of f, $f_x = 3x^2 - y$ and $f_y = 2y - x$, are zero at $\left(\frac{1}{6}, \frac{1}{12}\right)$ and $(0, 0)$. The second partials are $f_{xx} = 6x, f_{yy} = 2$, and $f_{xy} = 1$.

At $(0, 0), d < 0 \Rightarrow$ saddle point.

At $\left(\frac{1}{6}, \frac{1}{12}\right), f_{xx} > 0$ and $d > 0 \Rightarrow$ relative minimum.

69. The first partial derivatives of f, $f_x = 3x^2 - 3$ and $f_y = 3y^2 - 3$, are zero at $(\pm1, \pm1)$, $(\pm1, \mp1)$. The second partials are $f_{xx} = 6x, f_{yy} = 6y$, and $f_{xy} = 0$.

$(1, 1)$:	$f_{xx} > 0, d > 0$	Relative minimum
$(1, -1)$:	$f_{xx} > 0, d < 0$	Saddle point
$(-1, 1)$:	$f_{xx} < 0, d < 0$	Saddle point
$(-1, -1)$:	$f_{xx} < 0, d > 0$	Relative maximum

71. (a) $R = x_1 p_1 + x_2 p_2$

$= x_1(100 - x_1) + x_2(200 - 0.5x_2)$

$= -x_1^2 - \frac{1}{2}x_2^2 + 100x_1 + 200x_2$

(b) $R_{x_1} = -2x_1 + 100 = 0 \Rightarrow x_1 = 50$

$R_{x_2} = -x_2 + 200 = 0 \Rightarrow x_2 = 200$

By the Second-Partials Test, (50, 200) is a maximum.

(c) $R(50, 200) = \$22,500.00$

73. $F(x, y, \lambda) = x^2y - \lambda(x + 2y - 2)$

$\left.\begin{array}{l} F_x(x, y, \lambda) = 2xy - \lambda = 0 \\ F_y(x, y, \lambda) = x^2 - 2\lambda = 0 \end{array}\right\}$ $4xy = x^2$

$F_\lambda(x, y, \lambda) = -(x + 2y - 2) = 0, \quad y = \dfrac{2 - x}{2}$

Thus, $x = 0$ or $x = \frac{4}{3}$, and the corresponding y-values are $y = 1$ or $y = \frac{1}{3}$. This implies that the extrema occur at $(0, 1, 0)$ (relative minimum) and $\left(\frac{4}{3}, \frac{1}{3}, \frac{16}{27}\right)$ (relative minimum).

75. $F(x, y, z, \lambda) = xyz - \lambda(x + 2y + z - 4)$

$\left.\begin{array}{l} F_x = yz - \lambda = 0 \\ F_y = xz - 2\lambda = 0 \\ F_z = xy - \lambda = 0 \end{array}\right\}$ $xz = 2yz = 2xy \Rightarrow x = 2y$

$z = 2y$

$F_\lambda = -(x + 2y + z - 4) = 0$

$2y + 2y + 2y - 4 = 0 \Rightarrow y = \frac{2}{3}, x = \frac{4}{3}, z = \frac{4}{3}$

At $\left(\frac{4}{3}, \frac{2}{3}, \frac{4}{3}\right)$, the relative maximum value is $\frac{32}{27}$.

77. $F(x, y, z, \lambda, \mu) = x^2 + y^2 + z^2 - \lambda(x + z - 6) - \mu(y + z - 8)$

$F_x = 2x - \lambda = 0 \qquad\qquad x = \dfrac{\lambda}{2}$

$F_y = 2y - \mu = 0 \qquad\qquad y = \dfrac{\mu}{2}$

$F_z = 2z - \lambda - \mu = 0 \qquad z = \dfrac{\lambda + \mu}{2} = x + y$

$F_\lambda = -(x + z - 6) = 0 \qquad \left.\begin{array}{l} x + z = 6 \Rightarrow 2x + y = 6 \\ y + z = 8 \Rightarrow x + 2y = 8 \end{array}\right\}$

$F_\mu = -(y + z - 8) = 0$

$x = \dfrac{4}{3}, y = \dfrac{10}{3}, z = \dfrac{14}{3}$

$f\left(\frac{4}{3}, \frac{10}{3}, \frac{14}{3}\right) = 34\frac{2}{3}$ is a relative minimum.

79. Maximize $f(x, y) = 4x + xy + 2y$, subject to the constraint $20x + 4y = 2000$.

$F(x, y, \lambda) = 4x + xy + 2y - \lambda(20x + 4y - 2000)$

$\left.\begin{array}{l} F_x(x, y, \lambda) = 4 + y - 20\lambda = 0 \\ F_y(x, y, \lambda) = x + 2 - 4\lambda = 0 \end{array}\right\} \qquad \begin{array}{l} 4 + y = 5(x + 2) \\ y = 5x + 6 \end{array}$

$F_\lambda(x, y, \lambda) = -(20x + 4y - 2000) = 0, \qquad y - 500 - 5x$

Thus, $x = 49.4$ and $y = 5(49.4) + 6 = 253$ which implies that the maximum production level is $f(49.4, 253) \approx 13,202$.

81. (a) $\sum x_i = 1$

$\sum y_i = 0$

$\sum x_i^2 = 15$

$\sum x_i y_i = 15$

$a = \dfrac{4(15) - (1)(0)}{4(15) - (1)^2} = \dfrac{60}{59}$

$b = \dfrac{1}{4}\left(0 - \dfrac{60}{59}(1)\right) = -\dfrac{15}{59}$

$y = \dfrac{60}{59}x - \dfrac{15}{59}$

(b) $\left(\dfrac{60}{59}(-2) - \dfrac{15}{59} + 3\right)^2 + \left(\dfrac{60}{59}(-1) - \dfrac{15}{59} + 1\right)^2 + \left(\dfrac{60}{59}(1) - \dfrac{15}{59} - 2\right)^2 + \left(\dfrac{60}{59}(3) - \dfrac{15}{59} - 2\right)^2 \approx 2.746$

83. Using a graphing utility, $y = 14x + 19$. For 160 pounds per acre, $x = 1.6$ and $y = 14(1.6) + 19 = 41.4$ bushels per acre.

85. $\sum x_i = 6$

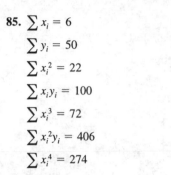

$\sum y_i = 50$

$\sum x_i^2 = 22$

$\sum x_i y_i = 100$

$\sum x_i^3 = 72$

$\sum x_i^2 y_i = 406$

$\sum x_i^4 = 274$

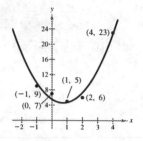

The system

$274a + 74b + 22c = 406$

$74a + 22b + 6c = 100$

$22a + 6b + 5c = 50$

has approximate solutions $a = 1.71$, $b = -2.57$, and $c = 5.56$. Therefore, the least squares quadratic is $y = 1.71x^2 - 2.57x + 5.56$.

87. $\displaystyle\int_0^1 \int_0^{1+x} (3x + 2y)\, dy\, dx = \int_0^1 \left[(3xy + y^2)\right]_0^{1+x} dx$

$= \displaystyle\int_0^1 [3x(1 + x) + (1 + x)^2]\, dx$

$= \displaystyle\int_0^1 (4x^2 + 5x + 1)\, dx$

$= \left[\dfrac{4x^3}{3} + \dfrac{5x^2}{2} + x\right]_0^1$

$= \dfrac{4}{3} + \dfrac{5}{2} + 1$

$= \dfrac{29}{6}$

89. $\displaystyle\int_1^2 \int_1^{2y} \dfrac{x}{y^2}\, dx\, dy = \int_1^2 \left[\dfrac{x^2}{2y^2}\right]_1^{2y} dy$

$= \displaystyle\int_1^2 \left[\dfrac{4y^2}{2y^2} - \dfrac{1}{2y^2}\right] dy$

$= \displaystyle\int_1^2 \left(2 - \dfrac{1}{2}y^{-2}\right) dy$

$= \left[2y + \dfrac{1}{2y}\right]_1^2$

$= \left(4 + \dfrac{1}{4}\right) - \left(2 + \dfrac{1}{2}\right)$

$= \dfrac{7}{4}$

91. $A = \displaystyle\int_{-2}^{2}\int_{5}^{9-x^2} dy\, dx = \int_{5}^{9}\int_{-\sqrt{9-y}}^{\sqrt{9-y}} dx\, dy$

$\displaystyle\int_{-2}^{2}\int_{5}^{9-x^2} dy\, dx = \int_{-2}^{2} [(9 - x^2) - 5]\, dx$

$\qquad = \displaystyle\int_{-2}^{2} (4 - x^2)\, dx$

$\qquad = \left[\left(4x - \dfrac{x^3}{3}\right)\right]_{-2}^{2}$

$\qquad = \left(8 - \dfrac{8}{3}\right) - \left(-8 + \dfrac{8}{3}\right)$

$\qquad = \dfrac{32}{3}$

93. $A = \displaystyle\int_{-3}^{6}\int_{1/3(x+3)}^{\sqrt{x+3}} -dy\, dx$

$\qquad = \displaystyle\int_{-3}^{6} \left(\sqrt{x+3} - \dfrac{1}{3}(x+3)\right) dx$

$\qquad = \left[\dfrac{2}{3}(x+3)^{3/2} - \dfrac{x^2}{6} - x\right]_{-3}^{6}$

$\qquad = (18 - 6 - 6) - \left(0 - \dfrac{3}{2} + 3\right)$

$\qquad = \dfrac{9}{2}$

95. $V = \displaystyle\int_{0}^{4}\int_{0}^{4} (xy)^2\, dy\, dx$

$\qquad = \displaystyle\int_{0}^{4}\int_{0}^{4} x^2 y^2\, dy\, dx$

$\qquad = \displaystyle\int_{0}^{4} \dfrac{x^2 y^3}{3}\Big]_{0}^{4}\, dx$

$\qquad = \displaystyle\int_{0}^{4} \dfrac{64x^2}{3}\, dx$

$\qquad = \dfrac{64x^3}{9}\Big]_{0}^{4}$

$\qquad = \dfrac{4096}{9}$

97. Average $= \dfrac{\displaystyle\int_{0}^{10}\int_{0}^{25-2.5x} (0.25 - 0.025x - 0.01y)\, dy\, dx}{\text{area}}$

$\qquad = \dfrac{10^{5/12}}{125}$

$\qquad = 0.083\overline{3}$ miles

Practice Test for Chapter 7

1. Find the distance between the points $(3, -7, 2)$ and $(5, 11, -6)$ and find the midpoint of the line segment joining the two points.

2. Find the standard form of the equation of the sphere whose center is $(1, -3, 0)$ and whose radius is $\sqrt{5}$.

3. Find the center and radius of the sphere whose equation is $x^2 + y^2 + z^2 - 4x + 2y + 8z = 0$.

4. Sketch the graph of the plane.

   (a) $3x + 8y + 6z = 24$ (b) $y = 2$

5. Identify the surface.

   (a) $\dfrac{x^2}{16} + \dfrac{y^2}{4} - \dfrac{z^2}{9} = 1$ (b) $z = \dfrac{x^2}{25} + y^2$

6. Find the domain of the function.

   (a) $f(x, y) = \ln(3 - x - y)$ (b) $f(x, y) = \dfrac{1}{x^2 + y^2}$

7. Find the first partial derivatives of $f(x, y) = 3x^2 + 9xy^2 + 4y^3 - 3x - 6y + 1$.

8. Find the first partial derivatives of $f(x, y) = \ln(x^2 + y^2 + 5)$.

9. Find the first partial derivatives of $f(x, y, z) = x^2 y^3 \sqrt{z}$.

10. Find the second partial derivatives of $z = \dfrac{x}{x^2 + y^2}$.

11. Find the relative extrema of $f(x, y) = 3x^2 + 4y^2 - 6x + 16y - 4$.

12. Find the relative extrema of $f(x, y) = 4xy - x^4 - y^4$.

13. Use Lagrange multipliers to find the minimum of $f(x, y) = xy$ subject to the constraint $4x - y = 16$.

14. Use Lagrange multipliers to find the minimum of $f(x, y) = x^2 - 16x + y^2 - 8y + 12$ subject to the constraint $x + y = 4$.

15. Find the least squares regression line for the points $(-3, 7)$, $(1, 5)$, $(8, -2)$, and $(4, 4)$.

16. Find the least squares regression quadratic for the points $(-5, 8)$, $(-1, 2)$, $(1, 3)$, and $(5, 5)$.

17. Evaluate $\displaystyle\int_0^3 \int_0^{\sqrt{x}} xy^3 \, dy \, dx$.

18. Evaluate $\displaystyle\int_{-1}^{2}\int_{0}^{3y} (x^2 - 4xy) \, dx \, dy$.

19. Set up a double integral to find the area of the indicated region.

(a)

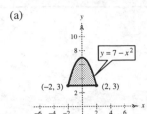

(b)

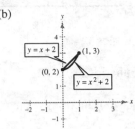

20. Find the volume of the solid bounded by the first octant and the plane $x + y + z = 4$.

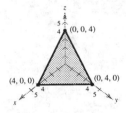

Graphing Calculator Required

21. Use a program similar to the one given on page ??? of the textbook to find the least squares regression line and the correlation coefficient for the given data points.

$(0, 20.4)$, $(1, 21.3)$, $(2, 22.9)$, $(3, 23.4)$, $(4, 24)$, $(5, 24.2)$, $(6, 24.9)$, $(7, 25.3)$, $(8, 26.2)$, $(9, 27.5)$, $(10, 29)$, $(11, 30.3)$, and $(12, 31.1)$

22. Use a graphing calculator or a computer algebra system to approximate $\displaystyle\int_{2}^{3}\int_{0}^{x+2} e^{x^2 y} \, dy \, dx$.

C H A P T E R 8
Trigonometric Functions

CHAPTER 8
Trigonometric Functions

Section 8.1   Radian Measure of Angles
Solutions to Odd-Numbered Exercises

1. (a) Positive: $45° + 360° = 405°$

    Negative: $45° - 360° = -315°$

    (b) Positive: $-41° + 360° = 319°$

    Negative: $-41° - 360° = -401°$

3. (a) Positive: $300° + 360° = 660°$

    Negative: $300° - 360° = -60°$

    (b) Positive: $740° - 2(360°) = 20°$

    Negative: $740° - 3(360°) = -340°$

5. (a) Positive: $\dfrac{\pi}{9} + 2\pi = \dfrac{19\pi}{9}$

    Negative: $\dfrac{\pi}{9} - 2\pi = -\dfrac{17\pi}{9}$

    (b) Positive: $\dfrac{2\pi}{3} + 2\pi = \dfrac{8\pi}{3}$

    Negative: $\dfrac{2\pi}{3} - 2\pi = -\dfrac{4\pi}{3}$

7. (a) Positive: $-\dfrac{9\pi}{4} + 2(2\pi) = \dfrac{7\pi}{4}$

    Negative: $-\dfrac{9\pi}{4} + 2\pi = -\dfrac{\pi}{4}$

    (b) Positive: $-\dfrac{2\pi}{15} + 2\pi = \dfrac{28\pi}{15}$

    Negative: $-\dfrac{2\pi}{15} - 2\pi = -\dfrac{32\pi}{15}$

9. $30°\left(\dfrac{\pi\,\text{radians}}{180°}\right) = \dfrac{\pi}{6}$ radians

11. $315°\left(\dfrac{\pi\,\text{radians}}{180°}\right) = \dfrac{7\pi}{4}$ radians

13. $-30°\left(\dfrac{\pi\,\text{radians}}{180°}\right) = -\dfrac{\pi}{6}$ radians

15. $\dfrac{3\pi}{2}\left(\dfrac{180°}{\pi}\right) = 270°$

17. $\dfrac{11\pi}{2}\left(\dfrac{180°}{\pi}\right) = 990°$

19. $-\dfrac{11\pi}{30}\left(\dfrac{180°}{\pi}\right) = -66°$

21. $-270°\left(\dfrac{\pi\,\text{radians}}{180°}\right) = -\dfrac{3\pi}{2}$ radians

23. $144°\left(\dfrac{\pi\,\text{radians}}{180°}\right) = \dfrac{4\pi}{5}$ radians

25. The angle θ is $\theta = 90° - 30° = 60°$. Since the vertical side is $c/2$, we can use the Pythagorean Theorem to find the length of the hypotenuse as follows.

$$c^2 = \left(5\sqrt{3}\right)^2 + \left(\dfrac{c}{2}\right)^2$$

$$\dfrac{3c^2}{4} = 75$$

$$c^2 = 100$$

$$c = 10$$

27. The angle θ is $\theta = 90° - 60° = 30°$. By the Pythagorean Theorem, the value of a is

$$a = \sqrt{8^2 - 4^2}$$
$$= \sqrt{64 - 16}$$
$$= \sqrt{48}$$
$$= 4\sqrt{3}.$$

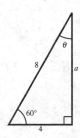

29. Since the triangle is isosceles, we have $\theta = 40°$.

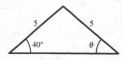

31. Since the large triangle is similar to the two smaller triangles, $\theta = 60°$. The hypotenuse of the large triangle is

$$\sqrt{2^2 + \left(2\sqrt{3}\right)^2} = \sqrt{16} = 4.$$

By similar triangles we have

$$\frac{s}{2} = \frac{2\sqrt{3}}{4} \Rightarrow s = \sqrt{3}.$$

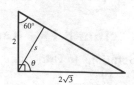

33. $h^2 + 3^2 = 6^2$

$\quad\quad h^2 = 27$

$\quad\quad\; h = 3\sqrt{3}$

$A = \dfrac{1}{2}bh$

$\quad = \dfrac{1}{2}(6)\left(3\sqrt{3}\right)$

$\quad = 9\sqrt{3}$ square inches

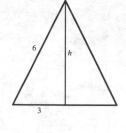

35. $h^2 + \left(\dfrac{5}{2}\right)^2 = 5^2$

$\quad\quad\quad h^2 = \dfrac{25}{2}$

$\quad\quad\quad\; h = \dfrac{5\sqrt{2}}{2}$

$A = \dfrac{1}{2}bh$

$\quad = \dfrac{1}{2}(5)\left(\dfrac{5\sqrt{2}}{2}\right)$

$\quad = \dfrac{25\sqrt{2}}{4}$ square feet

37. False. An obtuse angle is between 90° and 180°.

39. True. The angles would be 90°, 89°, and 1°.

41. Using similar triangles, we have

$$\frac{h}{24} = \frac{6}{8}$$

which implies that $h = 18$ feet.

43.

r	8 ft	15 in.	85 cm	24 in.	$\dfrac{12{,}963}{\pi}$ mi
s	12 ft	24 in.	200.28 cm	96 in.	8642 mi
θ	1.5	1.6	$\dfrac{3\pi}{4}$	4	$\dfrac{2\pi}{3}$

45. (a) The radian measure is

$$75°\left(\frac{\pi \text{ radians}}{180°}\right) = \frac{5\pi}{12} \text{ radians.}$$

(b) The distance moved is

$$S = \frac{5\pi}{12}(18.75) = 7.8125\pi \text{ in.} \approx 24.54 \text{ in.}$$

47. (a) Revolutions $= \dfrac{(3142 \text{ radians/minute})}{(2\pi \text{ radians/revolution})}$

$\quad\quad\quad\quad\quad\; \approx 500$ revolutions/minute

(b) Time $= \dfrac{(10{,}000 \text{ revolutions})}{(500 \text{ revolutions/minute})} = 20$ minutes

Section 8.2   The Trigonometric Functions

1. Since $x = 3$ and $y = 4$, it follows that $r = \sqrt{3^2 + 4^2} = 5$. Therefore, we have the following.

$\quad \sin \theta = \frac{4}{5} \quad\quad \csc \theta = \frac{5}{4}$

$\quad \cos \theta = \frac{3}{5} \quad\quad \sec \theta = \frac{5}{3}$

$\quad \tan \theta = \frac{4}{3} \quad\quad \cot \theta = \frac{3}{4}$

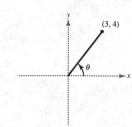

3. Since $x = -12$ and $y = -5$, it follows that $r = \sqrt{(-12)^2 + (-5)^2} = 13$.
Therefore, we have the following.

$$\sin \theta = -\frac{5}{13} \qquad \csc \theta = -\frac{13}{5}$$

$$\cos \theta = -\frac{12}{13} \qquad \sec \theta = -\frac{13}{12}$$

$$\tan \theta = \frac{5}{12} \qquad \cot \theta = \frac{12}{5}$$

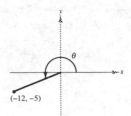

$(-12, -5)$

5. Since $x = -\sqrt{3}$ and $y = 1$, it follows that $r = \sqrt{3 + 1^2} = 2$.
Therefore, we have the following.

$$\sin \theta = \frac{1}{2} \qquad \csc \theta = 2$$

$$\cos \theta = -\frac{\sqrt{3}}{2} \qquad \sec \theta = -\frac{2\sqrt{3}}{3}$$

$$\tan \theta = -\frac{\sqrt{3}}{3} \qquad \cot \theta = -\sqrt{3}$$

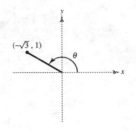

$(-\sqrt{3}, 1)$

7. $\csc \theta = \dfrac{1}{\sin \theta} = \dfrac{1}{1/2} = 2$

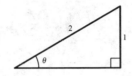

9. Since $x = 4$ and $r = 5$, the
length of the opposite side is
$y = \sqrt{5^2 - 4^2} = 3$. Therefore,
we have

$$\cot \theta = \frac{x}{y} = \frac{4}{3}.$$

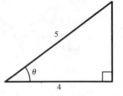

11. Since $x = 15$ and $y = 8$, the length of the hypotenuse is
$r = \sqrt{15^2 + 8^2} = 17$. Therefore, we have the following.

$$\sec \theta = \frac{r}{x} = \frac{17}{15}$$

13. The length of the third side of the triangle is

$$x^2 = 3^2 - 1^2 = 8$$

$$x = 2\sqrt{2}.$$

Therefore, we have the following.

$$\sin \theta = \frac{1}{3} \qquad \csc \theta = 3$$

$$\cos \theta = \frac{2\sqrt{2}}{3} \qquad \sec \theta = \frac{3}{2\sqrt{2}}$$

$$\tan \theta = \frac{1}{2\sqrt{2}} \qquad \cot \theta = 2\sqrt{2}$$

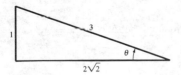

15. Since $x = 1$ and $r = 2$, the length of the opposite side is
$y = \sqrt{2^2 - 1^2} = \sqrt{3}$. Therefore, we have the following.

$$\sin \theta = \frac{\sqrt{3}}{2} \qquad \csc \theta = \frac{2\sqrt{3}}{3}$$

$$\cos \theta = \frac{1}{2} \qquad \sec \theta = 2$$

$$\tan \theta = \sqrt{3} \qquad \cot \theta = \frac{\sqrt{3}}{3}$$

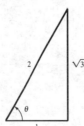

17. Since $x = 1$ and $y = 3$, the length of the hypotenuse is
$r = \sqrt{1^2 + 3^2} = \sqrt{10}$. Therefore, we have the following.

$$\sin \theta = \frac{3\sqrt{10}}{10} \qquad \csc \theta = \frac{\sqrt{10}}{3}$$

$$\cos \theta = \frac{\sqrt{10}}{10} \qquad \sec \theta = \sqrt{10}$$

$$\tan \theta = 3 \qquad \cot \theta = \frac{1}{3}$$

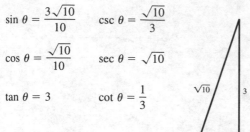

19. Since the sine is negative and the cosine is positive, θ must lie in Quadrant IV.

21. Since the sine is positive and the secant is positive, θ must lie in Quadrant I.

23. Since the cosecant is positive and the tangent is negative, θ must lie in Quadrant II.

25. (a) $\sin 60° = \dfrac{\sqrt{3}}{2}$ (b) $\sin\left(\dfrac{-2\pi}{3}\right) = -\dfrac{\sqrt{3}}{2}$

   $\cos 60° = \dfrac{1}{2}$ $\cos\left(\dfrac{-2\pi}{3}\right) = -\dfrac{1}{2}$

   $\tan 60° = \sqrt{3}$ $\tan\left(\dfrac{-2\pi}{3}\right) = \sqrt{3}$

27. (a) $\sin\left(-\dfrac{\pi}{6}\right) = -\dfrac{1}{2}$ (b) $\sin 150° = \dfrac{1}{2}$

   $\cos\left(-\dfrac{\pi}{6}\right) = \dfrac{\sqrt{3}}{2}$ $\cos 150° = -\dfrac{\sqrt{3}}{2}$

   $\tan\left(-\dfrac{\pi}{6}\right) = -\dfrac{\sqrt{3}}{3}$ $\tan 150° = -\dfrac{\sqrt{3}}{3}$

29. (a) $\sin 225° = -\dfrac{\sqrt{2}}{2}$ (b) $\sin(-225°) = \dfrac{\sqrt{2}}{2}$

   $\cos 225° = -\dfrac{\sqrt{2}}{2}$ $\cos(-225°) = -\dfrac{\sqrt{2}}{2}$

   $\tan 225° = 1$ $\tan(-225°) = -1$

31. (a) $\sin 750° = \dfrac{1}{2}$ (b) $\sin 510° = \dfrac{1}{2}$

   $\cos 750° = \dfrac{\sqrt{3}}{2}$ $\cos 510° = -\dfrac{\sqrt{3}}{2}$

   $\tan 750° = \dfrac{\sqrt{3}}{3}$ $\tan 510° = -\dfrac{\sqrt{3}}{3}$

33. (a) $\sin 12° \approx 0.2079$ (b) $\csc 12° \approx 4.8097$

35. (a) $\tan\left(\dfrac{\pi}{9}\right) \approx 0.3640$ (b) $\tan\left(\dfrac{10\pi}{9}\right) \approx 0.3640$

37. (a) $\cos(-110°) \approx -0.3420$

   (b) $\cos 250° \approx -0.3420$

39. (a) $\csc 2.62 = \dfrac{1}{\sin 2.62} \approx 2.0070$

   (b) $\csc 150° = \dfrac{1}{\sin 150°} = 2.0000$

41. (a) $\theta = \dfrac{\pi}{6}$ or $\theta = \dfrac{5\pi}{6}$ (b) $\theta = \dfrac{7\pi}{6}$ or $\theta = \dfrac{11\pi}{6}$

43. (a) $\theta = \dfrac{\pi}{3}$ or $\theta = \dfrac{2\pi}{3}$ (b) $\theta = \dfrac{3\pi}{4}$ or $\theta = \dfrac{7\pi}{4}$

45. (a) $\theta = \dfrac{3\pi}{4}$ or $\theta = \dfrac{7\pi}{4}$ (b) $\theta = \dfrac{5\pi}{6}$ or $\theta = \dfrac{11\pi}{6}$

47. Solving for $\sin \theta$ produces the following.

$$\sin \theta = \pm\frac{\sqrt{2}}{2}$$

$$\theta = \frac{\pi}{4}, \frac{3\pi}{4}, \frac{5\pi}{4}, \frac{7\pi}{4}$$

49. $\tan \theta(\tan \theta - 1) = 0$

$\qquad \tan \theta = 0 \qquad$ or $\quad \tan \theta = 1$

$\qquad \theta = 0, \pi, 2\pi \qquad \theta = \dfrac{\pi}{4}, \dfrac{5\pi}{4}$

51. $\qquad \sin 2\theta - \cos \theta = 0$

$2 \sin \theta \cos \theta - \cos \theta = 0$

$\cos \theta(2 \sin \theta - 1) = 0$

$\cos \theta = 0 \qquad$ or $\quad 2 \sin \theta = 1$

$\theta = \dfrac{\pi}{2}, \dfrac{3\pi}{2} \qquad \theta = \dfrac{\pi}{6}, \dfrac{5\pi}{6}$

53. Dividing both sides by $\cos \theta$ produces the following.

$\qquad \tan \theta = 1$

$\qquad \theta = \dfrac{\pi}{4}, \dfrac{5\pi}{4}$

55. Using the identity $\cos^2 \theta = 1 - \sin^2 \theta$ produces the following.

$\qquad 1 - \sin^2 \theta + \sin \theta = 1$

$\qquad \sin^2 \theta - \sin \theta = 0$

$\qquad \sin \theta(\sin \theta - 1) = 0$

$\sin \theta = 0 \qquad$ or $\quad \sin \theta = 1$

$\theta = 0, \pi, 2\pi \qquad \theta = \dfrac{\pi}{2}$

57. Since

$\qquad \tan 30° = \dfrac{1}{\sqrt{3}} = \dfrac{y}{100}$

it follows that

$\qquad y = \dfrac{100}{\sqrt{3}} = \dfrac{100\sqrt{3}}{3}.$

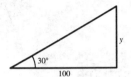

59. Since

$\qquad \cot 60° = \dfrac{1}{\sqrt{3}} = \dfrac{x}{25}$

it follows that

$\qquad x = \dfrac{25}{\sqrt{3}} = \dfrac{25\sqrt{3}}{3}.$

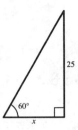

61. Since

$\qquad \sin 40° = \dfrac{10}{r} \approx 0.6428$

it follows that

$\qquad r = \dfrac{10}{0.6428} \approx 15.5572.$

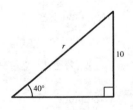

63. Let h be the height of the ladder. Then

$\qquad \sin 75° = \dfrac{h}{20} \approx 0.9659$

and $h = 20(0.9659) \approx 19.3185$ feet.

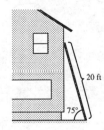

65. Let x be the distance from the shore. Then

$\qquad \cot 3° = \dfrac{x}{150} \approx 19.0811$

and

$\qquad x = 150(19.0811) \approx 2862.2$ feet.

67. (a) 10:00 P.M.: $T(0) = 98.6 + 4 \cos 0 = 98.6 + 4 = 102.6°$

(b) 4:00 A.M.: $T(6) = 98.6 + 4 \cos\left(\dfrac{6\pi}{36}\right) = 98.6 + 4\left(\dfrac{\sqrt{3}}{2}\right) \approx 102.1°$

(c) 10:00 A.M.: $T(12) = 98.6 + 4 \cos\left(\dfrac{12\pi}{36}\right) = 98.6 + 4\left(\dfrac{1}{2}\right) = 100.6°$

The temperature returns to normal when

$$98.6 = 98.6 + 4 \cos\left(\dfrac{\pi t}{36}\right)$$

$$0 = \cos\left(\dfrac{\pi t}{36}\right)$$

$$\dfrac{\pi}{2} = \dfrac{\pi t}{36}$$

$$t = 18 \text{ or } 4 \text{ P.M.}$$

69.

x	0	2	4	6	8	10
$f(x)$	0	2.7021	2.7756	1.2244	1.2979	4

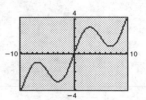

Section 8.3   Graphs of Trigonometric Functions

1. Period: $\dfrac{2\pi}{2} = \pi$

Amplitude: 2

3. Period: $\dfrac{2\pi}{1/2} = 4\pi$

Amplitude: $\dfrac{3}{2}$

5. Period: $\dfrac{2\pi}{\pi} = 2$

Amplitude: $\dfrac{1}{2}$

7. Period: $\dfrac{2\pi}{1} = 2\pi$

Amplitude: 2

9. Period: $\dfrac{2\pi}{10} = \dfrac{\pi}{5}$

Amplitude: 2

11. Period: $\dfrac{2\pi}{2/3} = 3\pi$

Amplitude: $\dfrac{1}{2}$

13. Period: $\dfrac{2\pi}{4\pi} = \dfrac{1}{2}$

Amplitude: 3

15. Period: $\dfrac{\pi}{2}$

17. Period: $\dfrac{2\pi}{5}$

19. Period: $\dfrac{\pi}{\pi/6} = 6$

21. The graph of this function has a period of π and matches graph (c).

23. The graph of this function has a period of 2 and matches graph (f).

25. The graph of this function has a period of 4π and matches graph (b).

27. Period: 4π

Amplitude: 1

x-intercepts: $(0, 0), (2\pi, 0), (4\pi, 0)$

Maximum: $(\pi, 1)$

Minimum: $(3\pi, -1)$

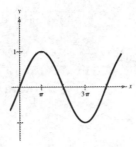

29. Period: 3π

Amplitude: 2

x-intercepts: $\left(\dfrac{3\pi}{4}, 0\right), \left(\dfrac{9\pi}{4}, 0\right), \left(\dfrac{15\pi}{4}, 0\right)$

Maxima: $(0, 2), (3\pi, 2)$

Minimum: $\left(\dfrac{3\pi}{2}, -2\right)$

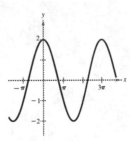

31. Period: $\dfrac{\pi}{3}$

Amplitude: 2

x-intercepts: $(0, 0), \left(\dfrac{\pi}{6}, 0\right), \left(\dfrac{\pi}{3}, 0\right), \left(\dfrac{\pi}{2}, 0\right), \left(\dfrac{2\pi}{3}, 0\right)$

Maxima: $\left(\dfrac{\pi}{4}, 2\right), \left(\dfrac{7\pi}{4}, 2\right)$

Minima: $\left(\dfrac{\pi}{12}, -2\right), \left(\dfrac{5\pi}{12}, -2\right)$

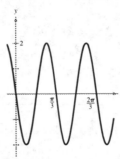

33. Period: 1

Amplitude: 1

x-intercepts: $\left(-\tfrac{3}{4}, 0\right), \left(-\tfrac{1}{4}, 0\right), \left(\tfrac{1}{4}, 0\right), \left(\tfrac{3}{4}, 0\right)$

Maxima: $(-1, 1), (0, 1), (1, 1)$

Minima: $\left(-\tfrac{1}{2}, -1\right), \left(\tfrac{1}{2}, 1\right)$

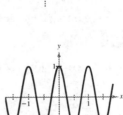

35. Period: 3

Amplitude: 1

x-intercepts: $(0, 0), \left(\tfrac{3}{2}, 0\right), (3, 0)$

Maximum: $\left(\tfrac{9}{4}, 1\right)$

Minimum: $\left(\tfrac{3}{4}, -1\right)$

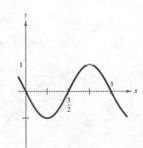

37. Period: π

x-intercepts: $(0, 0), (\pi, 0)$

Asymptotes: $x = -\dfrac{\pi}{2}, x = \dfrac{\pi}{2}, x = \dfrac{3\pi}{2}$

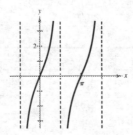

39. Period: $\dfrac{\pi}{2}$

x-intercepts: $\left(-\dfrac{3\pi}{4}, 0\right), \left(-\dfrac{\pi}{4}, 0\right), \left(\dfrac{\pi}{4}, 0\right), \left(\dfrac{3\pi}{4}, 0\right)$

Asymptotes: $x = -\dfrac{\pi}{2}, x = 0, x = \dfrac{\pi}{2}$

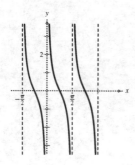

41. Period: 3π

Asymptotes: $x = 0, x = \dfrac{2\pi}{2}, 3\pi$

Relative minimum: $\left(\dfrac{3\pi}{4}, 1\right)$

Relative maximum: $\left(\dfrac{9\pi}{4}, -1\right)$

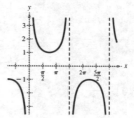

43. Period: π

Asymptotes: $x = -\dfrac{\pi}{4}, x = \dfrac{\pi}{4}, x = \dfrac{3\pi}{4}, x = \dfrac{5\pi}{4}, x = \dfrac{7\pi}{4}$

Relative minima: $(0, 2), (\pi, 2)$

Relative maxima: $\left(\dfrac{\pi}{2}, -2\right), \left(\dfrac{3\pi}{2}, -2\right)$

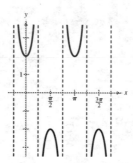

45. Period: 1

Asymptotes: $x = -\dfrac{3}{2}, x = -1, x = -\dfrac{1}{2}, x = 0, x = \dfrac{1}{2}, x = 1, x = \dfrac{3}{2}$

Relative minima: $\left(-\dfrac{7}{4}, 1\right), \left(-\dfrac{3}{4}, 1\right), \left(\dfrac{1}{4}, 1\right), \left(\dfrac{5}{4}, 1\right)$

Relative maxima: $\left(-\dfrac{5}{4}, -1\right), \left(-\dfrac{1}{4}, -1\right), \left(\dfrac{3}{4}, -1\right), \left(\dfrac{7}{4}, -1\right)$

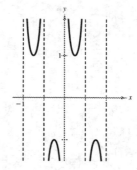

47.

x	-0.1	-0.01	-0.001	0.001	0.01	0.1
$f(x)$	-0.05	-0.005	-0.0005	0.0005	0.005	0.05

From this table, we estimate that $\displaystyle\lim_{x\to 0} \dfrac{1 - \cos x}{x} = 0$.

49.

x	-0.1	-0.01	-0.001	0.001	0.01	0.1
$f(x)$	0.2	0.20	0.200	0.200	0.20	0.2

From this table, we estimate that $\displaystyle\lim_{x\to 0} \dfrac{\sin x}{5x} = \dfrac{1}{5}$.

51. False, the amplitude is 3 (must be positive).

53. True, $\lim_{x \to 0} \dfrac{\sin 5x}{3x} - \dfrac{5}{3}\left(\lim_{x \to 0} \dfrac{\sin 5x}{5x}\right) = \dfrac{5}{3}(1) = \dfrac{5}{3}.$

55. (a) Period $= \dfrac{2\pi}{\pi/3} = 6$ seconds

    (b) Cycles per minute $= \dfrac{60}{6} = 10$

    (c)

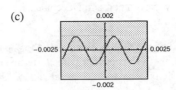

57. (a) Period $= \dfrac{2\pi}{880\pi} = \dfrac{1}{440}$

    (b) Frequency $= \dfrac{1}{\text{period}} = 440$

    (c)

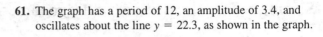

59. $P = 8000 + 2500 \sin \dfrac{2\pi t}{24}$

    $p = 12{,}000 + 4000 \cos \dfrac{2\pi t}{24}$

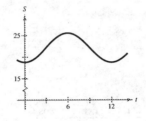

61. The graph has a period of 12, an amplitude of 3.4, and oscillates about the line $y = 22.3$, as shown in the graph.

63. December 25, 2000 is the 13,673rd day of this person's life.

Days in 1964:	165
9 leap years:	+ 9(366)
27 non-leap years:	+ 27(365)
Days in 2000:	+ 360
	13,674
	− 1
	13,673

(Do not count both first and last day.)

Therefore,

$$P(13{,}673) = \sin \frac{2\pi(13{,}673)}{23} \approx 0.1362$$

$$E(13{,}673) = \sin \frac{2\pi(13{,}673)}{28} \approx -0.9010$$

$$I(13{,}673) = \sin \frac{2\pi(13{,}673)}{33} \approx -0.8660.$$

65.

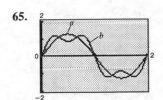

67. $\lim_{x \to 0} \dfrac{\sin x}{x} = 1$

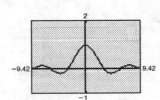

69. (a)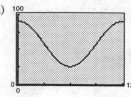

(b) Sales exceed 75,000 units during the months of January, February, November, and December.

71. Answers will vary.

73. When the population of hares (prey) increases, the population of lynxes (predator) increases as well, because there is more food. At some point, the lynxes devour hares faster than the hares can reproduce and the hare population decreases. As food becomes scarce, the lynx population decreases as well. At some point, there are only a few lynx, so the hares can begin to increase again.

Section 8.4   Derivatives of Trigonometric Functions

1. $y' = 2x + \sin x$

3. $y' = -3 \cos x$

5. $f'(x) = \dfrac{2}{\sqrt{x}} - 3 \sin x$

7. $f'(t) = -t^2 \sin t + 2t \cos t$

9. $g'(t) = \dfrac{t(-\sin t) - (\cos t)(1)}{t^2} = -\dfrac{t \sin t + \cos t}{t^2}$

11. $y' = \sec^2 x + 2x$

13. $y' = 5x(\sec x \tan x) + 5 \sec x$
$= 5 \sec x(x \tan x + 1)$

15. $y' = (-\sin 3x)(3) = -3 \sin 3x$

17. $y' = (\cos \pi x)\pi = \pi \cos \pi x$

19. $y' = x\left(\dfrac{1}{x^2}\right)\left(-\cos \dfrac{1}{x}\right) + \left(\sin \dfrac{1}{x}\right)(1)$
$= \sin \dfrac{1}{x} - \dfrac{1}{x} \cos \dfrac{1}{x}$

21. $y' = 12 \sec^2 4x$

23. $y = (\cos x)^2$
$y' = 2(\cos x)(-\sin x)$
$= -2 \cos x \sin x$
$= -\sin 2x$

25. $y' = -2 \cos x \sin x - 2 \cos x \sin x$
$= -4 \cos x \sin x$
$= -2 \sin 2x$

27. $y' = \dfrac{\cos x}{\sin x} = \cot x$

29. $y' = \dfrac{1}{\csc x^2 - \cot x^2}(-2x \csc x^2 \cot x^2 + 2x \csc^2 x^2)$
$= \dfrac{2x \csc x^2(\csc x^2 - \cot x^2)}{\csc x^2 - \cot x^2}$
$= 2x \csc x^2$

31. $y' = \sec^2 x - 1 = \tan^2 x$

33. $y' = \dfrac{1}{\sin^2 x} 2 \sin x \cos x$

$\qquad = \dfrac{2 \cos x}{\sin x}$

$\qquad = 2 \cot x$

35. $y' = \sec^2 x, \qquad y'\left(-\dfrac{\pi}{4}\right) = \left(\dfrac{2}{\sqrt{2}}\right)^2 = 2$

$\qquad y - (-1) = 2\left[x - \left(-\dfrac{\pi}{4}\right)\right]$

$\qquad\qquad y = 2x - 1 + \dfrac{\pi}{2}$

37. $y' = 4 \cos 4x, \qquad y'(\pi) = 4$

$\qquad y - 0 = 4(x - \pi)$

$\qquad\quad y = 4x - 4\pi$

39. $y = \dfrac{\cos x}{\sin x} = \cot x$

$\qquad y' = -\csc^2 x, \qquad y'\left(\dfrac{3\pi}{4}\right) = -2$

$\qquad\quad y + 1 = -2\left(x - \dfrac{3\pi}{4}\right)$

$\qquad\qquad\quad y = -2x - 1 + \dfrac{3\pi}{2}$

41. $y' = \dfrac{1}{\cot x}(-\csc^2 x), \qquad y'\left(\dfrac{\pi}{4}\right) = -2$

$\qquad y - 0 = -2\left(x - \dfrac{\pi}{4}\right)$

$\qquad\quad y = -2x + \dfrac{\pi}{2}$

43. $\qquad \sin x + \cos 2y = 1$

$\qquad \cos x - 2 \sin 2y \dfrac{dy}{dx} = 0$

$\qquad\qquad\qquad \dfrac{dy}{dx} = \dfrac{\cos x}{2 \sin 2y}$

At $\left(\dfrac{\pi}{2}, \dfrac{\pi}{4}\right)$, we have $\dfrac{dy}{dx} = 0$.

45. Since $y' = 2 \cos x - 3 \sin x$ and $y'' = -2 \sin x + 3 \cos x$, it follows that $y'' + y = 0$.

47. Since $y' = -2 \sin 2x + 2 \cos 2x$ and $y'' = -4 \cos 2x - 4 \sin 2x$, it follows that $y'' + 4y = 0$.

49. Since

$$y' = \dfrac{5}{4} \cos\left(\dfrac{5x}{4}\right),$$

the slope of the tangent line at $(0, 0)$ is $\dfrac{5}{4}$. There is one complete cycle of the graph in the interval $[0, 2\pi]$.

51. Since $y' = 2 \cos 2x$, the slope of the tangent line at $(0, 0)$ is 2. There are two complete cycles of the graph in the interval $[0, 2\pi]$.

53. Since $y' = \cos x$, the slope of the tangent line at $(0, 0)$ is 1. There is one complete cycle of the graph in the interval $[0, 2\pi]$.

55. The first derivative is zero when

$\qquad 2 \cos x + 2 \cos 2x = 0$

$\qquad 2[\cos x + 2 \cos^2 x - 1] = 0$

$\qquad 2 \cos^2 x + \cos x - 1 = 0$

$\qquad (2 \cos x - 1)(\cos x + 1) = 0$

$\qquad\qquad \cos x = \dfrac{1}{2}$ or $\cos x = -1$.

Critical numbers: $x = \dfrac{\pi}{3}, \dfrac{5\pi}{3}, \pi$

Relative maximum: $\left(\dfrac{\pi}{3}, \dfrac{3\sqrt{3}}{2}\right)$

Relative minimum: $\left(\dfrac{5\pi}{3}, -\dfrac{3\sqrt{3}}{2}\right)$

57. The first derivative is zero when

$\qquad 1 - 2 \cos x = 0$

$\qquad\qquad \cos x = \dfrac{1}{2}.$

Critical numbers: $x = \dfrac{\pi}{3}, \dfrac{5\pi}{3}, \dfrac{7\pi}{3}, \dfrac{11\pi}{3}$

Relative minimum: $\left(\dfrac{\pi}{3}, \dfrac{\pi}{3} - \sqrt{3}\right)$

Relative maximum: $\left(\dfrac{5\pi}{3}, \dfrac{5\pi}{3} + \sqrt{3}\right)$

59. The first derivative is zero when

$$f'(x) = e^{-x}(-\sin x) - e^{-x} \cos x = 0$$

$$e^{-x}(\sin x + \cos x) = 0$$

$$\tan x = -1$$

Critical numbers: $x = \dfrac{3\pi}{4}, \dfrac{7\pi}{4}$

Relative maximum: $\left(\dfrac{7\pi}{4}, 0.0029\right)$

Relative minimum: $\left(\dfrac{3\pi}{4}, -0.0670\right)$

61. $h(t) = 0.20t + 0.03 \sin 2\pi t$

$h'(t) = 0.20 + 0.06\pi \cos 2\pi t$

$h''(t) = -0.12\pi^2 \sin 2\pi t = 0$

$t = 0, \frac{1}{2}$

(a) $h'(t)$ is maximum when $t = 0$ (midnight).

(b) $h'(t)$ is minimum when $t = \frac{1}{2}$ (noon).

63. (a)

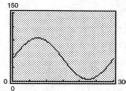

(b) $\dfrac{dh}{dx} = 45\left(\dfrac{\pi}{150}\right) \cos \dfrac{\pi x}{150} = \dfrac{3\pi}{10} \cos \dfrac{\pi x}{150}$

For $x = 50$, $h'(50) = 0.15\pi$.

For $x = 150$, $h'(150) = -0.3\pi$.

For $x = 200$, $h'(200) = -0.15\pi$.

For $x = 250$, $h'(250) = 0.15\pi$.

dh/dx is the rate the ride is rising or falling as it moves horizontally from its starting point.

(c) $\dfrac{3\pi}{10} \cos \dfrac{\pi x}{150} = 0 \Rightarrow \dfrac{\pi x}{150} = \dfrac{\pi}{2} \Rightarrow x = 75$ and $h(75) = 95$ ft

Lowest: $\dfrac{\pi x}{150} = \dfrac{3\pi}{2} \Rightarrow x = 225$ and $h(225) = 5$ ft

(d) dh/dx is greatest when

$$\cos \dfrac{\pi x}{150} = \pm 1 \text{ or } \dfrac{\pi x}{150} = \pi \Rightarrow x = 150 \text{ ft.}$$

65. $f'(x) = 2 \sec x(\sec x \tan x) = 2 \sec^2 x \tan x$

$g'(x) = 2 \tan x \sec^2 x$

67. (a) $f(t) = t^2 \sin t$

$f'(t) = t^2 \cos t + 2t \sin t$

$= t(t \cos t + 2 \sin t)$

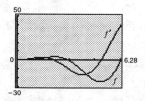

(b) $f'(t) = 0$ when $t = 0$, $t \approx 2.289$, and $t \approx 5.087$.

(c)

Interval	(0, 2.289)	(2.289, 5.087)	(5.087, 2π)
Sign of f'	+	−	+
Conclusion	f is increasing.	f is decreasing.	f is increasing.

69. (a) $f(x) = \sin x - \frac{1}{3}\sin 3x + \frac{1}{5}\sin 5x$

$f'(x) = \cos x - \cos 3x + \cos 5x$

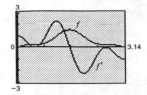

(b) $f'(x) = 0$ when $x \approx 0.524$, $x \approx 1.571$, and $x \approx 2.618$.

(c)

Interval	$(0, 0.524)$	$(0.524, 1.571)$	$(1.571, 2.618)$	$(2.618, \pi)$
Sign of f'	+	+	−	−
Conclusion	f is increasing.	f is increasing.	f is decreasing.	f is decreasing.

71. (a) $f(x) = \sqrt{2x}\,\sin x$

$f'(x) = \sqrt{2x}\cos x + \dfrac{\sin x}{\sqrt{2x}}$

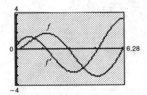

(b) $f'(x) = 0$ when $x \approx 1.837$ and $x \approx 4.816$.

(c)

Interval	$(0, 1.837)$	$(1.837, 4.816)$	$(4.816, 2\pi)$
Sign of f'	+	−	+
Conclusion	f is increasing.	f is decreasing.	f is increasing.

73. Relative maximum: $(4.49, -4.60)$

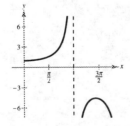

75. Relative maximum: $(1.27, 0.07)$

Relative minimum: $(3.38, -1.18)$

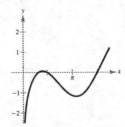

77. Relative maximum: $(3.96, 1)$

Section 8.5   Integrals of Trigonometric Functions

1. $\displaystyle\int (2 \sin x + 3 \cos x)\, dx = -2 \cos x + 3 \sin x + C$

3. $\displaystyle\int (1 - \csc t \cot t)\, dt = t + \csc t + C$

5. $\displaystyle\int (\csc^2 \theta - \cos \theta)\, d\theta = -\cot \theta - \sin \theta + C$

7. $\displaystyle\int \sin 2x\, dx = \frac{1}{2} \int 2 \sin 2x\, dx = -\frac{1}{2} \cos 2x + C$

9. $\displaystyle\int x \cos x^2\, dx = \frac{1}{2} \int 2x \cos x^2\, dx = \frac{1}{2} \sin x^2 + C$

11. $\displaystyle\int \sec^2 \frac{x}{2}\, dx = 2 \int \frac{1}{2} \sec^2 \frac{x}{2}\, dx = 2 \tan \frac{x}{2} + C$

13. $\displaystyle\int \tan 3x\, dx = \frac{1}{3} \int 3 \tan 3x\, dx = -\frac{1}{3} \ln|\cos 3x| + C$

15. $\displaystyle\int \tan^3 x \sec^2 x\, dx = \frac{\tan^4 x}{4} + C$

17. $\displaystyle\int \cot \pi x\, dx = \frac{1}{\pi} \int \pi \cot \pi x\, dx = \frac{1}{\pi} \ln|\sin \pi x| + C$

19. $\displaystyle\int \csc 2x\, dx = \frac{1}{2} \int 2 \csc 2x\, dx$

$$= \frac{1}{2} \ln|\csc 2x - \cot 2x| + C$$

21. $\displaystyle\int \frac{\sec^2 2x}{\tan 2x}\, dx = \frac{1}{2} \int \frac{1}{\tan 2x} (2 \sec^2 2x)\, dx$

$$= \frac{1}{2} \ln|\tan 2x| + C$$

23. $\displaystyle\int \frac{\sec x \tan x}{\sec x - 1}\, dx = \ln|\sec x - 1| + C$

25. $\displaystyle\int \frac{\sin x}{1 + \cos x}\, dx = -\ln|1 + \cos x| + C$

27. $\displaystyle\int \frac{\csc^2 x}{\cot^3 x}\, dx = -\int \cot^{-3} x (-\csc^2 x)\, dx$

$$= -\frac{\cot^{-2} x}{-2} + C$$

$$= \frac{1}{2} \tan^2 x + C$$

29. $\displaystyle\int e^x \sin e^x\, dx = -\cos e^x + C$

31. $\displaystyle\int e^{-x} \tan e^{-x}\, dx = -\int (-e^{-x}) \tan e^{-x}\, dx$

$$= \ln|\cos e^{-x}| + C$$

33. $\displaystyle\int (\sin 2x + \cos 2x)^2\, dx = \int (\sin^2 2x + 2 \sin 2x \cos 2x + \cos^2 2x)\, dx$

$$= \int (1 + \sin 4x)\, dx$$

$$= x - \frac{1}{4} \cos 4x + C$$

35. Using integration by parts, we let $u = x$ and $dv = \cos x\, dx$.
Then $du = dx$ and $v = \sin x$.

$$\int x \cos x\, dx = x \sin x - \int \sin x\, dx$$

$$= x \sin x + \cos x + C$$

37. Using integration by parts, we let $u = x$ and $dv = \sec^2 x\, dx$.
Then $du = dx$ and $v = \tan x$.

$$\int x \sec^2 x\, dx = x \tan x - \int \tan x\, dx$$

$$= x \tan x + \ln|\cos x| + C$$

39. $\displaystyle\int_0^{\pi/4} \cos\frac{4x}{3}\,dx = \frac{3}{4}\sin\frac{4x}{3}\Big]_0^{\pi/4}$

$$= \frac{3}{4}\left(\sin\frac{\pi}{3}\right)$$

$$= \frac{3\sqrt{3}}{8}$$

$$\approx 0.6495$$

41. $\displaystyle\int_{\pi/2}^{2\pi/3} \sec^2\frac{x}{2}\,dx = 2\tan\frac{x}{2}\Big]_{\pi/2}^{2\pi/3}$

$$= 2\left(\sqrt{3} - 1\right)$$

$$\approx 1.4641$$

43. $\displaystyle\int_{\pi/12}^{\pi/4} \csc 2x \cot 2x\,dx = -\frac{1}{2}\csc 2x\Big]_{\pi/12}^{\pi/4}$

$$= -\frac{1}{2}\left[\csc\frac{\pi}{2} - \csc\frac{\pi}{6}\right]$$

$$= -\frac{1}{2}[1 - 2]$$

$$= \frac{1}{2}$$

45. $\displaystyle\int_0^1 \tan(1-x)\,dx = \ln|\cos(1-x)|\,\Big]_0^1$

$$= \ln(\cos 0) - \ln(\cos 1)$$

$$\approx 0.6156$$

47. $\displaystyle\int_0^{2\pi} \cos\frac{x}{4}\,dx = 4\sin\frac{x}{4}\Big]_0^{2\pi} = 4$

49. Area $= \displaystyle\int_0^{\pi} (x + \sin x)\,dx$

$$= \left[\frac{x^2}{2} - \cos x\right]_0^{\pi}$$

$$= \frac{\pi^2}{2} + 2$$

$$\approx 6.9348 \text{ square units}$$

51. Area $= \displaystyle\int_0^{\pi} (\sin x + \cos 2x)\,dx$

$$= \left[-\cos x + \frac{1}{2}\sin 2x\right]_0^{\pi}$$

$$= 2 \text{ square units}$$

53. $V = \pi\displaystyle\int_0^{\pi/4} \sec^2 x\,dx$

$$= \pi\tan x\Big]_0^{\pi/4}$$

$$= \pi \text{ cubic units}$$

55. (a) $\dfrac{\pi/2}{8}\left[f(0) + 2f\left(\dfrac{\pi}{8}\right) + 2f\left(\dfrac{\pi}{4}\right) + 2f\left(\dfrac{3\pi}{8}\right) + f\left(\dfrac{\pi}{2}\right)\right] \approx 1.3655$

(b) $\dfrac{\pi/2}{12}\left[f(0) + 4f\left(\dfrac{\pi}{8}\right) + 2f\left(\dfrac{\pi}{4}\right) + 4f\left(\dfrac{3\pi}{8}\right) + f\left(\dfrac{\pi}{2}\right)\right] \approx 1.3708$

57. (a) $\dfrac{1}{3}\displaystyle\int_0^3 \left[217 + 13\cos\dfrac{\pi(t-3)}{6}\right]dt = \dfrac{1}{3}\left[217t + \dfrac{78}{\pi}\sin\dfrac{\pi(t-3)}{6}\right]_0^3 = \dfrac{1}{3}\left(651 + \dfrac{78}{\pi}\right) \approx 225.28 \text{ million barrels}$

(b) $\dfrac{1}{3}\displaystyle\int_3^6 \left[217 + 13\cos\dfrac{\pi(t-3)}{6}\right]dt = \dfrac{1}{3}\left[217t + \dfrac{78}{\pi}\sin\dfrac{\pi(t-3)}{6}\right]_3^6 = \dfrac{1}{3}\left(1302 + \dfrac{78}{\pi} - 651\right) \approx 225.28 \text{ million barrels}$

(c) $\dfrac{1}{12}\displaystyle\int_0^{12} \left[217 + 13\cos\dfrac{\pi(t-3)}{6}\right]dt = \dfrac{1}{12}\left[217t + \dfrac{78}{\pi}\sin\dfrac{\pi(t-3)}{6}\right]_0^{12} = \dfrac{1}{12}\left(2604 - \dfrac{78}{\pi} + \dfrac{78}{\pi}\right) \approx 217 \text{ million barrels}$

59. Total $= \displaystyle\int_0^{12} [4.65\sin(0.30x + 2.78) + 5.75]\,dt \approx 39.07 \text{ inches}$

61. (a) $C = 0.1 \int_8^{20} 12 \sin \frac{\pi(t-8)}{12} \, dt = -\frac{14.4}{\pi} \cos \frac{\pi(t-8)}{12} \Big]_8^{20} = -\frac{14.4}{\pi}(-1-1) \approx \$9.17$

(b) $C = 0.1 \int_{10}^{18} \left[12 \sin \frac{\pi(t-8)}{12} - 6 \right] dt = \left[-\frac{14.4}{\pi} \cos \frac{\pi(t-8)}{12} - 0.6t \right]_{10}^{18} \approx \$3.14$

Savings $\approx 9.17 - 3.14 = \$6.03$

63. $V = \int_0^2 1.75 \sin \frac{\pi t}{2} \, dt = -\frac{3.5}{\pi} \cos \frac{\pi t}{2} \Big]_0^2 \approx 2.2282$ liters   **65.** $\int_0^{\pi/2} \sqrt{x} \sin x \, dx \approx 0.9777$

Capacity increases by $2.2282 - 1.7189 = 0.5093$ liter.

67. $\int_0^{\pi} \sqrt{1 + \cos^2 x} \, dx \approx 3.8202$

Section 8.6   L'Hôpital's Rule

1. Yes, $\frac{0}{0}$.

3. No, $\frac{4}{\infty} \to 0$.

5. Yes, $\frac{\infty}{\infty}$.

7.

x	-0.1	-0.01	-0.001	0	0.001	0.01	0.1
$f(x)$	-0.35	-0.335	-0.3335		-0.3332	-0.332	-0.32

$\lim_{x \to 0} \frac{e^{-x} - 1}{3x} = \frac{-1}{3}$

9.

x	-0.1	-0.01	-0.001	0	0.001	0.01	0.1
$f(x)$	0.1997	0.2	0.2		0.2	0.2	0.1997

$\lim_{x \to 0} \frac{\sin x}{5x} = \frac{1}{5}$

11. $\lim_{x \to \infty} \frac{\ln x}{x^2} = \infty$

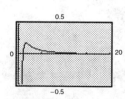

13. $\lim_{x \to 2} \frac{x^2 - x - 2}{x^2 - 5x + 6} = -3$

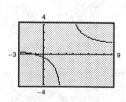

15. $\lim_{x \to 0} \frac{e^{-x} - 1}{x} = \lim_{x \to 0} \frac{-e^{-x}}{1} = -1$ (See Exercise 7.)

17. $\lim_{x \to 0} \frac{\sin x}{5x} = \lim_{x \to 0} \frac{\cos x}{5} = \frac{1}{5}$ (See Exercise 9.)

19. $\lim_{x \to \infty} \frac{\ln x}{x^2} = \lim_{x \to \infty} \frac{1/x}{2x} = 0$

21. $\lim_{x \to 2} \frac{x^2 - x - 2}{x^2 - 5x + 6} = \lim_{x \to 2} \frac{2x - 1}{2x - 5} = \frac{3}{-1} = -3$

(See Exercise 13.)

23. $\lim\limits_{x\to 0} \dfrac{2x + 1}{x} \dfrac{e^x}{} = \lim\limits_{x\to 0} \dfrac{2 - e^x}{1} = 1$

25. $\lim\limits_{x\to\infty} \dfrac{\ln x}{e^x} = \lim\limits_{x\to\infty} \dfrac{1/x}{e^x} = 0$

27. $\lim\limits_{x\to\infty} \dfrac{4x^2 + 2x - 1}{3x^2 - 7} = \lim\limits_{x\to\infty} \dfrac{8x + 2}{6x} = \lim\limits_{x\to\infty} \dfrac{8}{6} = \dfrac{4}{3}$

29. $\lim\limits_{x\to\infty} \dfrac{1 - x}{e^x} = \lim\limits_{x\to\infty} \dfrac{-1}{e^x} = 0$

31. $\lim\limits_{x\to 1} \dfrac{\ln x}{x^2 - 1} = \lim\limits_{x\to 1} \dfrac{1/x}{2x} = \dfrac{1}{2}$

33. $\lim\limits_{x\to 0} \dfrac{\sin 2x}{\sin 5x} = \lim\limits_{x\to 0} \dfrac{2 \cos 2x}{5 \cos 5x} = \dfrac{2}{5}$

35. $\lim\limits_{x\to 0} \dfrac{\sin x}{e^x - 1} = \lim\limits_{x\to 0} \dfrac{\cos x}{e^x} = 1$

37. $\lim\limits_{x\to\infty} \dfrac{x}{\sqrt{x + 1}} = \lim\limits_{x\to\infty} \dfrac{x/x}{\sqrt{(1/x) + (1/x^2)}}$

$\qquad = \lim\limits_{x\to\infty} \dfrac{1}{\sqrt{(1/x) + (1/x)}}$

$\qquad = \infty$

39. $\lim\limits_{x\to\infty} \dfrac{e^{3x}}{x^3} = \lim\limits_{x\to\infty} \dfrac{3e^{3x}}{3x^2} = \lim\limits_{x\to\infty} \dfrac{3e^{3x}}{2x} = \lim\limits_{x\to\infty} \dfrac{9e^{3x}}{2} = \infty$

41. $\lim\limits_{x\to\infty} \dfrac{x^2 + 2x + 1}{x^2 + 3} = \lim\limits_{x\to\infty} \dfrac{2x + 2}{2x} = \lim\limits_{x\to\infty} \dfrac{2}{2} = 1$

43. $\lim\limits_{x\to -1} \dfrac{x^3 + 3x^2 - 6x - 8}{2x^3 - 3x^2 - 5x + 6} = \dfrac{0}{6} = 0$

45. $\lim\limits_{x\to 3} \dfrac{\ln(x - 2)}{x - 2} = \dfrac{\ln(1)}{1} = 0$

47. $\lim\limits_{x\to 1} \dfrac{2 \ln x}{e^x} = \dfrac{2(0)}{e} = 0$

49. $\lim\limits_{x\to\infty} \dfrac{x^2}{e^{4x}} = \lim\limits_{x\to\infty} \dfrac{2x}{4e^{4x}} = \lim\limits_{x\to\infty} \dfrac{2}{16e^{4x}} = 0$

51. $\lim\limits_{x\to\infty} \dfrac{(\ln x)^4}{x} = \lim\limits_{x\to\infty} \dfrac{4(\ln x)^3(1/x)}{1}$

$\qquad = \lim\limits_{x\to\infty} \dfrac{12(\ln x)^2(1/x)}{1}$

$\qquad = \lim\limits_{x\to\infty} \dfrac{24(\ln x)(1/x)}{1}$

$\qquad = \lim\limits_{x\to\infty} \dfrac{24(1/x)}{1}$

$\qquad = 0$

53. $\lim\limits_{x\to\infty} \dfrac{(\ln x)^n}{x^m} = 0$

(The log term eventually disappears.)

55.

x	10	10^2	10^3	10^4	10^5	10^6
$\dfrac{(\ln x)^5}{x}$	6.47	20.71	15.73	6.63	2.02	0.503

$\lim\limits_{x\to\infty} \dfrac{(\ln x)^5}{x} = 0$

57. The limit of the denominator is not 0.

59. The limit of the numerator is not 0.

61. (a)

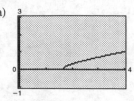

(b) $\lim\limits_{x\to 2} \dfrac{x-2}{\ln(3x-5)} = \dfrac{1}{3}$

63. (a)

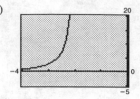

(b) $\lim\limits_{x\to -2} \dfrac{\sqrt{x^2-4}-5}{x+2}$ does not exist.

65. False, $\lim\limits_{x\to 0} \dfrac{x^2+3x-1}{x+1} = \dfrac{-1}{1} = -1.$

67. True

69. $\lim\limits_{x\to\infty} \dfrac{x}{\sqrt{x^2+1}} = \lim\limits_{x\to\infty} \dfrac{\sqrt{x^2+1}}{x} = \lim\limits_{x\to\infty} \dfrac{x}{\sqrt{x^2+1}}$

and repeated applications of L'Hôpital's Rule continue in this pattern.

$$\lim\limits_{x\infty} \dfrac{x}{\sqrt{x^2+1}} = \lim\limits_{x\infty} \dfrac{x/x}{\sqrt{(x^2/x^2)+(1/x^2)}}$$

$$= \lim\limits_{x\to\infty} \dfrac{1}{\sqrt{1+(1/x^2)}}$$

$$= 1$$

71. (a)

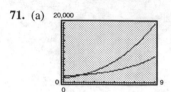

(b) For $0 \le t \le 9$, Home Depot had the larger rate of growth.

(c) Because of the term $0.1e^t$, Lowe's will ultimately have the largest rate of growth.

(d) The two curves intersect again at $t \approx 12.4$. Hence, Lowe's will surpass Home Depot in 1999.

Review Exercises for Chapter 8

1. Positive coterminal angle: $\dfrac{11\pi}{4} - 2\pi = \dfrac{3\pi}{4}$

Negative coterminal angle: $\dfrac{11\pi}{4} - 2(2\pi) = -\dfrac{5\pi}{4}$

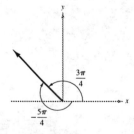

3. Positive coterminal angle: $110° + 360° = 470°$

Negative coterminal angle: $110° - 360° = -250°$

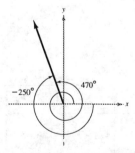

5. $210°\left(\dfrac{\pi \text{ radians}}{180°}\right) = \dfrac{7\pi}{6}$ radians

7. $-480°\left(\dfrac{\pi \text{ radians}}{180°}\right) = -\dfrac{8\pi}{3}$ radians ≈ -8.38

9. $\dfrac{7\pi}{3}\left(\dfrac{180°}{\pi}\right) = 420°$

11. $-\dfrac{3\pi}{5}\left(\dfrac{180°}{\pi}\right) = -108°$

13. $b = \sqrt{8^2 - 4^2} = \sqrt{48} = 4\sqrt{3}$

$\theta = 90° - 30° = 60°$

15. $C = 5$, $\theta = 60°$ (equilateral triangle)

$a = \sqrt{5^2 - \left(\frac{5}{2}\right)^2} = \frac{5}{2}\sqrt{3}$

17. $h = \sqrt{16^2 - 4.4^2} = \sqrt{236.64} \approx 15.38$ feet

19. The reference angle is 0.

21. The reference angle is $252° - 180° = 72°$.

23. $\cos 45° = \dfrac{\sqrt{2}}{2}$

25. $\tan \dfrac{\pi}{3} = \sqrt{3}$

27. $\sin\left(\dfrac{5\pi}{3}\right) = -\dfrac{\sqrt{3}}{2}$

29. $\sec(-180°) = 1$

31. $\tan 33° \approx 0.6494$

33. $\sin\left(-\dfrac{\pi}{9}\right) \approx -0.3420$

35. $\cos 70° = \dfrac{50}{r} \Rightarrow r = \dfrac{50}{\cos 70°} \approx 146.19$

37. $\tan 20° = \dfrac{25}{x} \Rightarrow x = \dfrac{25}{\tan 20°} \approx 68.69$

39. $2\cos x + 1 = 0$

$$\cos x = -\dfrac{1}{2}$$

$$x = \dfrac{2\pi}{3} + 2k\pi, \dfrac{4\pi}{3} + 2k\pi$$

41. $2\sin^2 x + 3\sin x + 1 = 0$

$(2\sin x + 1)(\sin x + 1) = 0$

$$\sin x = -\dfrac{1}{2} \text{ or } \sin x = -1$$

$$x = \dfrac{7\pi}{6} + 2k\pi, \dfrac{11\pi}{6} + 2k\pi, \text{ or } x = \dfrac{3\pi}{2} + 2k\pi$$

43. $\sec^2 x - \sec x - 2 = 0$

$(\sec x - 2)(\sec x + 1) = 0$

$$\sec x = 2 \text{ or } \sec x = -1$$

$$\cos x = \dfrac{1}{2} \text{ or } \cos x = -1$$

$$x = \dfrac{\pi}{3} + 2k\pi \text{ or } \dfrac{5\pi}{3} + 2k\pi \text{ or } x = \pi + 2k\pi$$

45. $h \approx 125 \tan 33° \approx 81.18$ feet

47. $y = 2 \cos 6x$

Period: $\dfrac{2\pi}{6} = \dfrac{\pi}{3}$

Amplitude: 2

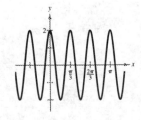

49. $y = \dfrac{1}{3} \tan x$

Period: π

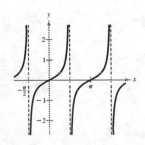

51. Period: $\dfrac{2\pi}{2/5} = 5\pi$

Amplitude: 3

x-intercepts: $(0, 0), \left(\dfrac{5\pi}{2}, 0\right), (5\pi, 0)$

Maximum: $\left(\dfrac{5\pi}{4}, 3\right)$

Minimum: $\left(\dfrac{15\pi}{4}, -3\right)$

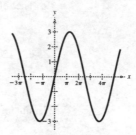

53. Period: 1

Asymptotes: $x = -\frac{1}{4}, x = \frac{1}{4}, x = \frac{3}{4}, x = \frac{5}{4}$

Maximum: $\left(\frac{1}{2}, -1\right), \left(\frac{3}{2}, -1\right)$

Minimum: $(0, 1), (1, 1)$

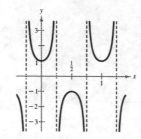

55. $S = 74 + \dfrac{3}{365}t + 40 \sin \dfrac{2\pi t}{365}$

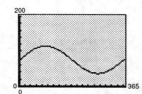

57. $y = \sin 5\pi x$

$y' = 5\pi \cos 5\pi x$

59. $y' = -x \sec^2 x - \tan x$

61. $y' = \dfrac{x^2(-\sin x) - \cos x(2x)}{(x^2)^2}$

$= \dfrac{-x \sin x - 2 \cos x}{x^3}$

63. $y' = 6 \sin 4x(4 \cos 4x) + 1$

$= 24 \sin 4x \cos 4x + 1$

$= 12 \sin 8x + 1$

65. $y' = 6 \csc^2 x(-\csc x \cot x)$

$= -6 \csc^3 x \cot x$

67. $y' = e^x(-\csc^2 x) + e^x \cot x$

$= e^x(\cot x - \csc^2 x)$

69. $y = 4 \sin 2x, \qquad (\pi, 0)$

$y' = 8 \cos 2x, \qquad y'(\pi) = 8$

$y - 0 = 8(x - \pi)$

$y = 8x - 8\pi$

71. $y = \dfrac{1}{4} \sin^2 2x, \qquad \left(\dfrac{\pi}{4}, \dfrac{1}{4}\right)$

$y' = \dfrac{1}{4} 2 \sin 2x \cdot \cos 2x(2) = \sin 2x \cos 2x, \quad y'\left(\dfrac{\pi}{4}\right) = 0$

$y - \dfrac{1}{4} = 0\left(x - \dfrac{\pi}{4}\right)$

$y = \dfrac{1}{4}$

73. $y = e^x \tan 2x, \qquad (0, 0)$

$y' = e^x(2 \sec^2 2x) + e^x \tan 2x, \qquad y'(0) = 2$

$y - 0 = 2(x - 0)$

$y = 2x$

75. Relative maximum: $(0.523, 1.128)$

Relative minimum: $(2.616, 0.443)$

77. $f(x) = \sin x(\sin x + 1)$

Relative maxima: $\left(\dfrac{\pi}{2}, 2\right), \left(\dfrac{3\pi}{2}, 0\right)$

Relative minima: $\left(\dfrac{7\pi}{6}, -\dfrac{1}{4}\right), \left(\dfrac{11\pi}{6}, -\dfrac{1}{4}\right)$

79. (a) Maximum daily sales of 114.75 thousand units occur on the 92nd day of the year.

(b) Minimum daily sales of 36.25 thousand units occur on the 273rd day of the year.

81. $\displaystyle\int (3 \sin x - 2 \cos x)\, dx = -3 \cos x - 2 \sin x + C$

83. $\displaystyle\int \sin^3 x \cos x\, dx = \dfrac{\sin^4 x}{4} + C$

85. $\displaystyle\int_0^{\pi} (1 + \sin x)\, dx = \left[x - \cos x \right]_0^{\pi}$

$$= [\pi - (-1)] - (-1)$$

$$= \pi + 2$$

87. $\displaystyle\int_{-\pi/3}^{\pi/6} 4 \sec x \tan x\, dx = 4 \sec x \Big]_{-\pi/3}^{\pi/3} = 0$

89. $\displaystyle\int_0^{\pi/2} \sin 2x\, dx = -\dfrac{1}{2} \cos 2x \Big]_0^{\pi/2} = -\dfrac{1}{2}(-1 - 1) = 1$

91. $\displaystyle\int_0^{\pi/2} (2 \sin x + \cos 3x)\, dx = \left[-2 \cos x + \dfrac{1}{3} \sin 3x \right]_0^{\pi/2}$

$$= \left(-\dfrac{1}{3}\right) - (-2)$$

$$= \dfrac{5}{3}$$

93. Average $= \dfrac{1}{12 - 0} \displaystyle\int_0^{12} \left[6.9 + \cos \dfrac{\pi(2t - 1)}{12} \right] dt$

$$= \dfrac{1}{12} \left[6.9t + \dfrac{6}{\pi} \sin \dfrac{\pi(2t - 1)}{12} \right]_0^{12}$$

$$= 6.9 \text{ quads}$$

95. $\displaystyle\int_0^{12} [3.45 \sin(0.31t + 2.51) + 3.45]\, dt \approx 21.3 \text{ inches}$

97. $\displaystyle\lim_{x \to 1} \dfrac{3x - 1}{5x + 5} = \dfrac{2}{10} = \dfrac{1}{5}$

99. $\displaystyle\lim_{x \to 1} \dfrac{x^3 - x^2 + 4x - 4}{x^3 - 6x^2 + 5x} = \lim_{x \to 1} \dfrac{3x^2 - 2x + 4}{3x^2 - 12x + 5}$

$$-\dfrac{5}{-4}$$

$$= -\dfrac{5}{4}$$

101. $\displaystyle\lim_{x \to 0} \dfrac{\sin \pi x}{\sin 2\pi x} = \lim_{x \to 0} \dfrac{\pi \cos \pi x}{2\pi \cos 2\pi x} = \dfrac{\pi}{2\pi} = \dfrac{1}{2}$

103. $\displaystyle\lim_{x \to 0} \dfrac{\sin^2 x}{e^x} = \dfrac{0}{1} = 0$

105. $g(x)$ grows faster than $f(x)$.

Practice Test for Chapter 8

1. (a) Express $\dfrac{12\pi}{23}$ in degree measure.   (b) Express $105°$ in radian measure.

2. Determine two coterminal angles (one positive and one negative) for the given angle.

   (a) $-220°$; give your answers in degrees.   (b) $\dfrac{7\pi}{9}$; give your answers in radians.

3. Find the six trigonometric functions of the angle θ if it is in standard position and the terminal side passes through the point $(12, -5)$.

4. Solve for θ, $(0 \le \theta \le 2\pi)$: $\sin^2 \theta + \cos \theta = 1$

5. Sketch the graph of the given function.

   (a) $y = 3 \sin \dfrac{x}{4}$ (b) $y = \tan 2\pi x$

6. Find the derivative of $y = 3x - 3 \cos x$.

7. Find the derivative of $f(x) = x^2 \tan x$.

8. Find the derivative of $g(x) = \sin^3 x$.

9. Find the derivative of $y = \dfrac{\sec x}{x^2}$.

10. Find the derivative of $y = \sin 5x \cos 5x$.

11. Find the derivative of $y = \sqrt{\csc x}$.

12. Find the derivative of $y = \ln|\sec x + \tan x|$.

13. Find the derivative of $f(x) = \cot e^{2x}$.

14. Find $\dfrac{dy}{dx}$: $\sin(x^2 + y) = 3x$

15. Find $\dfrac{dy}{dx}$: $\tan x - \cot 3y = 4$

16. Evaluate $\displaystyle\int \cos 4x \, dx$.

17. Evaluate $\displaystyle\int \csc^2 \dfrac{x}{8} \, dx$.

18. Evaluate $\displaystyle\int x \tan x^2 \, dx$.

19. Evaluate $\displaystyle\int \sin^5 x \cos x \, dx$.

20. Evaluate $\displaystyle\int \frac{\cos^2 x}{\sin x} \, dx$.

21. Evaluate $\displaystyle\int e^{\tan x} \sec^2 x \, dx$.

22. Evaluate $\displaystyle\int \frac{\sin x}{1 + \cos x} \, dx$.

23. Evaluate $\displaystyle\int (\sec x - \tan x)^2 \, dx$.

24. Evaluate $\displaystyle\int x \cos x \, dx$.

25. Evaluate $\displaystyle\int_0^{\pi/4} (2x - \cos x) \, dx$.

26. Use L'Hôpital's Rule to find the limit: $\displaystyle\lim_{x \to \infty} \frac{e^{2x}}{x^2}$

27. Use L'Hôpital's Rule to find the limit: $\displaystyle\lim_{x \to 0} \frac{\sin 7x}{3x}$

28. Use L'Hôpital's Rule to find the limit: $\displaystyle\lim_{x \to \infty} \frac{5x^3 + 7x^2 - 8}{9x^3 + 2x + 4}$

Graphing Calculator Required

29. Use a graphing utility to graph $f(x) = \sin x + \cos 2x$. What are the minimum and maximum values that $f(x)$ takes on?

30. Use a graphing utility to find $\displaystyle\lim_{x \to \infty} \frac{x}{\sqrt{x^2 + 4}}$. Try using L'Hôpital's Rule on this limit—what happens?

CHAPTER 9
Probability and Calculus

CHAPTER 9
Probability and Calculus

Section 9.1   Discrete Probability

Solutions to Odd-Numbered Exercises

1. (a) $S = \{HHH, HHT, HTH, THH, HTT, THT, TTH, TTT\}$

   (b) $A = \{HHH, HHT, HTH, THH\}$

   (c) $B = \{HTT, THT, TTH, TTT\}$

3. (a) $S = \{3, 6, 9, 12, 15, 18, 21, 24, 27, 30, 33, 36, 39, 42, 45, 48\}$

   (b) $A = \{12, 24, 36, 48\}$

   (c) $B = \{9, 36\}$

5. $1 - 0.29 - 0.47 = 0.24$

7. $1 - 0.9855 = 0.0145$

9.

Random variable	0	1	2
Frequency	1	2	1

11.

Random variable	0	1	2	3
Frequency	1	3	3	1

13. (a) $P(1 \leq x \leq 3) = P(1) + P(2) + P(3)$

$$= \frac{3}{20} + \frac{6}{20} + \frac{6}{20}$$

$$= \frac{15}{20}$$

$$= \frac{3}{4}$$

   (b) $P(x \geq 2) = P(2) + P(3) + P(4)$

$$= \frac{6}{20} + \frac{6}{20} + \frac{4}{20}$$

$$= \frac{16}{20}$$

$$= \frac{4}{5}$$

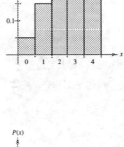

15. (a) $P(x \leq 3) = 0.041 + 0.189 + 0.247 + 0.326$

$$= 0.803$$

   (b) $P(x > 3) = 0.159 + 0.038$

$$= 0.197$$

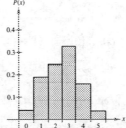

17. (a) $S = \{gggg, gggb, ggbg, gbgg, bggg, ggbb, gbbg, gbgb, bgbg, bbgg, bggb, gbbb, bgbb, bbgb, bbbg, bbbb\}$

(b)

x	0	1	2	3	4
$P(x)$	$\frac{1}{16}$	$\frac{4}{16}$	$\frac{6}{16}$	$\frac{4}{16}$	$\frac{1}{16}$

(c)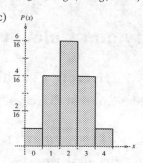

(d) Probability of at least one boy $= 1 -$ probability of all girls

$$P = 1 - \tfrac{1}{16} = \tfrac{15}{16}$$

19. There are 16 possibilities, as indicated in the following chart.

	RY	Ry	rY	ry
RY	RRYY	RRYy	RrYY	RrYy
Ry	RRYy	RRyy	RrYy	Rryy
rY	RrYY	RrYy	rrYY	rrYy
ry	RrYy	Rryy	rrYy	rryy

(a) Round, yellow seeds: can occur 9 ways

RRYY (1)

RRYy (2)

RrYY (2)

RrYy (4)

∴ Probability $= \frac{9}{16}$

(b) Wrinkled, yellow seeds: can occur 3 ways

rrYY (1)

rrYy (2)

∴ Probability $= \frac{3}{16}$

(c) Round, green seeds: can occur 3 ways

RRyy (1)

Rryy (2)

∴ Probability $= \frac{3}{16}$

(d) Wrinkled, green seeds: can occur 1 way

rryy (1)

∴ Probability $= \frac{1}{16}$

21. $E(x) = 1\left(\frac{1}{16}\right) + 2\left(\frac{3}{16}\right) + 3\left(\frac{8}{16}\right) + 4\left(\frac{3}{16}\right) + 5\left(\frac{1}{16}\right) = \frac{48}{16} = 3$

$V(x) = (1-3)^2\left(\frac{1}{16}\right) + (2-3)^2\left(\frac{3}{16}\right) + (3-3)^2\left(\frac{8}{16}\right) + (4-3)^2\left(\frac{3}{16}\right) + (5-3)^2\left(\frac{1}{16}\right) = 14\left(\frac{1}{16}\right) = \frac{7}{8} = 0.875$

$\sigma = \sqrt{V(x)} = 0.9354$

23. $E(x) = -3\left(\frac{1}{5}\right) + (-1)\left(\frac{1}{5}\right) + 0\left(\frac{1}{5}\right) + 3\left(\frac{1}{5}\right) + 5\left(\frac{1}{5}\right) = \frac{4}{5}$

$V(x) = \left(-3-\frac{4}{5}\right)^2\left(\frac{1}{5}\right) + \left(-1-\frac{4}{5}\right)^2\left(\frac{1}{5}\right) + \left(0-\frac{4}{5}\right)^2\left(\frac{1}{5}\right) + \left(3-\frac{4}{5}\right)^2\left(\frac{1}{5}\right) + \left(5-\frac{4}{5}\right)^2\left(\frac{1}{5}\right) = 8.16$

$\sigma = \sqrt{V(x)} \approx 2.857$

25. (a) $E(x) = 1\left(\frac{1}{4}\right) + 2\left(\frac{1}{4}\right) + 3\left(\frac{1}{4}\right) + 4\left(\frac{1}{4}\right) = \frac{10}{4} = 2.5$

$V(x) = (1-2.5)^2\left(\frac{1}{4}\right) + (2-2.5)^2\left(\frac{1}{4}\right) + (3-2.5)^2\left(\frac{1}{4}\right) + (4-2.5)^2\left(\frac{1}{4}\right) = 1.25$

(b) $E(x) = 2\left(\frac{1}{16}\right) + 3\left(\frac{2}{16}\right) + 4\left(\frac{3}{16}\right) + 5\left(\frac{4}{16}\right) + 6\left(\frac{3}{16}\right) + 7\left(\frac{2}{16}\right) + 8\left(\frac{1}{16}\right) = \frac{80}{16} = 5$

$V(x) = (2-5)^2\left(\frac{1}{16}\right) + (3-5)^2\left(\frac{2}{16}\right) + (4-5)^2\left(\frac{3}{16}\right) + (5-5)^2\left(\frac{4}{16}\right) + (6-5)^2\left(\frac{5}{36}\right) + (7-5)^2\left(\frac{6}{36}\right) + (8-5)^2\left(\frac{1}{16}\right)$

$= \frac{5}{2} = 2.5$

27. (a) $E(x) = 10(0.25) + 15(0.30) + 20(0.25) + 30(0.15) + 40(0.05) = 18.50$

$V(x) = (10 - 18.50)^2(0.25) + (15 - 18.50)^2(0.30) + (20 - 18.50)^2(0.25) + (30 - 18.50)^2(0.15) + (40 - 18.50)^2(0.05)$

$= 65.25$

$\sigma = \sqrt{V(x)} \approx 8.078$

(b) Expected revenue: $R = \$1.50(18.50)(1000) = \$27,750$

29. $E(x) = 0(0.995) + 30,000(0.0036) + 60,000(0.0011) + 90,000(0.0003) = 201$

Each customer should be charged \$201.

31. $E(x) = 35\left(\frac{1}{38}\right) + (-1)\left(\frac{37}{38}\right) = -\frac{2}{38} \approx -\$0.05$ or $E(x) = 36\left(\frac{1}{38}\right) + 0\left(\frac{37}{38}\right) = \frac{36}{38}$

$\frac{36}{38} - 1 = -\$0.05$

33. City 1: Expected value $= 0.3(20) + 0.7(-4) = 3.2$ million

City 2: Expected value $= 0.2(50) + 0.8(-9) = 2.8$ million

The company should open the store in City 1.

35. (a)

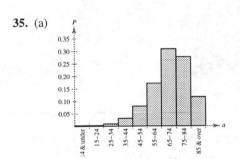

(b) 35–74: $0.0317 + 0.0800 + 0.1712 + 0.3082 = 0.5911$ or 59.11%

$[45, \infty)$: $0.08 + 0.1712 + 0.3082 + 0.2765 + 0.1164 = 0.9523$ or 95.23%

$[0, 34)$: $0.0032 + 0.0032 + 0.0096 = 0.016$ or 1.6%

37. (a) Total number $= 14 + 26 + 7 + 2 + 1 = 50$

(b)

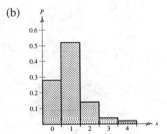

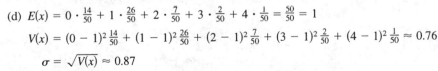

x	0	1	2	3	4
$P(x)$	$\frac{14}{50}$	$\frac{26}{50}$	$\frac{7}{50}$	$\frac{2}{50}$	$\frac{1}{50}$

(c) $P(1 \le x \le 3) = \frac{26}{50} + \frac{7}{50} + \frac{2}{50} = \frac{35}{50}$

(d) $E(x) = 0 \cdot \frac{14}{50} + 1 \cdot \frac{26}{50} + 2 \cdot \frac{7}{50} + 3 \cdot \frac{2}{50} + 4 \cdot \frac{1}{50} = \frac{50}{50} = 1$

$V(x) = (0 - 1)^2 \frac{14}{50} + (1 - 1)^2 \frac{26}{50} + (2 - 1)^2 \frac{7}{50} + (3 - 1)^2 \frac{2}{50} + (4 - 1)^2 \frac{1}{50} \approx 0.76$

$\sigma = \sqrt{V(x)} \approx 0.87$

39. Mean $= 4.4$

Standard deviation $= 2.816$

Section 9.2   Continuous Random Variables

1. $\int_0^8 \frac{1}{8}\, dx = \frac{1}{8}x \Big]_0^8 = 1$ and $f(x) = \frac{1}{8} \geq 0$ on $[0, 8]$.

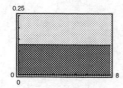

3. $\int_0^4 \frac{4-x}{8}\, dx = \frac{1}{8}\left(4x - \frac{x^2}{2}\right)\Big]_0^4 = 1$ and

$f(x) = \frac{4-x}{8} \geq 0$ on $[0, 4]$.

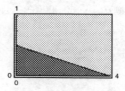

5. $\int_0^1 6x(1-x)\, dx = \int_0^1 (6x - 6x^2)\, dx$

$= \left[3x^2 - 2x^3\right]_0^1 = 1$

and $f(x) = 6x(1-x) \geq 0$ on $[0, 1]$.

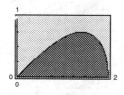

7. $\int_0^\infty \frac{1}{5}e^{-x/5}\, dx = \left[\lim_{b\to\infty} -e^{-x/5}\right]_0^b = 0 + 1 = 1$

and $f(x) = \frac{1}{5}e^{-x/5} \geq 0$ on $[0, \infty)$.

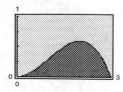

9. $\int_0^2 \frac{3}{8}x\sqrt{4-x^2}\, dx = -\frac{1}{2}\left(\frac{3}{8}\right)\left(\frac{2}{3}\right)(4 - x^2)^{3/2}\Big]_0^2 = 1$

and $f(x) = \frac{3}{8}x\sqrt{4-x^2} \geq 0$ on $[0, 2]$.

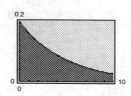

11. $\int_0^3 \frac{4}{27}x^2(3-x)\, dx = \frac{4}{27}\int_0^3 (3x^2 - x^3)\, dx$

$= \frac{4}{27}\left[x^3 - \frac{x^4}{4}\right]_0^3$

$= \frac{4}{27}\left(27 - \frac{81}{4}\right)$

$= 1$

and $f(x) = \frac{4}{27}x^2(3-x) \geq 0$ on $[0, 3]$.

13. $\int_0^\infty \frac{1}{3}e^{-x/3}\, dx = \lim_{b\to\infty}\left[-e^{-x/3}\right]_0^b = 0 - (-1) = 1$ and

$f(x) = \frac{1}{3}e^{-x/3} \geq 0$ on $[0, \infty)$.

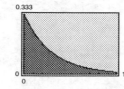

15. $\int_1^4 kx\, dx = \left[\frac{kx^2}{2}\right]_1^4 = \frac{15}{2}k = 1 \Rightarrow k = \frac{15}{2}$

17. $\int_{-2}^{2} k(4 - x^2)\,dx = k\left(4x - \dfrac{x^3}{3}\right)\Big]_{-2}^{2} = \dfrac{32k}{3} - 1 \Rightarrow k = \dfrac{3}{32}$

19. $\int_{0}^{\infty} ke^{-x/2}\,dx = \lim_{b \to \infty}\left[-2ke^{-x/2}\right]_{0}^{b} = 2k = 1 \Rightarrow k = \dfrac{1}{2}$

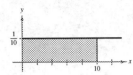

21. $\int_{a}^{b} \dfrac{1}{10}\,dx = \dfrac{x}{10}\Big]_{a}^{b} = \dfrac{b - a}{10}$

(a) $P(0 < x < 6) = \dfrac{6 - 0}{10} = \dfrac{3}{5}$

(b) $P(4 < x < 6) = \dfrac{6 - 4}{10} = \dfrac{1}{5}$

(c) $P(8 < x < 10) = \dfrac{10 - 8}{10} = \dfrac{1}{5}$

(d) $P(x \geq 2) = P(2 < x < 10) = \dfrac{10 - 2}{10} = \dfrac{4}{5}$

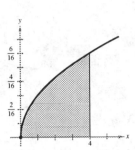

23. $\int_{a}^{b} \dfrac{3}{16}\sqrt{x}\,dx = \left(\dfrac{3}{16}\right)\dfrac{2}{3}x^{3/2}\Big]_{a}^{b} = \dfrac{1}{8}\left[b\sqrt{b} - a\sqrt{a}\right]$

(a) $P(0 < x < 2) = \dfrac{\sqrt{2}}{4} \approx 0.354$

(b) $P(2 < x < 4) = 1 - \dfrac{\sqrt{2}}{4} \approx 0.646$

(c) $P(1 < x < 3) = \dfrac{1}{8}(3\sqrt{3} - 1) \approx 0.525$

(d) $P(x \leq 3) = \dfrac{3\sqrt{3}}{8} \approx 0.650$

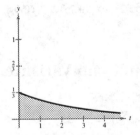

25. $\int_{a}^{b} \dfrac{1}{3}e^{-t/3}\,dt = e^{-t/3}\Big]_{a}^{b} = e^{-a/3} - e^{-b/3}$

(a) $P(t < 2) = e^{-0/3} - e^{-2/3} \approx 0.4866$

(b) $P(t \geq 2) = e^{-2/3} - 0 \approx 0.5134$

(c) $P(1 < t < 4) = e^{-1/3} - e^{-4/3} \approx 0.4529$

(d) $P(t = 3) = 0$

27. $P(a < t < b) = \int_{a}^{b} \dfrac{1}{30}\,dt = \dfrac{b - a}{30}$

(a) $P(0 \leq t \leq 5) = \dfrac{5 - 0}{30} = \dfrac{1}{6}$

(b) $P(18 < t < 30) = \dfrac{30 - 18}{30} = \dfrac{2}{5}$

29. $\int_{a}^{b} \dfrac{1}{3}e^{-t/3}\,dt = -e^{-t/3}\Big]_{a}^{b} = e^{-a/3} - e^{-b/3}$

(a) $P(0 < t < 2) = e^{-0/3} - e^{-2/3} = 1 - e^{-2/3} \approx 0.487$

(b) $P(2 < t < 4) = e^{-2/3} - e^{-4/3} \approx 0.250$

(c) $P(t > 2) = 1 - P(0 < t < 2) = e^{-2/3} \approx 0.513$

31. $P(0 < t < 1) = \int_{0}^{1} \dfrac{4}{3}e^{-4t/3}\,dt$

$= -e^{-4t/3}\Big]_{0}^{1}$

$= 1 - e^{-4/3}$

≈ 0.736

33. $\int_a^b \dfrac{1}{36} xe^{-x/e} \, dx = -\dfrac{1}{6} e^{-x/e}(x + 6) \Big]_a^b$

(a) $P(x < 6) = -\dfrac{1}{6} e^{-x/6}(x + 6) \Big]_0^6 = -\dfrac{1}{6}(12e^{-1} - 6) \approx 0.264$

(b) $P(6 < x < 12) = -\dfrac{1}{6} x^{-x/6}(x + 6) \Big]_6^{12} = 2e^{-1} - 3e^{-2} \approx 0.330$

(c) $P(x > 12) = 1 - P(x \leq 12) = 1 - (1 - 3e^{-2}) \approx 0.406$

35. Note that $\int_0^{15} f(x) \, dx = 1.$

(a) $P(0 \leq x \leq 10) = \int_0^{10} f(x) \, dx = 0.75$

75% probability of receiving up to 10 inches of rain

(b) $P(10 \leq x \leq 15) = \int_{10}^{15} f(x) \, dx = 1 - 0.75 = 0.25$

25% probability of receiving between 10 and 15 inches of rain

(c) $P(0 \leq x < 5) = \int_0^5 f(x) \, dx = 0.25$

25% probability of receiving less than 5 inches of rain

(d) $P(12 \leq x \leq 15) \approx 0.0955$

9.6% probability of receiving between 12 and 15 inches of rain

37. $P(49 \leq x \leq 51) \approx \int_{48.5}^{51.5} \dfrac{1}{5\sqrt{2\pi}} e^{-(x-50)^2/50} \, dx$

$= \dfrac{1}{5\sqrt{2\pi}} \int_{48.5}^{51.5} e^{-(x-50)^2/50} \, dx$

$\approx \dfrac{51.5 - 48.5}{3(12)} \left(\dfrac{1}{5\sqrt{2\pi}} \right) [f(48.50) + 4f(48.75) + 2f(49) + 4f(49.25) + 2f(49.50) + 4f(49.75)$

$\qquad\qquad + 2f(50) + 4f(50.25) + 2f(50.50) + 4f(50.75) + 2f(51) + 4f(51.25) + f(51.50)]$

$\approx 0.236 \text{ (where } f(x) = e^{-(x-50)^2/50}).$

Section 9.3   Expected Value and Variance

1. (a) $\mu = \int_a^b xf(x) \, dx = \int_0^8 x\left(\dfrac{1}{8}\right) dx = \dfrac{x^2}{16} \Big]_0^8 = 4$

(b) $\sigma^2 = \int_a^b x^2 f(x) \, dx - \mu^2 = \int_0^8 x^2\left(\dfrac{1}{8}\right) dx - (4)^2 = \dfrac{x^3}{24} \Big]_0^8 - 16 = \dfrac{64}{3} - 16 = \dfrac{16}{3}$

(c) $\sigma = \dfrac{4}{\sqrt{3}}$

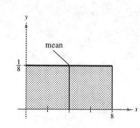

3. (a) $\mu = \int_a^b tf(t) \, dt = \int_0^6 t\left(\dfrac{t}{18}\right) dt = \dfrac{t^3}{54} \Big]_0^6 = 4$

(b) $\mu^2 = \int_a^b t^2 f(t) - \mu^2 = \int_0^6 t^2\left(\dfrac{t}{18}\right) dt - 4^2 = \dfrac{t^4}{72} \Big]_0^6 - 4^2 = 18 - 16 = 2$

(c) $\sigma = \sqrt{2}$

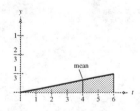

5. (a) $\mu = \displaystyle\int_a^b x f(x)\, dx$ **(b)** $\sigma^2 = \displaystyle\int_a^b x^2 f(x)\, dx - \mu^2$ **(c)** $\sigma = \dfrac{2\sqrt{5}}{21}$

$\qquad = \displaystyle\int_0^1 x\left(\frac{5}{2}x^{3/2}\right) dx$ $= \displaystyle\int_0^1 x^2\left(\frac{5}{2}x^{3/2}\right) dx - \left(\frac{5}{7}\right)^2$

$\qquad = \dfrac{5}{2}\displaystyle\int_0^1 x^{5/2}\, dx$ $= \dfrac{5}{2}\displaystyle\int_0^1 x^{7/2}\, dx - \dfrac{25}{49}$

$\qquad = \left(\dfrac{5}{2}\right)\dfrac{2}{7}x^{7/2}\Big]_0^1$ $= \left(\dfrac{5}{2}\right)\dfrac{2}{9}x^{9/2}\Big]_0^1 - \dfrac{25}{49}$

$\qquad = \dfrac{5}{7}$ $= \dfrac{5}{9} - \dfrac{25}{49}$

$\qquad \approx 0.714$ $= \dfrac{20}{441}$

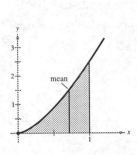

7. Mean $= \displaystyle\int_a^b x f(x)\, dx = \int_0^1 (x)6x(1-x)\, dx = \dfrac{1}{2}$

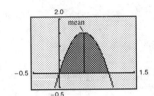

9. Mean $= \displaystyle\int_a^b x f(x)\, dx$

$\qquad = \displaystyle\int_0^3 x\,\dfrac{4}{3(x+1)^2}\, dx$

$\qquad = \dfrac{4}{3}\ln 4 - 1$

$\qquad \approx 0.848$

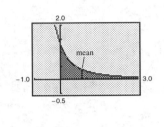

11. Median $= \displaystyle\int_0^m \dfrac{1}{9}e^{-t/9}\, dt = -e^{t/9}\Big]_0^m = 1 - e^{m/9} = \dfrac{1}{2}$

$\qquad e^{-m/9} = \dfrac{1}{2} \Longrightarrow -\dfrac{m}{9} = \ln\dfrac{1}{2}$

$\qquad\qquad\qquad\qquad m = -9\ln\dfrac{1}{2} \approx 6.238$

13. $f(x) = \frac{1}{10}$, $[0, 10]$ is a uniform density function.

Expected value (mean): $\dfrac{a+b}{2} = \dfrac{0+10}{2} = 5$

Variance: $\dfrac{(b-a)^2}{12} = \dfrac{(10-0)^2}{12} = \dfrac{100}{12} = \dfrac{25}{3}$

Standard deviation: $\dfrac{b-a}{\sqrt{12}} = \dfrac{10-0}{\sqrt{12}} \approx 2.887$

15. $f(x) = \frac{1}{8}e^{-x/8}$, $[0, \infty)$ is an exponential density function with $a = \frac{1}{8}$.

Expected value (mean): $\dfrac{1}{a} = 8$

Variance: $\dfrac{1}{a^2} = 64$

Standard deviation: $\dfrac{1}{a} = 8$

17. $f(x) = \dfrac{1}{11\sqrt{2\pi}}e^{-(x-100)^2/242}$,

$(-\infty, \infty)$ is a normal density function with $\mu = 100$ and $\sigma = 11$.

Expected value (mean): $\mu = 100$

Variance: $\sigma^2 = 121$

Standard deviation: $\sigma - 11$

19. Mean $= 0$

Standard deviation $- 1$

$P(0 \le x \le 0.85) \approx 0.3023$

21. Mean $= \frac{1}{6}$

Standard deviation $= \frac{1}{6}$

$P(x \ge 2.23) \approx 0.6896$

23. Mean $= 8$

Standard deviation $= 2$

$P(3 \leq x \leq 13) \approx 0.9876$

25. $\mu = 60, \sigma = 12$

(a) $P(x > 64) \approx 0.3694$

(b) $P(x > 70) \approx 0.2023$

(c) $P(x < 70) \approx 0.7977$

(d) $P(33 < x < 65) \approx 0.6493$

27. $f(t) = \frac{1}{10}$, $[0, 10]$ where $t = 0$ corresponds to 10:00 A.M.

(a) Mean $= \dfrac{10}{2} = 5$

The mean is 10:05 A.M.

Standard deviation $= \dfrac{10}{\sqrt{12}} \approx 2.887$ minutes

(b) $1 - \displaystyle\int_3^{10} \frac{1}{10}\, dx = 1 - \frac{7}{10} = \frac{3}{10} = 0.30$

29. (a) $f(t) = \frac{1}{2}e^{-t/2}$, since mean $= 2$.

(b) $P(0 < t < 1) = \displaystyle\int_0^1 \frac{1}{2}e^{-t/2}\, dt$

$= -e^{-t/2}\Big]_0^1$

$= 1 - e^{-1/2}$

≈ 0.3935

31. (a) Since $\mu = 5$, we have $f(t) = \frac{1}{5}e^{-t/5}$.

(b) $P(\mu - \sigma < t < \mu + \sigma) = P(0 < t < 10)$

$\displaystyle\int_0^{10} \frac{1}{5}e^{-t/5}\, dt = -e^{-t/5}\Big]_0^{10}$

$= 1 - e^{-2}$

≈ 0.865

$= 86.5\%$

33. (a) $\dfrac{174 - 150}{16} = \dfrac{3}{2} = 1.5$

Your score exceeded the national mean by 1.5 standard deviations.

(b) $P(x < 174) = 0.9332$

Thus, $0.9332 = 93.32\%$ of those who took the exam had scores lower than yours.

35. (a) $\mu = \displaystyle\int_0^6 \frac{1}{36}x^2(6 - x)\, dx = \frac{1}{36}\int_0^6 (6x^2 - x^3)\, dx = \frac{1}{36}\left[2x^3 - \frac{x^4}{4}\right]_0^6 = 3$

$\sigma^2 = \displaystyle\int_0^6 \frac{1}{36}x^3(6 - x)\, dx - (3)^2 = \frac{1}{36}\int_0^6 (6x^3 - x^4)\, dx - 9 = \frac{1}{36}\left[\frac{3x^4}{2} - \frac{x^5}{5}\right]_0^6 - 9 = \frac{54}{5} - 9 = \frac{9}{5}$

$\sigma = \sqrt{\dfrac{9}{5}} = \dfrac{3\sqrt{5}}{5} \approx 1.342$

(b) $\displaystyle\int_0^m \frac{1}{36}x(6 - x)\, dx = \frac{1}{36}\int_0^m (6x - x^2)\, dx = \frac{1}{36}\left[3x^2 - \frac{x^3}{3}\right]_0^m = \frac{1}{36}\left[3m^2 - \frac{m^3}{3}\right] = \frac{1}{2}$

$3m^2 - \dfrac{m^3}{3} = 18 \Rightarrow 0 = (m - 3)(m^2 - 6m - 18)$

$m = 3$ or $m = \dfrac{6 \pm \sqrt{108}}{2} = \dfrac{6 \pm 6\sqrt{3}}{2} = 3 \pm 3\sqrt{3}$

In the interval $[0, 6]$, $m = 3$.

(c) $P(\mu - \sigma < x < \mu + \sigma) = P\left(3 - \dfrac{3\sqrt{5}}{5} < x < 3 + \dfrac{3\sqrt{5}}{5}\right)$

$\approx P(1.6584 < x < 4.3416)$

$= \displaystyle\int_{1.6584}^{4.3416} \frac{1}{36}x(6 - x)\, dx$

$= \frac{1}{36}\left[3x^2 - \frac{x^3}{3}\right]_{1.6584}^{4.3416}$

≈ 0.626

$= 62.6\%$

37. $\mu = \displaystyle\int_0^1 \frac{15}{4}x^2\sqrt{1-x}\,dx$

$$= \frac{15}{4}\left(\frac{2}{-7}\right)\left[x^2(1-x)^{3/2} - 2\left(\frac{2}{-5}\right)\left[x(1-x)^{3/2} + \frac{2}{3}(1-x)^{3/2}\right]\right]_0^1 = -\frac{15}{14}\left[0 - \frac{4}{5}\left(\frac{2}{3}\right)\right] = \frac{4}{7}$$

$$\sigma^2 = \int_0^1 \frac{15}{4}x^3\sqrt{1-x}\,dx - \left(\frac{4}{7}\right)^2$$

$$= \frac{15}{4}\left(\frac{2}{-9}\right)\left[\left[x^3(1-x)^{3/2}\right]_0^1 - 3\int_0^1 x^2\sqrt{1-x}\,dx\right] - \frac{16}{49} = -\frac{5}{6}\left[0 - 3\left(\frac{4}{7}\right)\left(\frac{4}{15}\right)\right] - \frac{16}{49} = \frac{8}{21} - \frac{16}{49} = \frac{8}{147}$$

39. Using a graphing utility:

(a) $\mu = \displaystyle\int_0^\infty xf(x)\,dx = \int_0^\infty \frac{1}{25}x^2 e^{-x/5}\,dx = 10$

(b) $P(x \le 4) \approx 0.1912$

41. Mean $= \dfrac{11}{2} = $ median

43. Mean $= \displaystyle\int_0^{1/2} x(4)(1-2x)\,dx = \frac{1}{6}$

Median $= \displaystyle\int_0^m 4(1-2x)\,dx = \frac{1}{2}$

$$4x - 4x^2 \Big]_0^m = \frac{1}{2}$$

$$4m - 4m^2 = \frac{1}{2} \Longrightarrow m \approx 0.1465$$

$\left(m \approx 0.8536 \text{ is not in the interval } \left[0, \frac{1}{2}\right].\right)$

45. Mean $= 5$

Median $= 5\ln 2 \approx 3.4657$

47. $\displaystyle\int_0^m f(x)\,dx = \int_0^m 0.28e^{-0.28x}\,dx = 0.5$

$$-e^{-0.28m} + 1 = 0.5$$

$$e^{-0.28m} = 0.5$$

$$m = \frac{1}{-0.28}\ln 0.5 \approx 2.4755$$

49. (a) $\mu = \displaystyle\int_0^\infty \frac{1}{9}x^2 e^{-x/3}\,dx$ (Use integration by parts.)

$$= \lim_{b\to\infty} -3\left[\frac{x^2}{9}e^{-x/3} - 2\left(-\frac{x}{3} - 1\right)e^{-x/3}\right]_0^b$$

$$= 6$$

$$\sigma^2 = \int_0^\infty \frac{1}{9}x^3 e^{-x/3}\,dx - (6)^2 \quad \text{(Use integration by parts.)}$$

$$= \lim_{b\to\infty}\left[-\frac{x^3}{3}e^{-x/3}\right]_0^b + 9\int_0^\infty \frac{1}{9}x^2 e^{-x/3}\,dx - 36$$

$$= 0 + 9(6) - 36$$

$$= 18 \quad \text{(Use part (a).)}$$

$$\sigma = \sqrt{18} = 3\sqrt{2} \approx 4.243$$

(b) $P(x > 4) = 1 - P(x < 4) = 1 - \displaystyle\int_0^4 \frac{1}{9}xe^{-x/3}\,dx = 1 - \left[-\frac{1}{3}e^{-x/3}(x+3)\right]_0^4 \approx 0.615$

51.
$$P(x < 12) = 0.05$$

$$P\left(z < \frac{12 - \mu}{0.15}\right) = 0.05$$

$$0.5000 - P\left(\frac{12 - \mu}{0.15} < z < 0\right) = 0.05$$

$$P\left(\frac{12 - \mu}{0.15} < z < 0\right) = 0.4500$$

$$\frac{12 - \mu}{0.15} \approx -1.645$$

$$\mu \approx 12.25$$

53. $\mu = 12.30$, $\sigma = 1.50$

$$f(x) = \frac{1}{1.50\sqrt{2\pi}}e^{-(x - 12.30)^2/4.5}$$

$$P(9 < x < 12) = \frac{1}{1.50\sqrt{2\pi}}\int_9^{12} e^{-(x - 12.30)^2/4.5}\, dx$$

$$\approx 0.4068$$

$$= 40.68\%$$

55.
$$f(x) = \frac{1}{10\sqrt{2\pi}}e^{-(x - 110)^2/200} \text{ since } \mu = 110 \text{ and } \sigma = 10.$$

$$P(100 < x < 120) = \int_{100}^{120} \frac{1}{10\sqrt{2\pi}}e^{-(x - 110)^2/200}\, dx = \frac{1}{10\sqrt{2\pi}}\int_{100}^{120} e^{-(x - 110)^2/200}\, dx \approx 0.6828 = 68.28\%$$

Review Exercises for Chapter 9

1. The sample space consists of the twelve months of the year.

$S = \{$January, February, March, April, May, June, July, August, September, October, November, December$\}$

3. If the questions are numbered 1, 2, 3, and 4,

$S = \{123, 124, 134, 234\}$.

5. $S = \{0, 1, 2, 3\}$

7.

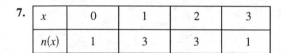

x	0	1	2	3
$n(x)$	1	3	3	1

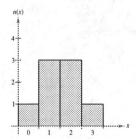

9. (a) $P(2 \le x \le 4) = P(2) + P(3) + P(4)$

$$= \frac{7}{18} + \frac{5}{18} + \frac{3}{18}$$

$$= \frac{15}{18}$$

$$= \frac{5}{6}$$

(b) $P(x \ge 3) = P(3) + P(4) + P(5)$

$$= \frac{5}{18} + \frac{3}{18} + \frac{2}{18}$$

$$= \frac{10}{18}$$

$$= \frac{5}{9}$$

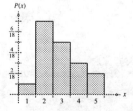

11.

x	2	3	4	5	6	7	8	9	10	11	12
$n(x)$	1	2	3	4	5	6	5	4	3	2	1

$n(S) = 36$

(a) $P(x = 8) = \frac{5}{36}$

(b) $P(x > 4) = 1 - P(x \le 4)$

$$= 1 - \frac{6}{36} = \frac{5}{6}$$

(c) $P(\text{doubles}) = \frac{6}{36} = \frac{1}{6}$

(d) $P(\text{double sixes}) = \frac{1}{36}$

13. Mean = 19.5

15. (a) $E(x) = 10(0.10) + 15(0.20) + 20(0.50) + 30(0.15) + 40(0.05) = 20.5$

   (b) $R = 20.5(1000)(0.75) = \$15,375$

17. $E(x) = \dfrac{24(1200) + 12(1500) + 35(2000) + 5(2200) + 4(3000)}{18} = \dfrac{139,800}{18} = 1747.50$

   $V(x) = (1200 - 1747.50)^2(24) + (1500 - 1747.50)^2(12) + (2000 - 1747.50)^2(35)$

   $\qquad + (2200 - 1747.50)^2(5) + \dfrac{(3000 - 1747.50)^2(4)}{80} = 218,243.7500$

   $\sigma = \sqrt{V(x)} \approx 467.1657$

19. $E(x) = 0(0.10) + 1(0.28) + 2(0.39) + 3(0.17) + 4(0.04) + 5(0.02) = 1.83$

   $V(x) = (0 - 1.83)^2(0.10) + (1 - 1.83)^2(0.28) + (2 - 1.83)^2(0.39) + (3 - 1.83)^2(0.17) + (4 - 1.83)^2(0.04)$

   $\qquad + (5 - 1.83)^2(0.02) \approx 1.1611$

   $\sigma = \sqrt{V(x)} \approx 1.0775$

21. $\displaystyle\int_0^4 \frac{1}{8}(4 - x)\,dx = \left[\frac{1}{2}x - \frac{x^2}{16}\right]_0^4 = 2 - 1 = 1$

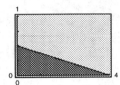

23. $\displaystyle\int_1^9 \frac{1}{4\sqrt{x}}\,dx = \frac{1}{4}\left[2\sqrt{x}\right]_1^9 = 1$

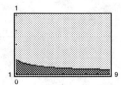

25. $P(0 < x < 2) = \displaystyle\int_0^2 \frac{1}{50}(10 - x)\,dx$

   $\qquad = \dfrac{1}{50}\left[10x - \dfrac{x^2}{2}\right]_0^2$

   $\qquad = \dfrac{9}{25}$

27. $P\left(0 < x < \dfrac{1}{2}\right) = \displaystyle\int_0^{1/2} \frac{2}{(x + 1)^2}\,dx$

   $\qquad = -\dfrac{2}{x + 1}\bigg]_0^{1/2}$

   $\qquad = -\dfrac{4}{3} + 2$

   $\qquad = \dfrac{2}{3}$

29. (a) $P(t \leq 10) = \displaystyle\int_0^{10} \frac{1}{20}\,dt = \frac{1}{20}t\bigg]_0^{10} = \frac{1}{2}$

   (b) $P(t \geq 15) = \displaystyle\int_{15}^{20} \frac{1}{20}\,dt = \frac{1}{20}t\bigg]_{15}^{20} = 1 - \frac{3}{4} = \frac{1}{4}$

31. $P(t > 8) = \displaystyle\int_8^{13} \frac{1}{4\sqrt{t - 4}}\,dt = \frac{1}{2}\sqrt{t - 4}\bigg]_8^{13} = \frac{1}{2}$

33. Mean $= \dfrac{5 - 0}{2} = 2.5$

35. Mean = 6

37. Mean $= \int_0^3 x \frac{2}{9} x(3-x)\, dx = \frac{2}{9}\left[x^3 - \frac{x^4}{4} \right]_0^3 = \frac{3}{2}$

Variance $= \int_0^3 \left(x - \frac{3}{2} \right)^2 \frac{2}{9} x(3-x)\, dx = \frac{9}{20}$

Standard deviation $= \sqrt{V(x)} = \frac{3}{2\sqrt{5}}$

39. Mean $= \int_0^\infty x \left(\frac{1}{2} e^{-x/2} \right) dx = 2$

Variance $= 4$

Standard deviation $= 2$

41. $\int_0^m 6x(1-x)\, dx = \frac{1}{2}$

$3x^2 - 2x^3 \Big]_0^m = \frac{1}{2}$

$3m^2 - 2m^3 = \frac{1}{2}$

$m = \frac{1}{2}$

43. $\int_0^m 0.25 e^{-x/4}\, dx = \frac{1}{2}$

$1 - e^{-m/4} = \frac{1}{2}$

$m \approx 2.7726$

45. $f(t) = \frac{1}{3} e^{-1/3}\, dt, \quad 0 \le t < \infty$

(a) $P(t < 2) = \int_0^2 f(t)\, dt \approx 0.4866$

(b) $P(2 < t < 4) = \int_2^4 f(t)\, dt \approx 0.2498$

47. $f(x) = \frac{1}{\sigma \sqrt{2\pi}} e^{-(x-\mu)^2/2\sigma^2}$

$P(x \ge 50) = \int_{50}^\infty \frac{1}{3\sqrt{2\pi}} e^{-(x-42)^2/[2(3)^2]}\, dx \approx 0.00383$

49. $\mu = 3.75, \, \omega = 0.5$

By Simpson's Rule with $n = 12$,

$P(3.5 < x < 4) \approx 0.3829$.

51. Draw a line dividing the area into two equal pieces.

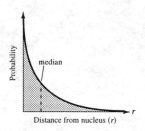

Distance from nucleus (r)

Practice Test for Chapter 9

1. A coin is tossed four times. What is the probability that at least two heads occur?

2. A card is chosen at random from a standard 52-card deck of playing cards. What is the probability that the card will be red and not a face card?

3. Find $E(x)$, $V(x)$, and σ for the given probability distribution.

x	-2	-1	0	3	4
$P(x)$	$\frac{2}{10}$	$\frac{1}{10}$	$\frac{4}{10}$	$\frac{2}{10}$	$\frac{1}{10}$

4. Find the constant k so that $f(x) = ke^{-x/4}$ is a probability density function over the interval $[0, \infty)$.

5. Find (a) $P(0 < x < 5)$ and (b) $P(x > 1)$ for the probability density function

   $$f(x) = \frac{x}{32}, \quad [0, 8].$$

6. Find (a) the mean, (b) the standard deviation, and (c) the median for the probability density function

   $$f(x) = \frac{3}{256}x(8 - x), \quad [0, 8].$$

7. Find (a) the mean, (b) the standard deviation, and (c) the median for the probability density function

   $$f(x) = \frac{6}{x^2}, \quad [2, 3].$$

8. Find the expected value, median, and standard deviation of the exponential density function $f(x) = 7e^{-7x}$, $[0, \infty)$.

Technology Required

9. Find $P(1.67 < x < 3.24)$ using the standard normal probability density function.

10. The monthly revenue x (in thousands of dollars) of a given shop is normally distributed with $\mu = 20$ and $\sigma = 4$. Approximate $P(19 < x < 24)$.

CHAPTER 10
Series and Taylor Polynomials

CHAPTER 10
Series and Taylor Polynomials

Section 10.1   Sequences

Solutions to Odd-Numbered Exercises

1. 2, 4, 8, 16, 32

3. $-\frac{1}{2}, \frac{1}{4}, -\frac{1}{8}, \frac{1}{16}, -\frac{1}{32}$

5. $3, \frac{9}{2}, \frac{27}{6}, \frac{81}{24}, \frac{243}{120}$

7. $-1, \frac{1}{4}, -\frac{1}{9}, \frac{1}{16}, -\frac{1}{25}$

9. This sequence converges since $\lim\limits_{n\to\infty} \dfrac{5}{n} = 0.$

11. This sequence converges since $\lim\limits_{n\to\infty} \dfrac{n+1}{n} = 1.$

13. This sequence converges since $\lim\limits_{n\to\infty} \dfrac{n^2 + 3n - 4}{2n^2 + n - 3} = \dfrac{1}{2}.$

15. This sequences diverges since

$$\lim_{n\to\infty} \frac{n^2 - 25}{n + 5} = \lim_{n\to\infty} (n - 5) = \infty.$$

17. This sequence converges since $\lim\limits_{n\to\infty} \dfrac{1 + (-1)^n}{n} = 0.$

19. This sequence converges since $\lim\limits_{n\to\infty} \left(3 - \dfrac{1}{2^n}\right) = 3.$

21. This sequence converges since

$$\lim_{n\to\infty} \frac{3^n}{4^n} = \lim_{n\to\infty} \left(\frac{3}{4}\right)^n = 0.$$

23. This sequence diverges since

$$\lim_{n\to\infty} \frac{(n+1)!}{n!} = \lim_{n\to\infty} \frac{(n+1)n!}{n!}$$

$$= \lim_{n\to\infty} (n + 1)$$

$$= \infty.$$

25. This sequence diverges since $\lim\limits_{n\to\infty} (-1)^n \dfrac{n}{n+1}$ does not exist.

27. The sequence $a_n = (-1)^n + 2$ oscillates between 1 and 3. Hence, $\lim\limits_{n\to\infty} a_n$ does not exist.

29. $a_n = 3n - 2$

31. $a_n = 5n - 6$

33. $a_n = \dfrac{n+1}{n+2}$

35. $a_n = \dfrac{(-1)^{n-1}}{2^{n-2}}$

37. $a_n = 1 + \dfrac{1}{n} - \dfrac{n+1}{n}$

39. $a_n = 2(-1)^n$

41. $a_n = \dfrac{1}{n!}$

43. Since $a_n = 3n - 1$, the next two terms are $a_5 = 14$ and $a_6 = 17$.

45. Since $a_n = \dfrac{1}{3} + \dfrac{2n}{3}$, we have $a_5 = \dfrac{11}{3}$ and $a_6 = \dfrac{13}{3}$.

47. Since $a_n = 3\left(\dfrac{1}{2}\right)^{n-1}$, the next two terms are $a_5 = \dfrac{3}{16}$ and $a_6 = -\dfrac{3}{32}$.

49. Since $a_n = 2(3^{n-1})$, the next two terms are 162 and 486.

51. Since $a_n = 20\left(\frac{1}{2}\right)^{n-1}$, the sequence is geometric.

53. Since $a_n = \frac{2}{3}n + 2$, the sequence is arithmetic.

55. One example is $a_n = \dfrac{3n + 1}{4n}$.

57. $A_n = P\left(1 + \dfrac{5}{12}\right)^n = 9000\left(1 + \dfrac{0.06}{12}\right)^n = 9000(1.005)^n$

The first 10 terms are 9045.00, 9090.23, 9135.68, 9181.35, 9227.26, 9273.40, 9319.76, 9366.36, 9413.20, 9460.26.

59. $A_n = 2000(11)[(1.1)^n - 1]$

(a) $A_1 = \$2200$ (b) $A_{20} \approx \$126{,}005.00$

    $A_2 = \$4620$ (c) $A_{40} \approx \$973{,}703.62$

    $A_3 = \$7282$

    $A_4 = \$10{,}210.20$

    $A_5 = \$13{,}431.22$

    $A_6 \approx \$16{,}974.34$

61. (a) $a_n = 59.14n + 519.71$

(Use the linear regression feature.)

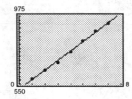

(b) For 2005, $n = 18$ and $a_{18} = 158{,}423$ dollars.

63. $A_n = 100(201)[(1.005)^n - 1]$

(a) $A_1 = \$100.50$ (b) $A_{60} \approx \$7011.89$

    $A_2 = \$201.50$ (c) $A_{240} \approx \$46{,}435.11$

    $A_3 \approx \$303.01$

    $A_4 \approx \$405.03$

    $A_5 \approx \$507.55$

    $A_6 \approx \$610.59$

65. (a) $S_1 = 1 = 1^2$

    $S_2 = 5 = 1^2 + 2^2$

    $S_3 = 14 = 1^2 + 2^2 + 3^2$

    $S_4 = 30 = 1^2 + 2^2 + 3^2 + 4^2$

    $S_5 = 55 = 1^2 + 2^2 + 3^2 + 4^2 + 5^2$

(b) $S_{20} = 2870$

67. (a) $A_1 = 1.3 - 0.15(1.3) = 1.3(0.85)$

    $A_2 = A_1 - 0.15A_1 = 0.85A_1 = 1.3(0.85)^2$

    $A_3 = 1.3(0.85)^3$

    $\vdots$

    $A_n = 1.3(0.85)^n$

(b) $A_1 = \$1.105$ billion

    $A_2 = \$0.939$ billion

    $A_3 = \$0.798$ billion

    $A_4 = \$0.679$ billion

(c) The sequence converges to 0.

$$\lim_{n \to \infty} 1.3(0.85)^n = 0$$

69. $a_n = \left(1 + \dfrac{1}{n}\right)^n$

$a_1 = 2$

$a_{10} \approx 2.593742460$

$a_{100} \approx 2.704813829$

$a_{1000} \approx 2.716923932$

$a_{10,000} \approx 2.718145927$

Note: $e \approx 2.718281828$

71. Year 1: $\$32{,}800$

Year 2: $(1.05)(32{,}800) = \$34{,}440$

Year 3: $(1.05)(34{,}400) = \$36{,}162$

Year 4: $(1.05)(36{,}162) = \$37{,}970.10$

73. First half hour: $2 = 2^1$ nth hour: 4^n

      Second half hour: $4 = 2^2$ 10 hours: $4^{10} = 1,048,576$

      Third half hour: $8 = 2^3$ 20 hours: $4^{20} \approx 1.0995 \times 10^{12}$ bacteria

      nth half hour: 2^n

Section 10.2   Series and Convergence

1. $S_1 = 1$

    $S_2 = \frac{5}{4} = 1.25$

    $S_3 = \frac{49}{36} \approx 1.361$

    $S_4 = \frac{205}{144} \approx 1.424$

    $S_5 = \frac{5269}{3600} \approx 1.464$

3. $S_1 = 3$

    $S_2 = \frac{9}{2} = 4.5$

    $S_3 = \frac{21}{4} = 5.25$

    $S_4 = \frac{45}{8} = 5.625$

    $S_5 = \frac{93}{16} = 5.8125$

5. This series diverges by the nth-Term Test since

$$\lim_{n \to \infty} \frac{n}{n+1} = 1 \neq 0.$$

7. This series diverges by the nth-Term Test since

$$\lim_{n \to \infty} \frac{n^2}{n^2+1} = 1 \neq 0.$$

9. This series diverges by the Test for Convergence of a Geometric Series since $|r| = \left|\frac{3}{2}\right| > 1$.

11. This series diverges by the Test for Convergence of a Geometric Series since $|r| = |1.055| > 1$.

13. This series converges by the Test for Convergence of a Geometric Series since $|r| = \left|\frac{3}{4}\right| < 1$.

15. This series converges by the Test for Convergence of a Geometric Series since $|r| = |0.9| < 1$.

17. Since $a = 1$ and $r = \frac{1}{2}$, we have

$$S = \frac{1}{1 - (1/2)} = 2.$$

19. Since $a = 1$ and $r = -\frac{1}{2}$, we have

$$S = \frac{1}{1 + (1/2)} = \frac{2}{3}.$$

21. Since $a = 2$ and $r = 1/\sqrt{2}$, we have

$$S = \frac{2}{1 - \left(1/\sqrt{2}\right)}$$

$$= \frac{2\sqrt{2}}{\sqrt{2} - 1}\left(\frac{\sqrt{2} + 1}{\sqrt{2} + 1}\right)$$

$$= 4 + 2\sqrt{2}$$

$$\approx 6.828.$$

23. Since $a = 1$ and $r = 0.1$, we have

$$S = \frac{1}{1 - 0.1} = \frac{1}{0.9} = \frac{10}{9}.$$

25. Since $a = 2$ and $r = -\frac{1}{3}$, we have

$$S = \frac{2}{1 + (1/3)} = \frac{3}{2}.$$

27. $\displaystyle\sum_{n=0}^{\infty} \left(\frac{1}{2^n} - \frac{1}{3^n}\right) = \sum_{n=0}^{\infty} \left(\frac{1}{2}\right)^n - \sum_{n=0}^{\infty} \left(\frac{1}{3}\right)^n$

$$= \frac{1}{1 - (1/2)} - \frac{1}{1 - (1/3)}$$

$$= 2 - \frac{3}{2}$$

$$= \frac{1}{2}$$

29. $\displaystyle\sum_{n=0}^{\infty}\left(\frac{1}{3^n}+\frac{1}{4^n}\right)=\sum_{n=0}^{\infty}\left(\frac{1}{3}\right)^n+\sum_{n=0}^{\infty}\left(\frac{1}{4}\right)^n$

$$=\frac{1}{1-(1/3)}+\frac{1}{1-(1/4)}$$

$$=\frac{3}{2}+\frac{4}{3}$$

$$=\frac{17}{6}$$

31. This series diverges by the nth-Term Test since

$$\lim_{n\to\infty}\frac{n+10}{10n+1}=\frac{1}{10}\neq 0.$$

33. This series diverges by the nth-Term Test since

$$\lim_{n\to\infty}\frac{n!+1}{n!}=1\neq 0.$$

35. This series diverges by the nth-Term Test since

$$\lim_{n\to\infty}\frac{3n-1}{2n+1}=\frac{3}{2}\neq 0.$$

37. This series diverges by the Test for Convergence of a Geometric Series since $r=1.075>1.$

39. This series converges by the Test for Convergence of a Geometric Series since $r=\frac{1}{4}<1.$

41. $0.666\overline{6}=\displaystyle\sum_{n=0}^{\infty}0.6(0.1)^n=\frac{0.6}{1-0.1}=\frac{0.6}{0.9}=\frac{2}{3}$

43. $0.81\overline{81}=\displaystyle\sum_{n=0}^{\infty}0.81(0.01)^n=\frac{0.81}{1-0.01}=\frac{0.81}{0.99}=\frac{9}{11}$

45. (a) $\displaystyle\sum_{i=0}^{n-1}8000(0.9)^i=\frac{8000[1-(0.9)^{(n-1)+1}]}{1-0.9}=80,000(1-0.9^n)$

(b) $\displaystyle\sum_{i=0}^{\infty}8000(0.9)^i=\frac{8000}{1-0.9}=80,000$

47. $D_1=16$

$D_2=0.64(16)+0.64(16)=32(0.64)$

$D_3=32(0.64)^2$

$\vdots$

$D=-16+\displaystyle\sum_{n=0}^{\infty}32(0.64)^n=-16+\frac{32}{1-0.64}=-16+\frac{32}{0.36}=\frac{2624}{36}\approx 72.89$ feet

49. $A=\displaystyle\sum_{n=1}^{60}100\left(1+\frac{0.10}{12}\right)^n=-100+\sum_{n=0}^{60}100\left(1+\frac{0.10}{12}\right)^n=-100+\frac{100\left[1-\left(1+\frac{0.10}{12}\right)^{61}\right]}{1-\left(1+\frac{0.10}{12}\right)}\approx\$7808.24$

51. $A=\displaystyle\sum_{n=0}^{\infty}100(0.75)^n=\frac{100}{1-0.75}=\$400$ million

53. $A=\displaystyle\sum_{n=0}^{19}0.01(2)^n=\frac{0.01(1-2^{20})}{1-2}=\$10,485.75$

55. $\displaystyle\sum_{n=1}^{\infty}n\left(\frac{1}{2}\right)^n=\frac{1}{2}+2\left(\frac{1}{4}\right)+3\left(\frac{1}{8}\right)+4\left(\frac{1}{16}\right)+5\left(\frac{1}{32}\right)+6\left(\frac{1}{64}\right)+\cdots$

$$=\frac{1}{2}+\frac{1}{2}+\frac{3}{8}+\frac{1}{4}+\frac{5}{32}+\frac{3}{32}+\cdots$$

$$\approx 2$$

57. At factory: 500

1 mile away: $(0.85)500$

2 miles away: $(0.85)^2 500$

12 miles away: $(0.85)^{12} 500 \approx 71.12$ ppm

61. $\displaystyle\sum_{n=1}^{\infty} n^2 \left(\frac{1}{2}\right)^n = 6$

65. $\displaystyle\sum_{n=1}^{\infty} e^2 \left(\frac{1}{e}\right)^n = \frac{e^2}{3-1} \approx 4.3003$

59. (a) $\displaystyle\sum_{i=1}^{10} 880 = \$8800$

(b) $\displaystyle\sum_{i=1}^{168} 880 = (168)(880) = \$147,840$

Hence, $147,840 - 100,000 = \$47,840$ more.

63. $\displaystyle\sum_{n=1}^{\infty} \frac{n!}{(n!)^2} = \sum_{n=1}^{\infty} \frac{1}{n!} = e - 1 \approx 1.7183$

Section 10.3   *p*-Series and the Ratio Test

1. The series

$$\sum_{n=1}^{\infty} \frac{1}{n^2}$$

is a *p*-series with $p = 2$.

3. The series

$$\sum_{n=1}^{\infty} \frac{1}{3^n} = \sum_{n=0}^{\infty} \frac{1}{3}\left(\frac{1}{3}\right)^n$$

is *not* a *p*-series. This series is geometric with $r = \frac{1}{3}$.

5. The series

$$\sum_{n=1}^{\infty} \frac{1}{n^n} = 1 + \frac{1}{2^2} + \frac{1}{3^3} + \frac{1}{4^4} + \cdots$$

is *not* a *p*-series. The exponent changes with each term.

7. This series converges since $p = 3 > 1$.

9. This series diverges since $p = \frac{1}{3} < 1$.

11. This series converges since $p = 1.03 > 1$.

13. $1 + \dfrac{1}{\sqrt{2}} + \dfrac{1}{\sqrt{3}} + \dfrac{1}{\sqrt{4}} + \cdots = \displaystyle\sum_{n=1}^{\infty} \frac{1}{\sqrt{n}} = \sum_{n=1}^{\infty} \frac{1}{n^{1/2}}$

Therefore, this series diverges since $p = \frac{1}{2} < 1$.

15. $1 + \dfrac{1}{2\sqrt{2}} + \dfrac{1}{3\sqrt{3}} + \dfrac{1}{4\sqrt{4}} + \cdots = \displaystyle\sum_{n-1}^{\infty} \frac{1}{n^{3/2}}$

Therefore, the series converges since $p = \frac{3}{2} > 1$.

17. Since $a_n = 3^n/n!$, we have

$$\lim_{n\to\infty} \left|\frac{a_{n+1}}{a_n}\right| = \lim_{n\to\infty} \left|\frac{3^{n+1}}{(n+1)!} \cdot \frac{n!}{3^n}\right|$$

$$= \lim_{n\to\infty} \frac{3}{n+1}$$

$$= 0$$

and the series converges.

19. Since $a_n = n!/3^n$, we have

$$\lim_{n\to\infty} \left|\frac{a_{n+1}}{a_n}\right| = \lim_{n\to\infty} \left|\frac{(n+1)!}{3^{n+1}} \cdot \frac{3^n}{n!}\right|$$

$$= \lim_{n\to\infty} \frac{n+1}{3}$$

$$= \infty$$

and the series diverges.

21. Since $a_n = n/4^n$, we have

$$\lim_{n\to\infty} \left|\frac{a_{n+1}}{a_n}\right| = \lim_{n\to\infty} \left|\frac{n+1}{4^{n+1}} \cdot \frac{4^n}{n}\right|$$

$$= \lim_{n\to\infty} \frac{n+1}{4n}$$

$$= \frac{1}{4}$$

and the series converges.

23. Since $a_n = 2^n/n^5$, we have

$$\lim_{n\to\infty} \left|\frac{a_{n+1}}{a_n}\right| = \lim_{n\to\infty} \left|\frac{2^{n+1}}{(n+1)^5} \cdot \frac{n^5}{2^n}\right|$$

$$= \lim_{n\to\infty} \frac{2n^5}{(n+1)^5}$$

$$= 2$$

and the series diverges.

25. Since

$$a_n = \frac{(-1)^n 2^n}{n!},$$

we have

$$\lim_{n \to \infty} \left| \frac{a_{n+1}}{a_n} \right| = \lim_{n \to \infty} \left| \frac{(-1)^{n+1} 2^{n+1}}{(n+1)!} \cdot \frac{n!}{(-1)^n 2^n} \right|$$

$$= \lim_{n \to \infty} \left| \frac{-2}{n+1} \right|$$

$$= 0$$

and the series converges.

27. Since

$$a_n = \frac{4^n}{3^n + 1},$$

we have

$$\lim_{n \to \infty} \left| \frac{a_{n+1}}{a_n} \right| = \lim_{n \to \infty} \left| \frac{4^{n+1}}{3^{n+1} + 1} \cdot \frac{3^n + 1}{4^n} \right|$$

$$= \frac{4}{3} > 1$$

and the series diverges.

29. Since

$$a_n = \frac{n 5^n}{n!},$$

we have

$$\lim_{n \to \infty} \left| \frac{a_{n+1}}{a_n} \right| = \lim_{n \to \infty} \left| \frac{(n+1) 5^{n+1}}{(n+1)!} \cdot \frac{n!}{n 5^n} \right|$$

$$= \lim_{n \to \infty} \left| \frac{5}{n} \right|$$

$$= 0 < 1$$

and the series converges.

31. $\displaystyle\sum_{n=1}^{\infty} \frac{1}{n^3} \approx \frac{1}{1} + \frac{1}{2^3} + \frac{1}{3^3} + \frac{1}{4^3} = \frac{2035}{1728} \approx 1.1777$

The error is less than

$$\frac{1}{(p-1)N^{p-1}} = \frac{1}{(3-1)4^{3-1}} = \frac{1}{32}.$$

33. $\displaystyle\sum_{n=1}^{\infty} \frac{1}{n^{3/2}} \approx 1 + \frac{1}{2\sqrt{2}} + \frac{1}{3\sqrt{3}} + \cdots + \frac{1}{10\sqrt{10}} \approx 1.995$

The error is less than

$$\frac{1}{(p-1)N^{p-1}} = \frac{1}{[(3/2)-1]10^{[(3/2)-1]}}$$

$$= \frac{2}{\sqrt{10}}$$

$$\approx 0.6325.$$

35. $\displaystyle\lim_{n \to \infty} \left| \frac{a_{n+1}}{A_n} \right| = \lim_{n \to \infty} \frac{1/[(n+1)^{3/2}]}{1/n^{3/2}}$

$$= \lim_{n \to \infty} \left(\frac{n}{n+1} \right)^{3/2}$$

$$= 1,$$

and the Ratio Test is inconclusive. (The series is a convergent *p*-series.)

37. $\displaystyle\sum_{n=1}^{\infty} \frac{2}{\sqrt[4]{n^3}} = \sum_{n=1}^{\infty} \frac{2}{n^{3/4}} = 2 + \frac{2}{2^{3/4}} + \cdots$

Diverges (*p*-series with $p = \frac{3}{4} < 1$)

Matches (a).

39. $\displaystyle\sum_{n=1}^{\infty} \frac{2}{n\sqrt{n}} = \sum_{n=1}^{\infty} \frac{2}{n^{3/2}} = 2 + \frac{2}{2^{3/2}} + \cdots$

Converges (*p*-series with $p = \frac{3}{2} > 1$)

Matches (b).

41. This series diverges by the *n*th-Term Test since

$$\lim_{n \to \infty} \frac{2n}{n+1} = 2 \neq 0.$$

43. This series converges by the *p*-series test since $p = \frac{4}{3} > 1$.

45. This series converges by the Geometric Series Test since $|r| = \left| -\frac{2}{3} \right| = \frac{2}{3} < 1$.

47. Since both series are convergent *p*-series, their difference is convergent.

49. This series diverges by the Geometric Series Test since $r = \frac{5}{4} > 1$.

51. This series diverges by the Ratio Test since

$$a_n = \frac{n!}{3^{n-1}}$$

and

$$\lim_{n \to \infty} \left| \frac{a_{n+1}}{a_n} \right| = \lim_{n \to \infty} \left| \frac{(n+1)!}{3^n} \cdot \frac{3^{n-1}}{n!} \right|$$

$$= \lim_{n \to \infty} \frac{n+1}{3}$$

$$= \infty.$$

53. $\displaystyle\sum_{n=1}^{100} \frac{1}{n^2} = \frac{1}{1} + \frac{1}{2^2} + \frac{1}{3^2} + \frac{1}{4^2} + \cdots + \frac{1}{100^2} \approx 1.635$

$\dfrac{\pi^2}{6} \approx 1.644934$

55. (a) Using a quadratic model with $k = 1$ corresponding to 1989, $a_k = 1.002 + 0.0073k + 0.0107k^2$. Hence,

$$\sum_{k=1}^{n} (0.0107k^2 + 0.0073k + 1.002).$$

(b) No, the Ratio Test yields a limit equal to 1.

Section 10.4   Power Series and Taylor's Theorem

1. $\displaystyle\sum_{n=0}^{\infty} \left(\frac{x}{4}\right)^n = 1 + \frac{x}{4} + \left(\frac{x}{4}\right)^2 + \left(\frac{x}{4}\right)^3 + \left(\frac{x}{4}\right)^4 + \cdots$

3. $\displaystyle\sum_{n=0}^{\infty} \frac{(-1)^{n+1}(x+1)^n}{n!} = -1 + (x+1) - \frac{(x+1)^2}{2} + \frac{(x+1)^3}{6} - \frac{(x+1)^4}{24} + \cdots$

5. $\displaystyle\lim_{n \to \infty} \left| \frac{a_{n+1}x^{n+1}}{a_n x^n} \right| = \lim_{n \to \infty} \left| \frac{(x/2)^{n+1}}{(x/2)^n} \right|$

$\qquad = \displaystyle\lim_{n \to \infty} \left| \frac{x}{2} \right| < 1$

$\Rightarrow |x| < 2$

Radius $= 2$

7. $\displaystyle\lim_{n \to \infty} \left| \frac{a_{n+1}x^{n+1}}{a_n x^n} \right| = \lim_{n \to \infty} \left| \frac{(-1)^{n+1}x^{n+1}/3(n+1)}{(-1)^n x^n/3n} \right|$

$\qquad = \displaystyle\lim_{n \to \infty} \left| \frac{n}{n+1}x \right|$

$\qquad = |x| < 1$

$\Rightarrow |x| < 1$

Radius $= 1$

9. $\displaystyle\lim_{n \to \infty} \left| \frac{a_{n+1}x^{n+1}}{a_n x^n} \right| = \lim_{n \to \infty} \left| \frac{(-1)^{n+1}x^{n+1}/(n+1)!}{(-1)^n x^n/n!} \right|$

$\qquad = \displaystyle\lim_{n \to \infty} \left| \frac{x}{n+1} \right|$

$\qquad = 0$

$\Rightarrow -\infty < x < \infty$

Radius $= \infty$

11. $\displaystyle\lim_{n \to \infty} \left| \frac{a_{n+1}x^{n+1}}{a_n x^n} \right| = \lim_{n \to \infty} \left| \frac{(n+1)!x^{n+1}/2^{n+1}}{n!x^n/2^n} \right|$

$\qquad = \displaystyle\lim_{n \to \infty} \left| \frac{(n+1)x}{2} \right|$

$\qquad = \infty$

This series converges only at $x = 0$.

Radius $= 0$

13. $\displaystyle\lim_{n \to \infty} \left| \frac{a_{n+1}x^{n+1}}{a_n x^n} \right| = \lim_{n \to \infty} \left| \frac{(-1)^{n+2}x^{n+1}/4^{n+1}}{(-1)^{n+1}x^n/4^n} \right|$

$\qquad = \displaystyle\lim_{n \to \infty} \left| \frac{x}{4} \right| < 1$

$\Rightarrow |x| < 4$

Radius $= 4$

15. $\lim\limits_{n\to\infty}\left|\dfrac{a_{n+1}(x-5)^{n+1}}{a_n(x-5)^n}\right| = \lim\limits_{n\to\infty}\left|\dfrac{(-1)^{n+2}(x-5)^{n+1}/(n+1)5^{n+1}}{(-1)^{n+1}(x-5)^n/n5^n}\right|$

$$= \lim_{n\to\infty}\left|\frac{x-5}{5}\cdot\frac{n+1}{n}\right|$$

$$= \left|\frac{x-5}{5}\right| < 1$$

$\Rightarrow |x-5| < 5$

Radius $= 5$

17. $\lim\limits_{n\to\infty}\left|\dfrac{a_{n+1}(x-1)^{n+1}}{a_n(x-1)^n}\right| = \lim\limits_{n\to\infty}\left|\dfrac{(-1)^{n+2}(x-1)^{n+2}/(n+2)}{(-1)^{n+1}(x-1)^{n+1}/(n+1)}\right|$

$$= \lim_{n\to\infty}\left|(x-1)\frac{n+1}{n+2}\right|$$

$$= \lim_{n\to\infty}|x-1|$$

$$= |x-1| < 1$$

$\Rightarrow |x-1| < 1$

Radius $= 1$

19. $\lim\limits_{n\to\infty}\left|\dfrac{a_{n+1}(x-c)^{n+1}}{a_n(x-c)^n}\right| = \lim\limits_{n\to\infty}\left|\dfrac{(x-c)^n/c^n}{(x-c)^{n-1}/c^{n-1}}\right|$

$$= \lim_{n\to\infty}\left|\frac{x-c}{c}\right| < 1$$

$\Rightarrow |x-c| < c$

Radius $= c$

21. $\lim\limits_{n\to\infty}\left|\dfrac{a_{n+1}x^{n+1}}{a_nx^n}\right| = \lim\limits_{n\to\infty}\left|\dfrac{(n+1)(-2)^nx^n/(n+2)!}{n(-2)^{n-1}x^{n-1}/(n+1)!}\right|$

$$= \lim_{n\to\infty}\left|\frac{n+1}{n}(-2)x\frac{1}{n+2}\right|$$

$$= 0$$

$\Rightarrow -\infty < x < \infty$

Radius $= \infty$

23. $\lim\limits_{n\to\infty}\left|\dfrac{a_{n+1}x^{n+1}}{a_nx^n}\right| = \lim\limits_{n\to\infty}\left|\dfrac{x^{2n+3}/(2n+3)!}{x^{2n+1}/(2n+1)!}\right| = \lim\limits_{n\to\infty}\left|\dfrac{x^2}{(2n+3)(2n+2)}\right| = 0$

$\Rightarrow -\infty < x < \infty$

Radius $= \infty$

25.

$f(x) = e^x$	$f(0) = 1$
$f'(x) = e^x$	$f'(0) = 1$
$f''(x) = e^x$	$f''(0) = 1$
$\vdots$	$\vdots$
$f^{(n)}(x) = e^x$	$f^{(n)}(0) = 1$

The power series for f is

$$e^x = f(0) + f'(0)x + \frac{f''(0)x^2}{2!} + \cdots = 1 + x + \frac{x^2}{2!} + \frac{x^3}{3!} + \cdots = \sum_{n=0}^{\infty}\frac{x^n}{n!}.$$

$$\lim_{n\to\infty}\left|\frac{x^{n+1}/(n+1)!}{x^n/n!}\right| = \lim_{n\to\infty}\left|\frac{x}{n+1}\right| = 0 \Rightarrow R = \infty$$

27. $f(x) = e^{2x}$ $f(0) = 1$

     $f'(x) = 2e^{2x}$ $f'(0) = 2$

     $f''(x) = 4e^{2x}$ $f''(0) = 4$

     $f'''(x) = 8e^{2x}$ $f'''(0) = 8$

$$\vdots$$

$$f^{(n)}(0) = 2^n$$

The power series for f is $e^{2x} = f(0) + f'(0)x + \dfrac{f''(0)x^2}{2!} + \cdots = 1 + 2x + \dfrac{4x^2}{2!} + \dfrac{8x^3}{3!} + \cdots = \displaystyle\sum_{n=0}^{\infty} \dfrac{(2x)^n}{n!}.$

$$\lim_{n\to\infty} \left| \frac{(2x)^{n+1}/(n+1)!}{(2x)^n/n!} \right| = \lim_{n\to\infty} \left| \frac{2x}{n+1} \right| = 0 \Rightarrow R = \infty$$

29. $f(x) = \dfrac{1}{x+1}$ $f(0) = 1$

     $f'(x) = \dfrac{-1}{(x+1)^2}$ $f'(0) = -1$

     $f''(x) = \dfrac{2}{(x+1)^3}$ $f''(0) = 2$

     $f'''(x) = \dfrac{-6}{(x+1)^4}$ $f'''(0) = -6$

$$\vdots$$

$$f^{(n)}(0) = (-1)^n n!$$

The power series for f is

$$\frac{1}{x+1} = f(0) + f'(0)x + \frac{f''(0)x^2}{2!} + \cdots$$

$$= 1 - x + \frac{2x^2}{2!} - \frac{6x^3}{3!} + \cdots + \frac{(-1)^n n! x^n}{n!} + \cdots = 1 - x + x^2 - x^3 + \cdots = \sum_{n=0}^{\infty} (-1)^n x^n.$$

$$\lim_{n\to\infty} \left| \frac{(-1)^{n+1} x^{n+1}}{(-1)^n x^n} \right| = \lim_{n\to\infty} |x| < 1 \Rightarrow R = 1$$

31. $f(x) = \sqrt{x}$ $f(2) = \sqrt{2}$

     $f'(x) = \dfrac{1}{2\sqrt{x}}$ $f'(2) = \dfrac{1}{2\sqrt{2}} = \dfrac{1}{4}\sqrt{2}$

     $f''(x) = -\dfrac{1}{4x\sqrt{x}}$ $f''(2) = -\dfrac{1}{8\sqrt{2}} = -\dfrac{1}{16}\sqrt{2}$

     $f'''(x) = \dfrac{3}{8x^2\sqrt{x}}$ $f'''(2) = \dfrac{3}{32\sqrt{2}} = \dfrac{3}{64}\sqrt{2}$

     $f^{(4)}(x) = -\dfrac{15}{16x^3\sqrt{x}}$ $f^{(4)}(2) = -\dfrac{15}{128\sqrt{2}} = -\dfrac{15}{256}\sqrt{2}$

The general pattern (for $n \geq 2$) is $f^{(n)}(2) = \dfrac{(-1)^{n+1} 1 \cdot 3 \cdot 5 \cdots (2n-3)}{2^{2n}}\sqrt{2}$.

The power series for $f(x) = \sqrt{x}$ is

$$\sqrt{x} = f(2) + f'(2)(x-2) + \frac{f''(2)(x-2)^2}{2!} + \cdots = \sqrt{2} + \frac{1}{4}\sqrt{2}(x-2) + \sum_{n=2}^{\infty} \frac{(-1)^{n+1} 1 \cdot 3 \cdots (2n-3)}{2^{2n}n!}\sqrt{2}(x-2)^n.$$

$$\lim_{n\to\infty} \left| \frac{1 \cdot 3 \cdots (2n-3)(2n-1)\sqrt{2}(x-2)^{n+1}/2^{2n+2}(n+1)!}{1 \cdot 3 \cdots (2n-3)\sqrt{2}(x-2)^n/2^{2n}n!} \right| = \lim_{n\to\infty} \left| \frac{(2n-1)(x-2)}{4(n+1)} \right| = \left| \frac{x-2}{2} \right| < 1 \Rightarrow R - 2$$

33. $f(x) = (1 + x)^{-3}$ $\qquad\qquad$ $f(0) = 1$

$\quad f'(x) = -3(1 + x)^{-4}$ $\qquad\qquad$ $f'(0) = -3$

$\quad f''(x) = 12(1 + x)^{-5}$ $\qquad\qquad$ $f''(0) = 12$

$\quad f'''(x) = -60(1 + x)^{-6}$ $\qquad\quad$ $f'''(0) = -60$

In general, $f^{(n)}(0) = (-1)^n \dfrac{(n + 2)!}{2}$. The power series is

$$\frac{1}{(1 + x)^3} = f(0) + f'(0)x + \frac{f''(0)x^2}{2!} + \cdots = 1 - 3x + 6x^2 - 10x^3 + \cdots = \sum_{n=0}^{\infty} (-1)^n \frac{(n + 2)(n + 1)}{2} x^n.$$

$R = 1$

35. $f(x) = (1 + x)^{-1/2}$ $\qquad\qquad$ $f(0) = 1$

$\quad f'(x) = -\dfrac{1}{2}(1 + x)^{-3/2}$ $\qquad\qquad$ $f'(0) = -\dfrac{1}{2}$

$\quad f''(x) = \dfrac{3}{4}(1 + x)^{-5/2}$ $\qquad\qquad$ $f''(0) = \dfrac{3}{4}$

$\quad f'''(x) = -\dfrac{15}{8}(1 + x)^{-7/2}$ $\qquad\qquad$ $f'''(0) = -\dfrac{15}{8}$

Thus, the general pattern is given by $f^{(n)}(0) = (-1)^n \left[\dfrac{1 \cdot 3 \cdot 5 \cdots (2n - 1)}{2^n}\right]$. The power series for f is

$$\frac{1}{\sqrt{1 + x}} = f(0) + f'(0)x + \frac{f''(0)x^2}{2!} + \frac{f'''(0)x^3}{3!} + \cdots$$

$$= 1 - \frac{1}{2}x + \frac{1 \cdot 3x^2}{2^2 2!} - \frac{1 \cdot 3 \cdot 5x^3}{2^3 3!} + \cdots$$

$$= 1 + \sum_{n=1}^{\infty} \frac{(-1)^n 1 \cdot 3 \cdot 5 \cdots (2n - 1)}{2^n n!} x^n.$$

$R = 1$

37. (a) $f(x) = \displaystyle\sum_{n=0}^{\infty} \left(\frac{x}{2}\right)^n = \sum_{n=0}^{\infty} \frac{x^n}{2^n}$

$\quad \displaystyle\lim_{n \to \infty} \left|\frac{(x/2)^{n+1}}{(x/2)^n}\right| = \left|\frac{x}{2}\right| < 1 \Rightarrow |x| < 2$

$\quad R = 2$

(b) $f'(x) = \displaystyle\sum_{n=1}^{\infty} \frac{nx^{n-1}}{2^n}$

$\quad \displaystyle\lim_{n \to \infty} \left|\frac{(n + 1)x/2^{n+1}}{nx^{n-1}/2^n}\right| = \lim_{n \to \infty} \left|\frac{n + 1}{2n}x\right| = \left|\frac{x}{2}\right| < 1 \Rightarrow R = 2$

(c) $f''(x) = \displaystyle\sum_{n=2}^{\infty} \frac{n(n - 1)x^{n-2}}{2^n}$

$\quad \displaystyle\lim_{n \to \infty} \left|\frac{(n + 1)nx^{n-1}/2^{n+1}}{n(n - 1)x^{n-2}/2^n}\right| = \lim_{n \to \infty} \left|\frac{n + 1}{n - 1} \cdot \frac{x}{2}\right| = \left|\frac{x}{2}\right| 1 \Rightarrow R = 2$

(d) $\displaystyle\int f(x)\,dx = \sum_{n=0}^{\infty} \frac{x^{n+1}}{(n + 1)2^n} + C$

$\quad \displaystyle\lim_{n \to \infty} \left|\frac{x^{n+2}/(n + 2)2^{n+1}}{x^{n+1}/(n + 1)2^n}\right| = \lim_{n \to \infty} \left|\frac{n + 1}{n + 2} \cdot \frac{x}{2}\right| = \left|\frac{x}{2}\right| < 1 \Rightarrow R = 2$

39. (a) $f(x) = \sum_{n=0}^{\infty} \frac{(x + 1)^{n+1}}{n + 1}$

$$\lim_{n\to\infty} \left| \frac{(x + 1)^{n+2}/(n + 2)}{(x + 1)^{n+1}/(n + 1)} \right| = \lim_{n\to\infty} \left| \frac{n + 1}{n + 2}(x + 1) \right| = |x + 1| < 1 \Rightarrow R = 1$$

(b) $f'(x) = \sum_{n=0}^{\infty} (x + 1)^n$

$$\lim_{n\to\infty} \left| \frac{(x + 1)^{n+1}}{(x + 1)^n} \right| = |x + 1|1 \Rightarrow R = 1$$

(c) $f''(x) = \sum_{n=1}^{\infty} n(x + 1)^{n-1}$

$$\lim_{n\infty} \left| \frac{(n + 1)(x + 1)^n}{n(x + 1)^{n-1}} \right| = \lim_{n\infty} \left| \frac{n + 1}{n}(x + 1) \right| = |x + 1| < 1 \Rightarrow R = 1$$

(d) $\int f(x)\, dx = \sum_{n=0}^{\infty} \frac{(x + 1)^{n+2}}{(n + 2)(n + 1)} + C$

$$\lim_{n\to\infty} \left| \frac{(x + 1)^{n+3}/(n + 3)(n + 2)}{(x + 1)^{n+2}/(n + 2)(n + 1)} \right| = |x + 1| < 1 \Rightarrow R = 1$$

41. Since the power series for e^x is

$$e^x = \sum_{n=0}^{\infty} \frac{x^n}{n!}$$

it follows that the power series for e^{x^3} is

$$e^{x^3} = \sum_{n=0}^{\infty} \frac{(x^3)^n}{n!} = \sum_{n=0}^{\infty} \frac{x^{3n}}{n!}.$$

43. $3x^2 e^x = \frac{d}{dx}[e^{x^3}]$

$$= \sum_{n=1}^{\infty} \frac{3nx^{3n-1}}{n!}$$

$$= 3 \sum_{n=1}^{\infty} \frac{x^{3n-1}}{(n - 1)!}$$

$$= 2 \sum_{n=0}^{\infty} \frac{x^{3n+2}}{n!}$$

45. Since the power series for $1/(1 + x)$ is

$$f(x) = \frac{1}{1 + x} = \sum_{n=0}^{\infty} (-1)^n x^n$$

it follows that the power series for $1/(1 + x^4)$ is

$$f(x^4) = \frac{1}{1 + x^4} = \sum_{n=0}^{\infty} (-1)^n x^{4n}.$$

47. $\frac{1}{1 + x^2} = \sum_{n=0}^{\infty} (-1)^n x^{2n}$

$$\frac{2x}{1 + x^2} = \sum_{n=0}^{\infty} (-1)^n (2x)x^{2n} = 2 \sum_{n=0}^{\infty} (-1)^n x^{2n+1}$$

$$\ln(1 + x^2) = \int \frac{2x}{1 + x^2}\, dx$$

$$= 2 \sum_{n=0}^{\infty} \frac{(-1)^n x^{2n+2}}{2n + 2} = \sum_{n=0}^{\infty} \frac{(-1)^n x^{2n+2}}{n + 1}$$

49. $\frac{1}{x} = \sum_{n=0}^{\infty} (-1)^n (x - 1)^n$

$$\ln x = \int \frac{1}{x}\, dx = \sum_{n=0}^{\infty} \frac{(-1)^n (x - 1)^{n+1}}{n + 1}$$

51. $-\frac{1}{x} = \sum_{n=0}^{\infty} (-1)^{n+1}(x - 1)^n$

Differentiating, $\frac{1}{x^2} = \sum_{n=1}^{\infty} (-1)^{n+1} n(x - 1)^{n-1}.$

53. $e^x = 1 + x + \frac{x^2}{2!} + \frac{x^3}{3!} + \frac{x^4}{4!} + \frac{x^5}{5!} + \cdots$

$$e^{1/2} = 1 + \frac{1}{2} + \frac{1}{2!(2^2)} + \frac{1}{3!(2^3)} + \frac{1}{4!(2^4)} + \frac{1}{5!(2^5)} + \cdots$$

Since $1/[6!(2^6)] \approx 0.00002$, the first six terms are sufficient to approximate $e^{1/2}$ to four decimal places.

$$e^{1/2} \approx 1 + \frac{1}{2} + \frac{1}{8} + \frac{1}{48} + \frac{1}{384} + \frac{1}{3840} \approx 1.6487$$

55. $f(0.5) = \sum_{n=1}^{\infty} \frac{(-1)^{n+1}(0.5-1)^n}{n}$

$= \sum_{n=1}^{\infty} \frac{(-1)^{n+1}(-1)^n(1/2)^n}{n}$

$= \sum_{n=1}^{\infty} \frac{(-1)^{2n+1}}{2^n n}$

$= -\sum_{n=1}^{\infty} \frac{1}{2^n n}$

≈ 0.6931

57. $f(0.1) = \sum_{n=1}^{\infty} \frac{(-1)^{n+1}(0.1-1)^n}{n}$

$= \sum_{n=1}^{\infty} \frac{(-1)^{n+1}(-0.9)^n}{n}$

$= \sum_{n=1}^{\infty} \frac{(-1)^{2n+1}(0.9)^n}{n}$

$= -\sum_{n=1}^{\infty} \frac{(0.9)^n}{n}$

≈ -2.3018

Section 10.5   Taylor Polynomials

1. $e^x = \sum_{n=0}^{\infty} \frac{1}{n!} x^n$

  (a) $S_1(x) = 1 + x$

  (b) $S_2(x) = 1 + x + \frac{x^2}{2}$

  (c) $S_3(x) = 1 + x + \frac{x^2}{2} + \frac{x^3}{6}$

  (d) $S_4(x) = 1 + x + \frac{x^2}{2} + \frac{x^3}{6} + \frac{x^4}{24}$

3. $\frac{1}{\sqrt{x+1}} = 1 + \frac{x}{2^1 \cdot 1!} - \frac{x^2}{2^2 \cdot 2!} + \frac{3x^3}{2^3 \cdot 3!} - \frac{3 \cdot 5 x^4}{2^4 \cdot 4!}$

  (a) $S_1(x) = 1 + \frac{x}{2}$

  (b) $S_2(x) = 1 + \frac{x}{2} - \frac{x^2}{8}$

  (c) $S_3(x) = 1 + \frac{x}{2} - \frac{x^2}{8} + \frac{3x^3}{48}$

  (d) $S_4(x) = 1 + \frac{x}{2} - \frac{x^2}{8} + \frac{3x^3}{48} - \frac{15x^4}{384}$

5. $f(x) = \frac{x}{x+1} = 1 - \frac{1}{x+1}$

  (a) $S_1(x) = x$

  (b) $S_2(x) = x - x^2$

  (c) $S_3(x) = x - x^2 + x^3$

  (d) $S_4(x) = x - x^2 + x^3 - x^4$

7. $S_1(x) = 1 + \frac{x}{2}$

$S_2(x) = 1 + \frac{x}{2} + \frac{x^2}{8}$

$S_3(x) = 1 + \frac{x}{2} + \frac{x^2}{8} + \frac{x^3}{48}$

$S_4(x) = 1 + \frac{x}{2} + \frac{x^2}{8} + \frac{x^3}{48} + \frac{x^4}{384}$

x	0	0.25	0.50	0.75	1.0
$f(x)$	1.0000	1.1331	1.2840	1.4550	1.6487
$S_1(x)$	1.0000	1.1250	1.2500	1.3750	1.5000
$S_2(x)$	1.0000	1.1328	1.2813	1.4453	1.6250
$S_3(x)$	1.0000	1.1331	1.2839	1.4541	1.6458
$S_4(x)$	1.0000	1.1331	1.2840	1.4549	1.6484

9. $\frac{1}{1+x^2} = \sum_{n=0}^{\infty} (-1)^n x^{2n}$

  (a) $S_2(x) = 1 - x^2$

  (b) $S_4(x) = 1 - x^2 + x^4$

  (c) $S_6(x) = 1 - x^2 + x^4 - x^6$

  (d) $S_8(x) = 1 - x^2 + x^4 - x^6 + x^8$

11. $S_4(x) = 1 - x^2 + x^4$

13. $y = -\frac{1}{2}x^2 + 1$ is a parabola through $(0, 1)$; matches (d).

15. $y = e^{-1/2}[(x + 1) + 1]$ is a line; matches (a).

17. $S_6(x) = 1 - x + \dfrac{x^2}{2} - \dfrac{x^3}{6} + \dfrac{x^4}{24} - \dfrac{x^5}{120} + \dfrac{x^6}{720}$

$$f\left(\frac{1}{2}\right) \approx 1 - \frac{1}{2} + \frac{1}{8} - \frac{1}{48} + \frac{1}{384} - \frac{1}{3840} + \frac{1}{46{,}080}$$

$$\approx 0.607$$

19. $f(x) = \ln x, c = 2$

$$S_6(x) = \ln 2 + \frac{1}{2}(x - 2) - \frac{1}{8}(x - 2)^2 + \frac{1}{24}(x - 2)^3 - \frac{1}{64}(x - 2)^4 + \frac{1}{160}(x - 2)^5 - \frac{1}{384}(x - 2)^6 f\left(\frac{3}{2}\right)$$

$$\approx \ln 2 + \frac{1}{2}\left(-\frac{1}{2}\right) - \frac{1}{8}\left(\frac{1}{4}\right) + \frac{1}{24}\left(-\frac{1}{8}\right) - \frac{1}{64}\left(\frac{1}{16}\right) - \frac{1}{160}\left(\frac{1}{32}\right) - \frac{1}{384}\left(\frac{1}{64}\right)$$

$$= 0.4055$$

21. $\quad S_6(x) = 1 - x^2 + \dfrac{x^4}{2} - \dfrac{x^6}{6}$

$$\int_0^1 e^{-x^2}\, dx \approx \int_0^1 \left(1 - x^2 + \frac{x^4}{2} - \frac{x^6}{6}\right) dx$$

$$= \left[x - \frac{x^3}{3} + \frac{x^5}{10} - \frac{x^7}{42}\right]_0^1$$

$$\approx 0.74286$$

23. $\quad S_6(x) = 1 - \dfrac{1}{2}x^2 + \dfrac{3}{8}x^4 - \dfrac{5}{16}x^6$

$$\int_0^{1/2} \frac{1}{\sqrt{1 + x^2}}\, dx \approx \int_0^{1/2} \left(1 - \frac{1}{2}x^2 + \frac{3}{8}x^4 - \frac{5}{16}x^6\right) dx$$

$$= \left[x - \frac{x^3}{6} + \frac{3x^5}{40} - \frac{5x^7}{112}\right]_0^{1/2}$$

$$= \frac{1}{2} - \frac{1}{48} + \frac{3}{1280} - \frac{5}{14{,}336}$$

$$\approx 0.481$$

25. Since the $(n + 1)$ derivative of $f(x) = e^x$ is e^x, the maximum value of $|f^{n+1}(x)|$ on the interval $[0, 2]$ is $e^2 < 8$. Therefore, the nth remainder is bounded by

$$|R_n| \le \left|\frac{8}{(n + 1)!}(x - 1)^{n+1}\right|, \quad 0 \le x \le 2$$

$$|R_n| \le \frac{8}{(n + 1)!}(1)$$

with $n = 7$,

$$\frac{8}{(7 + 1)!} = 1.98 \times 10^{-4} < 0.001.$$

Thus, $n = 7$ will approximate e^x with an error less than 0.001.

27. $|R_5| \le \dfrac{f^{(6)}(z)}{6!}x^6 = \dfrac{e^{-z}}{6!}x^6$

Since $e^{-z} \le 1$ in the interval $[0, 1]$, it follows that

$$R_5 \le \frac{1}{6!} \approx 0.00139.$$

29. (a) $\displaystyle\sum_{n=0}^{\infty} P(n) = \sum_{n=0}^{\infty}\left(\frac{1}{2}\right)^{n+1} = \sum_{n=0}^{\infty}\frac{1}{2}\left(\frac{1}{2}\right)^n = \frac{1/2}{1 - (1/2)} = 1$

(b) Expected value $= \displaystyle\sum_{n=0}^{\infty} nP(n) = \sum_{n=0}^{\infty} n\left(\frac{1}{2}\right)^{n+1} = 1$ (See Example 5.)

(c) The expected daily profit is $\$10(1) = \$10$.

Section 10.6   Newton's Method

1. $x_2 = x_1 - \dfrac{f(x_1)}{f'(x_1)} = 2.2 - \dfrac{(2.2)^2 - 5}{2(2.2)} \approx 2.2364$

3. $f'(x) = 3x^2 + 1$

n	x_n	$f(x_n)$	$f'(x_n)$	$\dfrac{f(x_n)}{f'(x_n)}$	$x_n - \dfrac{f(x_n)}{f'(x_n)}$
1	0.5000	-0.3750	1.7500	-0.2143	0.7143
2	0.7143	0.0787	2.5306	0.0311	0.6832
3	0.6832	0.0021	2.4002	0.0009	0.6823

Approximation: $x \approx 0.682$

5. $f'(x) = \dfrac{5}{2\sqrt{x-1}} - 2$

n	x_n	$f(x_n)$	$f'(x_n)$	$\dfrac{f(x_n)}{f'(x_n)}$	$x_n - \dfrac{f(x_n)}{f'(x_n)}$
1	1.2	-0.1639	3.5902	-0.0457	1.2457
2	1.2457	-0.0131	3.0440	-0.0043	1.2500
3	1.2500	-0.000094	3.0003	-0.00003	1.25

Approximation: $x \approx 1.25$ (exact!)

7. $f'(x) = \dfrac{1}{x} + 1$

n	x_n	$f(x_n)$	$f'(x_n)$	$\dfrac{f(x_n)}{f'(x_n)}$	$x_n - \dfrac{f(x_n)}{f'(x_n)}$
1	0.6000	0.0892	2.1667	0.4120	0.5588
2	0.5588	-0.0231	2.7895	-0.0083	0.5671
3	0.5671	-0.0002	3.7634	-0.0001	0.5672

Approximation: $x \approx 0.567$

9. $f'(x) = -2xe^{-x^2} - 2x = -2x(e^{-x^2} + 1)$

n	x_n	$f(x_n)$	$f'(x_n)$	$\dfrac{f(x_n)}{f'(x_n)}$	$x_n - \dfrac{f(x_n)}{f'(x_n)}$
1	0.8000	-0.1127	-2.4437	0.0461	0.7539
2	0.7539	-0.0019	-2.3619	0.0008	0.7531

Approximations: $x \approx \pm 0.753$

11. $f'(x) = 3x^2 - 27$

n	x_n	$f(x_n)$	$f'(x_n)$	$\dfrac{f(x_n)}{f'(x_n)}$	$x_n - \dfrac{f(x_n)}{f'(x_n)}$
1	-5.0000	-17.0000	48.0000	-0.3542	-4.6458
2	-4.6458	-1.8371	37.7513	-0.0487	-4.5972
3	-4.5972	-0.0329	36.4019	-0.0009	-4.5963

n	x_n	$f(x_n)$	$f'(x_n)$	$\dfrac{f(x_n)}{f'(x_n)}$	$x_n - \dfrac{f(x_n)}{f'(x_n)}$
1	-1.0000	-1.0000	-24.0000	0.0417	-1.0417
2	-1.0417	-0.0053	-23.7448	0.0002	-1.0419

n	x_n	$f(x_n)$	$f'(x_n)$	$\dfrac{f(x_n)}{f'(x_n)}$	$x_n - \dfrac{f(x_n)}{f'(x_n)}$
1	6.0000	27.0000	81.0000	0.3333	5.6667
2	5.6667	1.9630	69.3333	0.0283	5.6384
3	5.6384	0.0136	68.3731	0.0002	5.6382

Approximations: $x \approx -4.596, -1.042, 5.638$

13. Let $h(x) = f(x) - g(x) = 4 - x - \ln x$. Then $h'(x) = -1 - (1/x)$. Starting with $x_1 = 3$, you obtain

$x_1 = 3$

$x_2 = 2.926$

$x_3 = 2.92627$

$x \approx 2.926.$

15. Let $3 - x = 1/(x^2 + 1)$ and define

$$h(x) = \frac{1}{x^2 + 1} + x - 3, \text{ then } h'(x) = -\frac{2x}{(x^2 + 1)^2} + 1.$$

n	x_n	$h(x_n)$	$h'(x_n)$	$\dfrac{h(x_n)}{h'(x_n)}$	$x_n - \dfrac{h(x_n)}{h'(x_n)}$
1	3.0000	0.1000	0.9400	0.1064	2.8936
2	2.8936	0.0003	0.9341	0.0003	2.8933

Approximation: $x \approx 2.893$

17. From the graph we see that the function has one zero and it is in the interval $(11, 12)$. Let $x_1 = 12$.

$$f(x) = \frac{1}{4}x^3 - 3x^2 + \frac{3}{4}x - 2$$

$$f'(x) = \frac{3}{4}x^2 - 6x + \frac{3}{4}$$

$$x_{n+1} = x_n - \frac{f(x_n)}{f'(x_n)}$$

$$= x_n - \frac{(1/4)x_n{}^3 - 3x_n{}^2 + (3/4)x_n - 2}{(3/4)x_n{}^2 - 6x_n + (3/4)}$$

$$= x_n - \frac{x_n{}^3 - 12x_n{}^2 + 3x_n - 8}{3x_n{}^2 - 24x_n + 3}$$

$$= \frac{2x_n{}^3 - 12x_n{}^2 + 8}{2x_n{}^2 - 24x_n + 3}$$

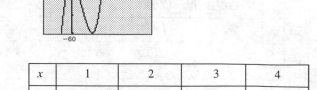

x	1	2	3	4
x_n	12.0000	11.8095	11.8033	11.8033

Zero: $x \approx 11.8033$

19. $f(x) = -x^4 + 5x^2 - 5$

$x \approx \pm 1.9021, \pm 1.1756$

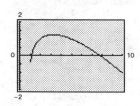

21. $f(x) = x^2 - \ln x - \frac{3}{2}$

$x \approx 1.3385, 0.2359$

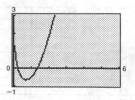

23. 1.1459, 7.8541

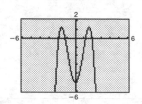

25. 0.8655

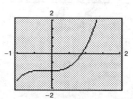

27. Newton's Method fails because $f'(x_1) = 0$.

29. Newton's Method fails because

$$\lim_{x \to \infty} x_n = \begin{cases} 1 = x_1 = x_3 = \ldots \\ 0 = x_2 = x_4 = \ldots \end{cases}.$$

Therefore, the limit does not exist.

31. Let $f(x) = x^2 - a$, then $f'(x) = 2x$.

$$x_{n+1} = x_n - \frac{x_n{}^2 - a}{2x_n} = \frac{x_n{}^2 + a}{2x_n}$$

33. $x_{i+1} = \dfrac{x_i{}^2 + 7}{2x_i}$

i	1	2	3	4	5
x_i	2.0000	2.7500	2.6477	2.6458	2.6458

Approximation: $\sqrt{7} \approx 2.646$

35. Let $f(x) = x^4 - 6$, then $f'(x) = 4x^3$.

$$x_{i+1} = x_i - \frac{x_i{}^4 - 6}{4x_i{}^3} = \frac{3x_i{}^4 + 6}{4x_i{}^3}$$

i	1	2	3	4	5
x_i	2.0000	1.6875	1.5778	1.5652	1.5651

Approximation: $\sqrt[4]{6} \approx 1.565$

37. Let $f(x) = (1/x) - a$, then $f'(x) = -1/x^2$.

$$x_{n+1} = x_n - \frac{(1/x_n) - a}{-1/(x_n{}^2)}$$

$$= x_n + (x_n - ax_n{}^2) = x_n(2 - ax_n)$$

39. The time is given by

$$T = \frac{\sqrt{x^2 + 4}}{3} + \frac{\sqrt{x^2 - 6x + 10}}{4}.$$

To minimize the time, we set dT/dx equal to zero and solve for x. This produces the equation

$$7x^4 - 42x^3 + 43x^2 + 216x - 324 = 0.$$

Let $f(x) = 7x^4 - 42x^3 + 43x^2 + 216x - 324$. Since $f(1) = -100$ and $f(2) = 56$, the solution is in the interval $(1, 2)$.

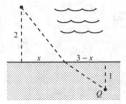

n	x_n	$f(x_n)$	$f'(x_n)$	$\dfrac{f(x_n)}{f'(x_n)}$	$x_n - \dfrac{f(x_n)}{f'(x_n)}$
1	1.7000	19.5887	135.6240	0.1444	1.5556
2	1.5556	−1.0414	150.2782	−0.0069	1.5625
3	1.5629	−0.0092	149.5693	−0.0001	1.5626

Approximation: $x \approx 1.563$ miles

41. To maximize C, we set dC/dt equal to zero and solve for t. This produces

$$C' = -\frac{3t^4 - 2t^3 + 300t + 50}{(50 + t^3)^2} = 0.$$

Let $f(t) = 3t^4 + 2t^3 - 300t - 50$. Since $f(4) = -354$ and $f(5) = 575$, the solution is in the interval $(4, 5)$.

n	t_n	$f(t_n)$	$f'(t_n)$	$\dfrac{f(t_n)}{f'(t_n)}$	$t_n - \dfrac{f(t_n)}{f'(t_n)}$
1	4.5000	12.4375	915.0000	0.0136	4.4864
2	4.4864	0.0658	904.3822	0.0001	4.4863

Approximation: $t \approx 4.486$ hours

43. $P(t) = A(t) - C(t) = 100{,}000e^{0.8\sqrt{t}}e^{-0.10t} - \displaystyle\int_0^t 1000e^{-0.10u}\, du = 100{,}000e^{0.8\sqrt{t}-0.10t} - 1000\displaystyle\int_0^t e^{-0.10u}\, du$

$P'(t) = 100{,}000\left(\dfrac{0.4}{\sqrt{t}} - 0.10\right)e^{0.8\sqrt{t}-0.10t} - 1000e^{-0.10t} = 1000e^{-0.10t}\left[100\left(\dfrac{0.4}{\sqrt{t}} - 0.10\right)e^{0.8\sqrt{t}} - 1\right] = 0$

Let $f(t) = \left(\dfrac{40}{\sqrt{t}} - 10\right)e^{0.8\sqrt{t}} - 1$, then:

$f'(t) = \left(\dfrac{40}{\sqrt{t}} - 10\right)\left(\dfrac{0.4}{\sqrt{t}}e^{0.8\sqrt{t}}\right) + e^{0.8\sqrt{t}}\left(-\dfrac{20}{t\sqrt{t}}\right) = e^{0.8\sqrt{t}}\left(\dfrac{16}{t} - \dfrac{4}{\sqrt{t}} - \dfrac{20}{t\sqrt{t}}\right)$

$t_{n+1} = t_n - \dfrac{f(t_n)}{f'(t_n)}$

n	1	2	3	4
t_n	16	15.8696	15.8686	15.8686

The timber should be harvested in 15.8686 years.

Review Exercises for Chapter 10

1. $a_n = \left(-\dfrac{1}{3}\right)^n$: $-\dfrac{1}{3}, \dfrac{1}{9}, -\dfrac{1}{27}, \dfrac{1}{81}, -\dfrac{1}{243}$

3. $a_n = \dfrac{4^n}{n!}$: $4, 8, 10\frac{2}{3}, 10\frac{2}{3}, \dfrac{128}{15}$

5. The sequence converges since

$$\lim_{n\to\infty} \frac{2n+3}{n^2} = 0.$$

7. The sequence diverges since

$$\lim_{n\to\infty} \frac{n^3}{n^2+1} = \infty.$$

9. The sequence converges since

$$\lim_{n\to\infty} \left(5 + \frac{1}{3^n}\right) = 5 + 0 = 5.$$

11. The sequence converges since

$$\lim_{n\to\infty} \frac{1}{n^{4/3}} = 0.$$

13. $a_n = \dfrac{n}{3n}$ or $a_n = \dfrac{1}{3}$

15. $a_n = (-1)^{n-1}\left(\dfrac{2^{n-1}}{3^n}\right)$, $n = 1, 2, 3, \ldots$

OR

$$a_n = (-1)^n\left(\frac{2^n}{3^{n+1}}\right), \quad n = 0, 1, 2, \ldots$$

17. (a) $a_1 = 15{,}000$

$a_2 = 15{,}000 + 10{,}000$

$\vdots$

$a_n = 15{,}000 + 10{,}000(n-1)$

(b) $a_1 + a_2 + a_3 + a_4 + a_5 = \displaystyle\sum_{n=1}^{15} [15{,}000 + 10{,}000(n-1)] = (15{,}000)5 + 100{,}000 = \$175{,}000$

19. $S_0 = 1$

$S_1 = 1 + \dfrac{3}{2} = \dfrac{5}{2} = 2.5$

$S_2 = 1 + \dfrac{3}{2} + \dfrac{9}{4} = \dfrac{19}{4} = 4.75$

$S_3 = 1 + \dfrac{3}{2} + \dfrac{9}{4} + \dfrac{27}{8} = \dfrac{65}{8} = 8.125$

$S_4 = 1 + \dfrac{3}{2} + \dfrac{9}{4} + \dfrac{27}{8} + \dfrac{81}{16} = \dfrac{211}{16} = 13.1875$

21. $S_1 = \dfrac{1}{2!} = \dfrac{1}{2} = 0.5$

$S_2 = \dfrac{1}{2!} - \dfrac{1}{4!} = \dfrac{11}{24} \approx 0.4583$

$S_3 = \dfrac{1}{2!} - \dfrac{1}{4!} + \dfrac{1}{6!} = \dfrac{331}{720} \approx 0.4597$

$S_4 = \dfrac{1}{2!} - \dfrac{1}{4!} + \dfrac{1}{6!} - \dfrac{1}{8!} = \dfrac{18{,}535}{40{,}320} \approx 0.4597$

$S_5 = \dfrac{1}{2!} - \dfrac{1}{4!} + \dfrac{1}{6!} - \dfrac{1}{8!} + \dfrac{1}{10!} = \dfrac{1{,}668{,}151}{3{,}628{,}800} \approx 0.4597$

23. This series diverges by the nth-Term Test since

$$\lim_{n\to\infty} \frac{n^2+1}{n(n+1)} = 1 \neq 0.$$

25. This geometric series converges since $r = 0.25 < 1$.

27. $\displaystyle\lim_{n\to\infty} \frac{2n}{n+5} = 2 \neq 0$, and the series diverges.

29. $\displaystyle\lim_{n\to\infty} \left(\frac{5}{4}\right)^n = \infty \neq 0$, and the series diverges.

31. $S_n = \sum_{k=0}^{n} \left(\frac{1}{5}\right)^k = \frac{1 - (1/5)^{n+1}}{1 - (1/5)} = \frac{5}{4}\left[1 - \left(\frac{1}{5}\right)^{n+1}\right]$

33. $S_n = \sum_{k=0}^{n} \frac{1}{2^n} + \sum_{k=0}^{n} \frac{1}{4^n}$

$= \frac{1 - (1/2)^{n+1}}{1 - (1/2)} + \frac{1 - (1/4)^n}{1 - (1/4)}$

$= 2\left[1 - \left(\frac{1}{2}\right)^{n+4}\right] + \frac{4}{3}\left[1 - \left(\frac{1}{4}\right)^{n+1}\right]$

35. $\sum_{n=0}^{\infty} \frac{1}{4}(4^n)$ diverges since $\lim_{n\to\infty} a_n \neq 0$.

37. $\sum_{n=0}^{\infty} [(0.5)^n + (0.2)^n] = \frac{1}{1 - 0.5} + \frac{1}{1 - 0.2}$

$= 2 + \frac{5}{4}$

$= \frac{13}{4}$

39. (a) $D_1 = 8$

$D_2 = 0.7(8) + 0.7(8) = 16(0.7)$

$D_3 = 16(0.7)^2$

$D = -8 + 16 + 16(0.7) + 16(0.7)^2 + \cdots$

(b) $D = -8 + \sum_{n=0}^{\infty} 16(0.7)^n$

$= -8 + \frac{16}{1 - 0.7}$

$= -8 + \frac{160}{3}$

$= \frac{136}{3}$ feet

41. $\sum_{n=1}^{\infty} \frac{1}{n^4}$ converges by the p-series test since $p = 4 > 1$.

43. $\sum_{n=1}^{\infty} \frac{1}{n\sqrt[4]{n}} = \sum_{n=1}^{\infty} \frac{1}{n^{5/4}}$ converges by the p-series test since $p = \frac{5}{4} > 1$.

45. $6, 5, 4\frac{2}{3}, \ldots$

Matches (a).

47. $10, 3, \ldots$

Matches (d).

49. $\sum_{n=1}^{\infty} \frac{1}{n^6} \approx 1 + \frac{1}{2^6} + \frac{1}{3^6} + \frac{1}{4^6} \approx 1.0172$

error $< \frac{1}{(5)4^5} \approx 1.9531 \times 10^{-4}$

51. $\sum_{n=1}^{\infty} \frac{1}{n^{5/4}} \approx 1 + \frac{1}{2^{5/4}} + \cdots + \frac{1}{7^{5/4}} \approx 2.17857$

error $< \frac{1}{(1/4)(7)^{1/4}} < 2.4592$

53. This series converges by the Ratio Test since

$\lim_{n\to\infty} \left| \frac{(n+1)4^{n+1}}{(n+1)!} \cdot \frac{n!}{n4^n} \right| = \lim_{n\to\infty} \frac{4}{n} = 0 < 1.$

55. This series diverges by the Ratio Test since

$\lim_{n\to\infty} \left| \frac{(-1)^{n+1}3^{n+1}}{n+1} \cdot \frac{n}{(-1)^n 3^n} \right| = \lim_{n\to\infty} \left| 3\frac{n}{n+1} \right| = 3 > 1.$

57. This series converges by the Ratio Test since $\lim_{n\to\infty} \left| \frac{(n+1)2^{n+1}}{(n+1)!} \cdot \frac{n!}{n2^n} \right| = \lim_{n\to\infty} \frac{2}{n} = 0 < 1.$

59. $\lim_{n\to\infty} \left| \frac{(-1)^{n+1}(x-2)^{n+1}/(n+2)^2}{(-1)^n(x-2)^n/(n+1)^2} \right| = \lim_{n\to\infty} \left| \frac{(n+1)^2}{(n+2)^2}(x-2) \right| = |x - 2| < 1 \Rightarrow R = 1$

61. $\lim\limits_{n\infty} \left| \dfrac{(n+1)!(x-3)^{n+1}}{n!(x-3)^n} \right| = \lim\limits_{n\infty} |(n+1)(x-3)| = \infty \Rightarrow R = 0$

63. $f(x) = e^{-0.5x}$ $\qquad\qquad f(0) = 1$

$\qquad f'(x) = -\dfrac{1}{2}e^{-1/2x}$ $\qquad f'(0) = -\dfrac{1}{2}$

$\qquad f''(x) = \dfrac{1}{4}e^{-1/2x}$ $\qquad f''(0) = \dfrac{1}{4}$

$$\vdots$$

$$f^{(n)}(0) = \left(-\dfrac{1}{2}\right)^n$$

$e^{-0.5x} = 1 - \dfrac{1}{2}x + \dfrac{1}{4}\cdot\dfrac{x^2}{2} - \dfrac{1}{8}\cdot\dfrac{x^3}{3!} + \cdots = \sum\limits_{n=0}^{\infty} \left(-\dfrac{1}{2}\right)^n \dfrac{x^n}{n!}$

65. $f(x) = \dfrac{1}{x}$ $\qquad\qquad f(-1) = -1$

$\qquad f'(x) = -\dfrac{1}{x^2}$ $\qquad\qquad f'(-1) = -1$

$\qquad f''(x) = \dfrac{2}{x^3}$ $\qquad\qquad f''(-1) = -2$

$\qquad f'''(x) = -\dfrac{6}{x^4}$ $\qquad\qquad f'''(-1) = -6$

$$\vdots$$

$$f^{(n)}(-1) = -(n!)$$

The power series for f is

$$\dfrac{1}{x} = f(-1) + f'(-1)(x+1) + \dfrac{f''(-1)(x+1)^2}{2!} + \cdots$$

$$= -1 - (x+1) - \dfrac{2(x+1)^2}{2!} - \dfrac{6(x+1)^3}{3!} - \cdots$$

$$= -[1 + (x+1) + (x+1)^2 + (x+1)^3 + \cdots]$$

$$= -\sum\limits_{n=0}^{\infty} (x+1)^n.$$

67. $\ln(x+2) = \ln\left[2\left(\dfrac{x}{2}+1\right)\right]$

$$= \ln 2 + \ln\left(\dfrac{x}{2}+1\right)$$

$$= \ln 2 + \dfrac{1}{2}x - \dfrac{1}{8}x^2 + \dfrac{1}{24}x^3 - \dfrac{1}{64}x^4 + \cdots$$

$$= \ln 2 + \sum\limits_{n=1}^{\infty} (-1)^{n+1} \dfrac{(x/2)^2}{n}$$

69. $(1+x^2)^2 = 1 + 2x^2 + \dfrac{2(1)x^4}{2!} + \cdots$

$$= 1 + 2x^2 + x^4 + \cdots$$

71. $\dfrac{1}{(x+3)^2} \approx \dfrac{1}{9} - \dfrac{2}{27}x + \dfrac{1}{27}x^2 - \dfrac{4}{243}x^3 + \dfrac{5}{729}x^4 - \dfrac{2}{729}x^5 + \dfrac{7}{6561}x^6$

73. $\ln(x + 2) \approx \ln 3 + \frac{1}{3}(x - 1) - \frac{1}{18}(x - 1)^2 + \frac{1}{81}(x - 1)^3 - \frac{1}{324}(x - 1)^4 + \frac{1}{1215}(x - 1)^5 - \frac{1}{4374}(x - 1)^6$

75. $f(1.25) \approx 4.770479903$

77. $f(1.5) \approx 0.9162835738$

79. error $= R_n = \dfrac{f^{(n+1)}(z)}{(n + 1)!}(x - c)^{n+1}$

$f(x) = \dfrac{2}{x} \Longrightarrow f^{(6)}(x) = \dfrac{1440}{x^7} \leq 1440$ on $\left[1, \dfrac{3}{2}\right]$.

$R_n \leq \dfrac{1440}{6!}(x - 1)^6 < 2\left(\dfrac{1}{2}\right)^6 = \dfrac{1}{32}$

81. $\sqrt{1 + x^3} = 1 + \dfrac{x^3}{2} - \dfrac{x^6}{8} + \cdots$

$\displaystyle\int_0^{0.3} \sqrt{1 + x^3}\, dx \approx \left[x + \dfrac{x^4}{8} - \dfrac{x^7}{56}\right]_0^{0.3}$

$= 0.3 + \dfrac{(0.3)^4}{8} - \dfrac{(0.3)^7}{56}$

≈ 0.301

83. $\ln(x^2 + 1) = x^2 - \dfrac{1}{2}x^4 + \dfrac{1}{3}x^6 - \cdots$

$\displaystyle\sum_0^{0.75} \ln(x^2 + 1)\, dx \approx \sum_0^{0.75}\left(x^2 - \dfrac{1}{2}x^4 + \dfrac{1}{3}x^6\right) dx = \left[\dfrac{x^3}{3} - \dfrac{x^5}{10} + \dfrac{x^7}{21}\right]_0^{0.75} \approx 0.12325$

85. Expected value $= \displaystyle\sum_{n=0}^{\infty} nP(n) = \sum_{n=0}^{\infty} n\left(\dfrac{1}{3}\right)^{n+1} = 0\left(\dfrac{1}{3}\right) + 1\left(\dfrac{1}{3}\right)^2 + 2\left(\dfrac{1}{3}\right)^3 + \cdots$

Since $(1 - x)^{-2} = 1 + 2x + 3x^2 + 4x^3 + \cdots$ (binomial series),

$\left(1 - \dfrac{1}{3}\right)^{-2} = 1 + 2\left(\dfrac{1}{3}\right) + 3\left(\dfrac{1}{3}\right)^2 + \cdots = \dfrac{9}{4}$

and

Expected value $= 1\left(\dfrac{1}{3}\right)^2 + 2\left(\dfrac{1}{3}\right)^3 + 3\left(\dfrac{1}{3}\right)^4 + \cdots$

$= \left(\dfrac{1}{3}\right)^2\left[1 + 2\left(\dfrac{1}{3}\right) + 3\left(\dfrac{1}{3}\right)^2 + \cdots\right]$

$= \left(\dfrac{1}{3}\right)^2\left(\dfrac{9}{4}\right) = \dfrac{1}{4} = 0.25$ units.

Expected production cost $= (23.00)(0.25) = \$5.75$

87. $f(x) = 2x^3 + 3x - 1$

$f'(x) = 6x^2 + 3$

$x_{n+1} = x_n - \dfrac{f(x_n)}{f'(x_n)} = x_n - \dfrac{2x_n^3 + 3x_n - 1}{6x_n^3 + 3}$

$x_1 = 0$ (initial guess)

$x_2 = \dfrac{1}{3}$

$x_3 = 0.31\overline{31}$

$x_4 = 0.3129$

$x \approx 0.313$

89. $f(x) = \ln 3x + x$

$f'(x) = \dfrac{1}{x} + 1$

$x_{n+1} = x_n - \dfrac{f(x_n)}{f'(x_n)}$

$x_1 = 0.5$ (initial guess)

$x_2 = 0.1982$

$x_3 = 0.2514$

$x_4 = 0.2576$

$x \approx 0.258$

91. Let

$$h(x) = f(x) - g(x) = x^5 - (x + 3)$$
$$h'(x) = 5x^4 - 1.$$

$x_1 = 1.5$ (initial guess)

$x_2 = 1.37275$

$x_3 = 1.34279$

$x_4 = 1.3413$

$x \approx 1.341$

95. Beginning: 1

Year 1: $1 + 0.07 = 1.07$

Year 2: $1.07 + 1.07(0.07) = 1.1449$

Year 3: 1.2250

Year 4: 1.3108

Year 5: 1.4026

Year 6: 1.5007

Year 7: 1.6058

Year 8: 1.7182

Year 9: 1.8385

Year 10: 1.9672

93. Let

$$h(x) = f(x) - g(x) = x^3 - e^{-x}$$
$$h'(x) = 3x^2 + e^{-x}.$$

$x_1 = 1$ (initial guess)

$x_2 = 0.8123$

$x_3 = 0.7743$

$x_4 = 0.7729$

$x \approx 0.773$

97. (a) $V_n = 120,000(0.7)^n$

(b) $V_5 = 120,000(0.7)^5 = \$20,168.40$

Practice Test for Chapter 10

1. Find the general term of the sequence $\frac{1}{2}, \frac{2}{5}, \frac{3}{10}, \frac{4}{17}, \frac{5}{26}, \ldots$.

2. Find the general term of the sequence $5, -7, 9, -11, 13, \ldots$.

3. Determine the convergence or divergence of the sequence whose general term is $a_n = \dfrac{n^2}{3n^2 + 4}$.

4. Determine the convergence or divergence of the sequence whose general term is $a_n = \dfrac{4n}{\sqrt{n^2 + 1}}$.

5. Find the sum of the series $\displaystyle\sum_{n=0}^{\infty} \left(\frac{1}{5^n} - \frac{1}{7^n} \right)$.

6. Determine the convergence or divergence of the series $\displaystyle\sum_{n=1}^{\infty} \frac{3^n}{n!}$.

7. Determine the convergence or divergence of the series $\displaystyle\sum_{n=1}^{\infty} \frac{1}{n\sqrt[3]{n}}$.

8. Determine the convergence or divergence of the series $\displaystyle\sum_{n=1}^{\infty} \frac{n}{2n + 3}$.

9. Determine the convergence or divergence of the series $\displaystyle\sum_{n=0}^{\infty} \frac{(-1)^n 6^n}{5^n}$.

10. Determine the convergence or divergence of the series $\displaystyle\sum_{n=1}^{\infty} \frac{\sqrt[3]{n}}{\sqrt{n}}$.

11. Determine the convergence or divergence of the series $\displaystyle\sum_{n=1}^{\infty} \frac{5^n n!}{(n + 1)!}$.

12. Determine the convergence or divergence of the series $\displaystyle\sum_{n=0}^{\infty} 4(0.27)^n$.

13. Determine the convergence or divergence of the series $\displaystyle\sum_{n=1}^{\infty} \left(1 + \frac{1}{3^n} \right)$.

14. Find the radius of convergence of the power series $\displaystyle\sum_{n=0}^{\infty} \frac{(-1)^n (x - 3)^n}{(n + 4)^2}$.

15. Find the radius of convergence of the power series $\displaystyle\sum_{n=0}^{\infty} \frac{x^n}{(n + 1)!}$.

16. Apply Taylor's Theorem to find the power series (centered at 0) for $f(x) = e^{-4x}$.

17. Apply Taylor's Theorem to find the power series (centered at 1) for $f(x) = \dfrac{1}{\sqrt[3]{x}}$.

18. Use the ninth-degree Taylor polynomial for e^{x^3} to approximate the value of $\displaystyle\int_0^{0.213} e^{x^3}dx$.

19. Use Newton's Method to approximate the zero of the function $f(x) = x^3 + x - 3$. (Make your approximation good to three decimal places.)

20. Use Newton's Method to approximate $\sqrt[4]{10}$ to three decimal places.

Graphing Calculator Required

21. Use SUM SEQ to evaluate the following sums.

(a) $\displaystyle\sum_{n=0}^{10} \dfrac{4}{2^n}$ (b) $\displaystyle\sum_{n=1}^{8} 3n!$

22. Graph the function $y = e^{2x}$ as well as the Taylor polynomials of degree 2, 4, and 6 on the same set of axes.

23. Use a program similar to the one on page 702 in the textbook to approximate the real roots of $3x^4 - 2x^3 + 5x^2 + 6x - 10 = 0$.

A P P E N D I C E S

A P P E N D I X   A
Alternate Introduction to the Fundamental Theorem of Calculus

Solutions to Odd-Numbered Exercises

1. Left Riemann sum: 0.518
   Right Riemann sum: 0.768

3. Left Riemann sum: 0.746
   Right Riemann sum: 0.646

5. Left Riemann sum: 0.859
   Right Riemann sum: 0.659

7. Midpoint Rule: 0.673

9. (a)

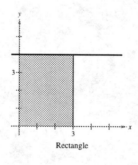

(b)

n	5	10	50	100
Left sum, S_L	1.6	1.8	1.96	1.98
Right sum, S_R	2.4	2.2	2.04	2.02

11. $\displaystyle\int_0^5 3\,dx$

13. $\displaystyle\int_{-4}^4 (4 - |x|)\,dx = \int_{-4}^0 (4 + x)\,dx + \int_0^4 (4 - x)\,dx$

15. $\displaystyle\int_{-2}^2 (4 - x^2)\,dx$

17. $A = 12$

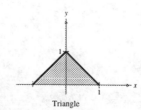

Rectangle

19. $A = 8$

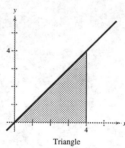

Triangle

21. $A = 14$

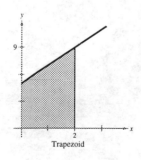

Trapezoid

23. $A = 1$

Triangle

25. $A = \dfrac{9\pi}{2}$

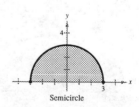

Semicircle

A P P E N D I X   D
Differential Equations

Section D.1   Solutions of Differential Equations

Solutions to Odd-Numbered Exercises

1. $y' = 3x^2$

3. $y' = -2e^{-2x}$ and $y' + 2y = -2e^{-2x} + 2(e^{-2x}) = 0$

5. $y' = 6x^2$ and $y' - \dfrac{3}{x}y = 6x^2 - \dfrac{3}{x}(2x^3) = 0$

7. $y'' = 2$ and $x^2y'' - 2y = x^2(2) - 2(x^2) = 0$

9. $y' = ye^{2x}$

$y'' = 8e^{2x}$ and $y'' - y' - 2y = 8e^{2x} - 4e^{2x} - 2(2x^{2x}) = 0$

11. By differentiation, we have

$$\frac{dy}{dx} = -\frac{1}{x^2}.$$

13. By differentiating, we have

$$\frac{dy}{dx} = 4Ce^{4x} = 4y.$$

15. Since $\dfrac{dy}{dt} = -\left(\dfrac{1}{3}\right)Ce^{-t/3}$, we have $3\dfrac{dy}{dt} + y - 7 = 3\left(-\dfrac{1}{3}Ce^{-t/3}\right) + (Ce^{-t/3} + 7) - 7 = 0$.

17. Since $y' = 2Cx - 3$, we have $xy' - 3x - 2y = x(2Cx - 3) - 3x - 2(Cx^2 - 3x) = 0$.

19. Since $y' = 2x + 2 - (C/x^2)$, we have $xy' + y = \left(2x^2 + 2x - \dfrac{C}{x}\right) + \left(x^2 + 2x + \dfrac{C}{x}\right) = 3x^2 + 4x = x(3x + 4)$.

21. Since $y' = \frac{1}{2}C_1e^{x/2} - 2C_2e^{-2x}$, we have $y'' = \frac{1}{4}C_1e^{x/2} + 4C_2e^{-2x}$, and it follows that

$2y'' + 3y' - 2y = \frac{1}{2}C_1e^{x/2} + 8C_2e^{-2x} + \frac{3}{2}C_1e^{x/2} - 6C_2e^{-2x} - 2C_1e^{x/2} - 2C_2e^{-2x} = 0$.

23. Since $y' = 4bx^3/(4 - a) + aCx^{a-1}$, we have

$$y' - \frac{ay}{x} = \left[\frac{4bx^3}{4 - a} + aCx^{a-1}\right] - \frac{a}{x}\left[\frac{bx^4}{4 - a} + Cx^a\right]$$

$$= \frac{4bx^3}{4 - a} + aCx^{a-1} - \frac{abx^3}{4 - a} - aCx^{a-1} = \frac{bx^3(4 - a)}{4 - a} = bx^3.$$

25. Since

$$y' = -2(1 + Ce^{x^2})^{-2}2xCe^{x^2} = -\frac{4xCe^{x^2}}{(1 + Ce^{x^2})^2}$$

we have

$$y' + 2xy = -\frac{4xCe^{x^2}}{(1 + Ce^{x^2})^2} + \frac{4x}{1 + Ce^{x^2}}$$

$$= \frac{-4xCe^{x^2} + 4x + 4xCe^{x^2}}{(1 + Ce^{x^2})^2} = x\left(\frac{2}{1 + Ce^{x^2}}\right)^2 = xy^2$$

27. Since $y' = \ln x + 1 + C$, we have $x(y' - 1) - (y - 4) = x(\ln x + 1 + C - 1) - (x \ln x + Cx + 4 - 4) = 0$.

29. By implicit differentiation, we have $2x + 2yy' = Cy'$, which implies that $2x = y'(C - 2y)$ and

$$y' = \frac{2x}{C - 2y} = \frac{2xy}{Cy - 2y^2} = \frac{2xy}{(x^2 + y^2) - 2y^2} = \frac{2xy}{x^2 - y^2}.$$

31. $2x + y + xy' = 0 \Rightarrow y' = \frac{-2x - y}{x} = -2 - \frac{y}{x}$

$$y'' = -\frac{y'}{x} + \frac{y}{x^2} = \frac{2}{x} + \frac{2y}{x^2}$$

$$x^2 y'' - 2(x + y) = x^2\left(\frac{2}{x} + \frac{2y}{x^2}\right) - 2(x + y) = 0$$

33. $y' = -2e^{-2x}$

$y'' = 4e^{-2x}$

$y''' = -8e^{-2x}$

$y^{(4)} = 16e^{-2x}$

Therefore, we have $y^{(4)} - 16y = 16e^{-2x} - 16(e^{-2x}) = 0$.

35. $y = 4x^{-1}$

$y' = -4x^{-2}$

$y'' = 8x^{-3}$

$y''' = -24x^{-4}$

$y^{(4)} = 96x^{-5}$

Therefore, we have $y^{(4)} - 16y = 96x^{-5} - 16(4x^{-1}) \neq 0$ and y is not a solution of the given differential equation.

37. $y = \frac{2}{9}xe^{-2x}$

$y' = -\frac{4}{9}xe^{-2x} + \frac{2}{9}e^{-2x}$

$y'' = \frac{8}{9}xe^{-2x} - \frac{8}{9}e^{-2x}$

$y''' = -\frac{16}{9}xe^{-2x} + \frac{24}{9}e^{-2x}$

Therefore,

$$y''' - 3y' + 2y = \left(-\tfrac{16}{9}xe^{-2x} + \tfrac{24}{9}e^{-2x}\right) - 3\left(-\tfrac{4}{9}xe^{-2x} + \tfrac{2}{9}e^{-2x}\right) + 2\left(\tfrac{2}{9}xe^{-2x}\right) = 2e^{-2x}.$$

This is *not* a solution to $y''' - 3y' + 2y = 0$.

39. $y = xe^x$

$y' = xe^x + e^x$

$y'' = xe^x + 2e^x$

$y''' = xe^x + 3e^x$

Therefore,

$$y''' - 3y' + 2y = (xe^x + 3e^x) - 3(xe^x + e^x) + 2(xe^x) = 0.$$

This *is* a solution to $y''' - 3y' + 2y = 0$.

41. Since $y' = -2Ce^{-2x} = -2y$, it follows that $y' + 2y = 0$. To find the particular solution, we use the fact that $y = 3$ when $x = 0$. That is, $3 = Ce^0 = C$. Thus, $C = 3$ and the particular solution is $y = 3e^{-2x}$.

43. Since $y' = C_2(1/x)$ and $y'' = -C_2(1/x^2)$, it follows that $xy'' + y' = 0$. To find the particular solution, we use the fact that $y = 5$ and $y' = 1/2$ when $x = 1$. That is,

$$\tfrac{1}{2} = C_2 \tfrac{1}{1} \quad \Rightarrow \quad C_2 = \tfrac{1}{2}$$

$$5 = C_1 + \tfrac{1}{2}(0) \quad \Rightarrow \quad C_1 = 5$$

Thus, the particular solution is
$y = 5 + \tfrac{1}{2}\ln|x| = 5 + \ln\sqrt{|x|}$.

45. Since $y' = 4C_1e^{4x} - 3C_2e^{-3x}$, and $y'' = 16C_1e^{4x} + 9C_2e^{-3x}$, it follows that $y'' - y' - 12y = 0$. To find the particular solution, we use the fact that $y = 5$ and $y' = 6$ when $x = 0$. That is,

$$C_1 + C_2 = 5$$
$$4C_1 - 3C_2 = 6$$

which implies that $C_1 = 3$ and $C_2 = 2$. The particular solution is

$$y = 3e^{4x} + 2e^{-3x}$$

47. Since

$$y' = e^{2x/3}\left(\tfrac{2}{3}C_1 + \tfrac{2}{3}C_2x + C_2\right)$$
$$y'' = e^{2x/3}\left(\tfrac{4}{9}C_1 + \tfrac{4}{9}C_2x + \tfrac{4}{3}C_2\right)$$

it follows that $9y'' - 12y' + 4y = 0$. To find the particular solution, we use the fact that $y = 4$ when $x = 0$, and $y = 0$ when $x = 3$. That is,

$$4 = e^0[C_1 + C_2(0)] \quad \Rightarrow \quad C_1 = 4$$
$$0 = e^2[4 + C_2(3)] \quad \Rightarrow \quad C_2 = -\tfrac{4}{3}$$

Therefore, the particular solution is $y = e^{2x/3}\left(4 - \tfrac{4}{3}x\right) = \tfrac{4}{3}e^{2x/3}(3 - x)$.

49. When $C = 1$, the graph is a parabola $y = x^2$.

When $C = 2$, the graph is a parabola $y = 2x^2$.

When $C = 4$, the graph is a parabola $y = 4x^2$.

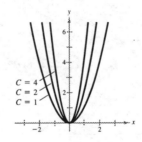

51. When $C = 0$, the graph is a straight line.

When $C = 1$, the graph is a parabola opening upward with a vertex at $(-2, 0)$.

When $C = -1$, the graph is a parabola opening downward with a vertex at $(-2, 0)$.

When $C = 2$, the graph is a parabola opening upward with a vertex at $(-2, 0)$.

When $C = -2$, the graph is a parabola opening downward with a vertex at $(-2, 0)$.

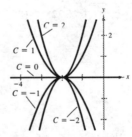

53. $y = \displaystyle\int 3x^2\, dx = x^3 + C$

55. $y = \displaystyle\int \frac{x + 3}{x}\, dx = \int\left(1 + \frac{3}{x}\right) dx = x + 3\ln|x| + C$

57. Letting $u = x - 3$, we have the following.

$$y = \int x\sqrt{x - 3}\, dx$$

$$= \int (u + 3)u^{1/2}\, du$$

$$= \int (u^{3/2} + 3u^{1/2})\, du = \tfrac{2}{5}u^{5/2} + 2u^{3/2} + C = \tfrac{2}{5}u^{3/2}(u + 5) + C$$

$$= \tfrac{2}{5}(x - 3)^{3/2}(x + 2) + C$$

59. Since $y = 4$ when $x = 4$, we have $4^2 = C4^3$ which implies that $C = \frac{1}{4}$ and the particular solutions is $y^2 = x^3/4$.

61. Since $y = 3$ when $x = 0$, we have $3 = Ce^0$ which implies that $C = 3$ and the particular solution is $y = 3e^x$.

63. (a) Since $N = 100$ when $t = 0$, it follows that $C = 650$. Therefore, the population function is $N = 750 - 650e^{-kt}$. Moreover, since $N = 160$ when $t = 2$, it follows that

$$160 = 750 - 650e^{-2k}$$

$$e^{-2k} = \frac{59}{65}$$

$$k = -\frac{1}{2}\ln\frac{59}{65} \approx -0.0484$$

Thus, the population function is $N = 750 - 650e^{-0.0484t}$.

(b) See accompanying graph.

(c) When $t = 4$, $N \approx 214$.

65. $x = 30,000 - Ce^{-kt}$. Since the product is new, $x = 0$ when $t = 0 \Rightarrow C = 30,000$. When $t = 1$, $x = 2000 \Rightarrow$

$$2000 = 30,000 - 30,000e^{-k}$$

$$\frac{1}{15} = 1 - e^{-k}$$

$$e^{-k} = \frac{14}{15}$$

$$k = -\ln\left(\frac{14}{15}\right) \approx 0.06899$$

$$x = 30,000 - 30,000e^{-0.06899t}$$

Year, t	2	4	6	8	10
Units, x	3867	7235	10,169	12,725	14,951

67. True

69.
$$y = a + Ce^{k(1-b)t}$$

$$\frac{dy}{dt} = Ck(1 - b)e^{k(1-b)t}$$

$$a + b(y - a) + \frac{1}{k}\frac{dy}{dt} = a + b[(a + Ce^{k(1-b)t}) - a] + \frac{1}{k}[Ck(1 - b)e^{k(1-b)t}]$$

$$= a + bCe^{k(1-b)t} + C(1 - b)e^{k(1-b)t}$$

$$= a + Ce^{k(1-b)t}[b + (1 - b)] = a + Ce^{k(1-b)t} = y$$

71. Since

$$\frac{ds}{dh} = -\frac{13}{\ln 3}\left(\frac{1/2}{h/2}\right) = -\frac{13}{\ln 3}\frac{1}{h}$$

and $-13/\ln 3$ is a constant, we can conclude that the equation is a solution to

$$\frac{ds}{dh} = \frac{k}{h} \text{ where } k = -\frac{13}{\ln 3}.$$

Section D.2   Separation of Variables

1. Yes, $\dfrac{dy}{dx} = \dfrac{x}{y+3}$

$(y+3)\,dy = x\,dx$

3. Yes, $\dfrac{dy}{dx} = \dfrac{1}{x} + 1$

$dy = \left(\dfrac{1}{x} + 1\right)dx$

5. No, the variables cannot be separated.

7. $\dfrac{dy}{dx} = 2x$

$\displaystyle\int dy = \int 2x\,dx$

$y = x^2 + C$

9. $\dfrac{dy}{dx} = \dfrac{1}{x}$

$\displaystyle\int dy = \int \dfrac{1}{x}\,dx$

$y = \ln|x| + C$

11. $3y^2\dfrac{dy}{dx} = 1$

$\displaystyle\int 3y^2\,dy = \int dx$

$y^3 = x + C$

$y = \sqrt[3]{x + C}$

13. $\dfrac{dy}{dx} = xy$

$\displaystyle\int \dfrac{1}{y}\,dy = \int x\,dx$

$\ln|y| = \dfrac{1}{2}x^2 + C_1$

$y - e^{(x^2/2) + C_1} = e^{C_1}e^{x^2/2} = Ce^{x^2/2}$

15. $\dfrac{dy}{dt} = \dfrac{e^t}{4y}$

$\displaystyle\int 4y\,dy = \int e^t\,dt$

$2y^2 = e^t + C_1$

$y^2 = \dfrac{1}{2}e^t + C$

17. $e^y\dfrac{dy}{dt} = 3t^2 + 1$

$\displaystyle\int e^y\,dy = \int (3t^2 + 1)\,dt$

$e^y = t^3 + t + C$

$y = \ln|t^3 + t + C|$

19. $(2 + x)\dfrac{dy}{dx} = 2y$

$\displaystyle\int \dfrac{1}{2y}\,dy = \int \dfrac{1}{2 + x}\,dx$

$\dfrac{1}{2}\ln|y| = \ln|C_1(2 + x)|$

$\sqrt{y} = C_1(2 + x)$

$y = C(2 + x)^2$

21.
$$\frac{dy}{dx} = \sqrt{1 - y}$$

$$\int (1 - y)^{-1/2}\, dy = \int dx$$

$$-2(1 - y)^{1/2} = x + C_1$$

$$\sqrt{1 - y} = \frac{-x}{2} + C$$

$$1 - y = \left(C - \frac{x}{2}\right)^2$$

$$y = 1 - \left(C - \frac{x}{2}\right)^2$$

23.
$$y' = \frac{dy}{dx} = (2x - 1)(y + 3)$$

$$\int \frac{1}{y + 3}\, dy = \int (2x + 1)\, dx$$

$$\ln|y + 3| = x^2 - x + C_1$$

$$y + 3 = e^{x^2 - x + C_1}$$

$$y = -3 + Ce^{x^2 - x}$$

25.
$$y' = \frac{dy}{dx} = \frac{x}{y} - \frac{x}{1 + y} = x\left(\frac{1}{y + y^2}\right)$$

$$\int (y + y^2)\, dy = \int x\, dx$$

$$\frac{y^2}{2} + \frac{y^3}{3} = \frac{x^2}{2} + C_1$$

$$3y^2 + 2y^3 = 3x^2 + C$$

27.
$$y\frac{dy}{dx} = e^x$$

$$\int y\, dy = \int e^x\, dx$$

$$\frac{y^2}{2} = e^x + C$$

When $x = 0$, $y = 4$. Therefore, $C = 7$ and the particular solution is $y^2 = 2e^x + 14$.

29.
$$\frac{dy}{dx} = -x(y + 4)$$

$$\int \frac{1}{y + 4}\, dy = \int -x\, dx$$

$$\ln|y + 4| = -\frac{x^2}{2} + C_1$$

$$|y + 4| = e^{-x^2/2 + C_1}$$

$$y = -4 + Ce^{-x^2/2}$$

When $x = 0$, $y = -5 \Rightarrow C = -1$ and $y = -4 - e^{-x^2/2}$.

31.
$$\frac{dy}{dx} = x^2(1 + y)$$

$$\frac{dy}{1 + y} = x^2\, dx$$

$$\ln(1 + y) = x^2 + C_1$$

$$y + 1 = e^{x^2 + C_1} = Ce^{x^2}$$

$$y = Ce^{x^2} - 1$$

$$y = 4e^{x^2} - 1$$

33.
$$\frac{dy}{dx} = \frac{6x}{5y}$$

$$\int 5y\, dy = \int 6x\, dx$$

$$\frac{5}{2}y^2 = 3x^2 + C_1 \implies 5y^2 = 6x^2 + C$$

When $x = -1$, $y = 1 \implies 5 = 6 + C \implies C = -1$.
Therefore, $5y^2 = 6x^2 - 1$ or $6x^2 - 5y^2 = 1$ (hyperbola).

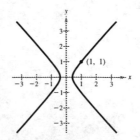

35. $$\frac{dv}{dt} = 3.456 - 0.1v$$

$$\int \frac{dv}{3.456 - 0.1v} = \int dt$$

$$-10 \ln|3.456 - 0.1v| = t + C_1$$

$$(3.456 - 0.1v)^{-10} = C_2 e^t$$

$$3.456 - 0.1v = Ce^{-0.1t}$$

$$v = -10Ce^{-0.1t} + 34.56$$

When $t = 0$, $v = 0$. Therefore, $C = 3.456$ and the solution is $v = 34.56(1 - e^{-0.1t})$.

37. From the differential equation, we have $T = Ce^{kt} + T_0$. We have $T_0 = 90$, and when $t = 0$, $T = 1500$. Thus, $1500 = Ce^0 + 90$ which implies that $C = 1410$. When $t = 1$, $T = 1120$ which implies that

$$1120 = 1410e^k + 90$$

$$k = \ln \frac{1030}{1410} = \ln \frac{103}{141}.$$

Therefore, $T = 1410e^{[\ln(103/141)]t} + 90$. When $t = 5$, we have $T \approx 383.298°$.

39. From the differential equation, we have $T = Ce^{kt} + T_0$. Since $T_0 = 0$ and $T = 70$ when $t = 0$, we have $T = 70e^{kt}$. When $t = 1$, $T = 48$ and we have $48 = 70e^k$ which implics that $k = \ln \frac{48}{70} = \ln \frac{24}{35}$. Therefore, $T = 70e^{[\ln(24/35)]t}$.

(a) When $t = 6$, we have $T \approx 7.277°$.

(b) When $T = 10$, we have

$$10 = 70e^{[\ln(24/35)]t}$$

$$\ln\left(\frac{1}{7}\right) = \ln\left(\frac{24}{35}\right)t$$

$$t = \frac{\ln(1/7)}{\ln(24/35)} \approx 5.158 \text{ hours.}$$

41. $$\int \frac{dN}{30 - N} = \int k \, dt$$

$$-\ln|30 - N| = kt + C_1$$

$$30 - N = e^{-(kt + C_1)}$$

$$30 - N = C_2 e^{-kt}$$

$$N = 30 + Ce^{-kt}$$

43. $$\frac{dy}{dx} = -k\frac{y}{x}$$

$$\int \frac{dy}{y} = \int \frac{-k}{x} \, dx$$

$$\ln|y| = -k \ln|x| + C_1 - \ln|x|^{-k} + \ln C = \ln(C|x|^{-k})$$

$$y = cx^{-k}$$

Section D.3   First-Order Linear Differential Equations

1. $x^3 - 2x^2y' + 3y = 0$

$$-2x^2y' + 3y = -x^3$$

$$y' + \frac{-3}{2x^2}y = \frac{x}{2}$$

3. $xy' + y = xe^x$

$$y' + \frac{1}{x}y = e^\lambda$$

5. $y + 1 = (x - 1)y'$

$(1 - x)y' + y = -1$

$y' + \dfrac{1}{1 - x}y = \dfrac{1}{x - 1}$

7. For this linear differential equation, we have $P(x) = 3x$ and $Q(x) = 6x$. Therefore, the integrating factor is $u(x) = e^{\int 3x\,dx} = e^{3x}$ and the general solution is

$$y = \frac{1}{u(x)} \int Q(x)u(x)\,dx = e^{-3x} \int 6e^{3x}\,dx = e^{-3x}(2e^{3x} + C) = 2 + Ce^{-3x}.$$

9. For this linear differential equation, we have $P(x) = 1$ and $Q(x) = e^{-x}$. Therefore, the integrating factor is $u(x) = e^{\int dx} = e^{x}$ and the general solution is

$$y = \frac{1}{u(x)} \int Q(x)u(x)\,dx = e^{-x} \int e^{-x} e^{x}\,dx = e^{-x}(x + C).$$

11. For this linear differential equation, we have $P(x) = 1/x$ and $Q(x) = 3x + 4$. Therefore, the integrating factor is

$$u(x) = e^{\int 1/x\,dx} = e^{\ln x} = x$$

and the general solution is

$$y = \frac{1}{u(x)} \int Q(x)u(x)\,dx = \frac{1}{x} \int (3x + 4)x\,dx = \frac{1}{x}(x^3 + 2x^2 + C) = x^2 + 2x + \frac{C}{x}.$$

13. For this linear differential equation, we have $P(x) = 5x$ and $Q(x) = x$. Therefore, the integrating factor is

$$u(x) = e^{\int 5x\,dx} = e^{(5/2)x^2}$$

and the general solution is

$$y = \frac{1}{u(x)} \int Q(x)u(x)\,dx = \frac{1}{e^{(5/2)x^2}} \int xe^{(5/2)x^2}\,dx = \frac{1}{e^{(5/2)x^2}}\left(\frac{1}{5}e^{(5/2)x^2} + C\right) = \frac{1}{5} + Ce^{-(5/2)x^2}$$

15. For this linear differential equation

$$y' + y\left(\frac{1}{x - 1}\right) = x + 1$$

we have $P(x) = 1/(x - 1)$ and $Q(x) = x + 1$. Therefore, the integrating factor is $u(x) = e^{\int 1/(x-1)\,dx} = e^{\ln(x-1)} = x - 1$ and the general solution is

$$y = \frac{1}{u(x)} \int Q(x)u(x)\,dx = \frac{1}{x - 1} \int (x + 1)(x - 1)\,dx = \frac{1}{x - 1}\left(\frac{x^3}{3} - x + C_1\right)$$

$$= \frac{x^3 - 3x + C}{3(x - 1)}.$$

17. For this linear differential equation $y' + \dfrac{1}{x}y = x + \dfrac{1}{x}$, we have $P(x) = \dfrac{1}{x}$ and $Q(x) = x + \dfrac{1}{x}$. Therefore, the integrating factor is

$$u(x) = e^{\int 1/x\,dx} = e^{\ln x} = x$$

and the general solution is

$$y = \frac{1}{u(x)} \int Q(x)u(x)\,dx = \frac{1}{x} \int \left(x + \frac{1}{x}\right)x\,dx$$

$$= \frac{1}{x}\left[\frac{x^3}{3} + x + C\right] = \frac{x^2}{3} + 1 + \frac{C}{x}$$

19. Separation of Variables:

$$\frac{dy}{dx} = 4 - y$$

$$\int \frac{dy}{4 - y} = \int dx$$

$$-\ln|4 - y| = x + C_1$$

$$4 - y = e^{-(x + C_1)}$$

$$4 - y = C_2 e^{-x}$$

$$y = 4 - C_2 e^x = 4 + Ce^{-x}$$

First-Order Linear: $P(x) = 1, \quad Q(x) = 4$

$$u(x) = e^{\int 1\, dx} = e^x$$

$$y = \frac{1}{e^x} \int 4e^x\, dx = \frac{1}{e^x}[4e^x + C] = 4 + Ce^{-x}$$

21. Separation of Variables:

$$\frac{dy}{dx} = 2x(1 + y)$$

$$\int \frac{dy}{1 + y} = \int 2x\, dx$$

$$\ln|1 + y| = x^2 + C_1$$

$$1 + y = e^{x^2 + C_1}$$

$$y = Ce^{x^2} - 1$$

First-Order Linear: $P(x) = -2x, \quad Q(x) = 2x$

$$u(x) = e^{\int -2x\, dx} = e^{-x^2}$$

$$y = \frac{1}{e^{-x^2}} \int 2xe^{-x^2}\, dx = e^{x^2}[-e^{-x^2} + C] = Ce^{x^2} - 1$$

23. Since $P(x) = 1$ and $Q(x) = 6e^x$, the integrating factor is

$$u(x) = e^{\int dx} = e^x$$

and the general solution is

$$y = \frac{1}{e^x} \int 6e^x(e^x)\, dx = \frac{1}{e^x}[3e^{2x} + C] = 3e^x + Ce^{-x}.$$

Since $y = 3$ when $x = 0$, it follows that $C = 0$, and the particular solution is $y = 3e^x$.

25. Since $P(x) = 1/x$ and $Q(x) = 0$, the integrating factor is $u(x) = e^{\int 1/x\, dx} = e^{\ln x} = x$ and the general solution is

$$y = \frac{1}{x} \int 0\, dx = \frac{C}{x}.$$

Since $y = 2$ when $x = 2$, it follows that $C = 4$, and the particular solution is $y = 4/x$ or $xy = 4$.

27. Since $P(x) = 1$ and $Q(x) = x$, the integrating factor is $u(x) = e^{\int dx} = e^x$ and the general solution is

$$y = e^{-x} \int xe^x\, dx = e^{-x}(xe^x - e^x + C) = x - 1 + Ce^{-x}.$$

Since $y = 4$ when $x = 0$, it follows that $C = 5$, and the particular solution is $y = x - 1 + 5e^{-x}$.

29. Since $P(x) = -2/x$ and $Q(x) = -x$, the integrating factor is

$$u(x) = e^{\int -2/x\, dx} = e^{-2\ln x} = \frac{1}{x^2}$$

and the general solution is

$$y = x^2 \int (-x)\left(\frac{1}{x^2}\right) dx = x^2(-\ln|x| + C).$$

Since $y = 5$ when $x = 1$, it follows that $C = 5$, and the particular solution is $y = x^2(5 - \ln|x|)$.

31. Since $P(t) = 0.2$ and $Q(t) = 20 + 0.2t$, the integrating factor is $u(t) = e^{\int 0.2\, dt} = e^{t/5}$ and the general solution is

$$S = e^{-t/5} \int e^{t/5}\left(20 + \frac{t}{5}\right) dt.$$

Using integration by parts, the integral is

$$S = e^{-t/5}(100e^{t/5} + te^{t/5} - 5e^{t/5} + C) = 100 + t - 5 + Ce^{-t/5} = 95 + t + Ce^{-t/5}.$$

Since $S = 0$ when $t = 0$, it follows that $C = -95$, and the particular solution is $S = t + 95(1 - e^{-t/5})$. During the first 10 years, the sales are as follows.

t	0	1	2	3	4	5	6	7	8	9	10
s	0	18.22	33.32	45.86	56.31	65.05	72.39	78.57	83.82	88.30	92.14

33. $\dfrac{dp}{dx}\left(1 - \dfrac{400}{3x}\right) = \dfrac{p}{x}$

$$\frac{dp}{dx}\left(x - \frac{400}{3}\right) = p$$

$$\int \frac{dp}{p} = \int \frac{dx}{x - (400/3)}$$

$$\ln|p| = \ln\left|x - \frac{400}{3}\right| + \ln|C|$$

$$p = C\left(x - \frac{400}{3}\right)$$

$$340 = C\left(20 - \frac{400}{3}\right) \implies C = -3$$

$$p = -3\left(x - \frac{400}{3}\right) = 400 - 3x$$

35. $\qquad\qquad D(t) = S(t)$

$$480 + 5p(t) - 2p'(t) = 300 + 8p(t) + p'(t)$$

$$180 = 3p'(t) + 3p(t)$$

$$60 = p'(t) + p(t)$$

$$P(t) = 1, \quad Q(t) = 60$$

$$u(t) = e^{\int 1\, dt} = e^t$$

$$p(t) = \frac{1}{e^t} \int 60e^t\, dt = \frac{1}{e^t}[60e^t + C] = 60 + Ce^{-t}$$

$$p(0) = 60 + C = 75 \implies C = 15$$

$$p(t) = 60 + 15e^{-t} = 15(4 + e^{-t})$$

37. (a) Since $P(t) = -r$ and $Q(t) = Pt$, the integrating factor is $u(t) = e^{\int -r\,dt} = e^{-rt}$ and the general solution is

$$A = e^{rt}\int Pte^{-rt}\,dt = Pe^{rt}\left(-\frac{t}{r}e^{-rt} - \frac{1}{r^2}e^{-rt} + C_1\right) = \frac{P}{r^2}(-rt - 1 + Ce^{rt}).$$

Since $A = 0$ when $t = 0$, it follows that $C = 1$, and the particular solution is

$$A = \frac{P}{r^2}(e^{rt} - rt - 1).$$

(b) When $t = 10$, $P = 500{,}000$ and $r = 0.09$, $A \approx \$34{,}543{,}402$.

39. (a)
$$\frac{dA}{dt} = rA - P$$

$$\frac{dA}{dt} - rA = -P$$

$$u(x) = e^{-\int r\,dt} = e^{-rt} \qquad \text{integrating factor}$$

$$A = e^{rt}\int -Pe^{-rt}\,dt = e^{rt}\left(\frac{P}{r}e^{-rt} + C\right) = \frac{P}{r} + Ce^{rt}$$

Since $A = A_0$ when $t = 0$, we have

$$C = A_0 - \frac{P}{r} \implies A = \frac{P}{r} + \left(A_0 - \frac{P}{r}\right)e^{rt}.$$

(b) If $A_0 = 2{,}000{,}000$, $r = 0.07$, $P = 250{,}000$ and $t = 5$,

$$A = \frac{250{,}000}{0.07} + \left(2{,}000{,}000 - \frac{250{,}000}{0.07}\right)e^{0.07(5)}$$

$$\approx \$1{,}341{,}465.28$$

(c) For $P = 40{,}000$, $t = 20$, $r = 0.08$ and $A = 0$,

$$0 = \frac{40{,}000}{0.08} + \left(A_0 - \frac{40{,}000}{0.08}\right)e^{0.08(20)}$$

$$-A_0e^{1.6} = 500{,}000 - 500{,}000e^{1.6}$$

$$A_0 = \$399{,}051.74.$$

Section D.4   Applications of Differential Equations

1. The general solution is $y = Ce^{kx}$. Since $y = 1$ when $x = 0$, it follows that $C = 1$. Thus, $y = e^{kx}$. Since $y = 2$ when $x = 3$, it follows that $2 = e^{3k}$ which implies that

$$k = \frac{\ln 2}{3} \approx 0.2310.$$

Thus, the particular solution is $y \approx e^{0.2310x}$

3. The general solution is $y = Ce^{kx}$. Since $y = 4$ when $x = 0$, it follows that $C = 4$. Thus, $y = 4e^{kx}$. Since $y = 1$ when $x = 4$, it follows that $\frac{1}{4} - e^{4k}$ which implies that

$$k = \frac{1}{4}\ln\frac{1}{4} \approx -0.3466.$$

Thus, the particular solution is $y \approx 4e^{-0.3466x}$.

5. The general solution is $y = Ce^{kx}$. Since $y = 2$ when $x = 2$ and $y = 4$ when $x = 3$, it follows that $2 = Ce^{2k}$ and $4 = Ce^{3k}$. By equating C-values from these two equations, we have the following.

$$2e^{-2k} = 4e^{-3k}$$

$$\tfrac{1}{2} = e^{-k} \implies k = \ln 2 \approx 0.6931$$

This implies that

$$C = 2e^{-2\ln 2} = 2e^{\ln(1/4)} = 2\left(\tfrac{1}{4}\right) = \tfrac{1}{2}.$$

Thus, the particular solution is

$$y = \tfrac{1}{2}e^{x\ln 2} \approx \tfrac{1}{2}e^{0.6931x}.$$

7. The general solution is $y = Ae^{kt}$ with $A = 2000$. Since $y = 2983.65$ when $t = 5$, we have

$$2983.65 = 2000e^{5k}$$

$$k = \frac{\ln(1.491825)}{5} \approx 0.08.$$

Thus, the particular solution is $y = 2000e^{0.08t}$. When $t = 10$, $y = 2000e^{0.08(10)} \approx \$4451.08$.

9.
$$\frac{dS}{dt} = k(L - S)$$

$$\int \frac{dS}{L - S} = \int k\,dt$$

$$-\ln|L - S| = kt + C_1$$

$$L - S = e^{-kt - C_1}$$

$$S = L + Ce^{-kt}$$

Since $S = 0$ when $t = 0$, we have $0 = L + C \implies C = -L$. Thus, $S = L(1 - e^{-kt})$.

11. The general solution is $y = Ce^{20kx}(20 - y)$. Since $y = 1$ when $x = 0$, it follows that $C = \tfrac{1}{19}$. Thus,

$$y = \frac{1}{19}e^{20kx}(20 - y).$$

Since $y = 10$ when $x = 5$, it follows that

$$19 = e^{100k}$$

$$20k = \frac{\ln 19}{5} \approx 0.5889.$$

Thus, the particular solution is

$$y = \frac{1}{19}e^{0.5889x}(20 - y)$$

$$y(19 + e^{0.5889x}) = 20e^{0.5889x}$$

$$y = \frac{20e^{0.5889x}}{19 + e^{0.5889x}} = \frac{20}{1 + 19e^{-0.5889x}}.$$

13. The general solution is $y = Ce^{5000kx}(5000 - y)$. Since $y = 250$ when $x = 0$, it follows that $C = \frac{1}{19}$. Thus,

$$y = \frac{1}{19}e^{5000kx}(5000 - y).$$

Since $y = 2000$ when $x = 25$, it follows that

$$\frac{38}{3} = e^{125,000k}$$

$$5000k = \frac{\ln(38/3)}{25} \approx 0.10156.$$

Thus, the particular solution is

$$y = \frac{1}{19}e^{0.10156x}(5000 - y)$$

$$y(19 + e^{0.10156x}) = 5000e^{0.10156x}$$

$$y = \frac{5000e^{0.10156x}}{19 + e^{0.10156x}} = \frac{5000}{1 + 19e^{-0.10156x}}.$$

15.
$$\frac{dN}{dt} = kN(500 - N)$$

$$\int \frac{dN}{N(500 - N)} = \int k\,dt$$

$$\frac{1}{500}\int \left[\frac{1}{N} + \frac{1}{500 - N}\right] dN = \int k\,dt$$

$$\ln|N| - \ln|500 - N| = 500(kt + C_1)$$

$$\frac{N}{500 - N} = e^{500kt + C_2} = Ce^{500kt}$$

$$N = \frac{500Ce^{500kt}}{1 + Ce^{500kt}}$$

When $t = 0$, $N = 100$. Thus, $100 = \frac{500C}{1 + C} \Rightarrow C = 0.25$. Thus, $N = \frac{125e^{500kt}}{1 + 0.25e^{500kt}}$.

When $t = 4$, $N = 200$. Thus, $200 = \frac{125e^{2000k}}{1 + 0.25e^{2000k}} \Rightarrow k = \frac{\ln(8/3)}{2000} \approx 0.00049$. Therefore,

$$N = \frac{125e^{0.2452t}}{1 + 0.25e^{0.2452t}} = \frac{500}{1 + 4e^{-0.2452t}}.$$

17. The differential equation is given by the following.

$$\frac{dP}{dn} = kP(L - P)$$

$$\int \frac{1}{P(L - P)}\,dP = \int k\,dn$$

$$\frac{1}{L}[\ln|P| - \ln|L - P|] = kn + C_1$$

$$\frac{P}{L - P} = Ce^{Lkn}$$

$$P = \frac{CLe^{Lkn}}{1 + Ce^{Lkn}} = \frac{CL}{e^{-Lkn} + C}$$

19. The general solution is $y = \dfrac{-1}{kt + C}$. Since $y = 45$ when $t = 0$, it

follows that $45 = \dfrac{-1}{C}$ and $C = \dfrac{-1}{45}$. Therefore,

$$y = -\frac{1}{kt - (1/45)} = \frac{45}{1 - 45kt}$$

Since $y = 4$ when $t = 2$, we have $4 = \dfrac{45}{1 - 45k(2)} \implies k = -\dfrac{45}{360}$.

Thus,

$$y = \frac{45}{1 + (41/8)t} = \frac{360}{8 + 41t}.$$

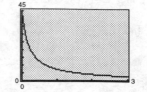

21. Since $y = 100$ when $t = 0$, it follows that $100 = 500e^{-C}$, which implies that $C = \ln 5$. Therefore, we have $y = 500e^{(-\ln 5)e^{-kt}}$. Since $y = 150$ when $t = 2$, it follows that

$$150 = 500e^{(-\ln 5)e^{-2k}}$$

$$e^{-2k} = \frac{\ln 0.3}{\ln 0.2}$$

$$k = -\frac{1}{2}\ln\frac{\ln 0.3}{\ln 0.2} \approx 0.1452.$$

Therefore, y is given by $y = 500e^{-1.6904e^{-0.1451t}}$

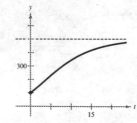

23. From Example 3, the general solution is

$$y = 60e^{-Ce^{-kt}}$$

Since $y = 8$ when $t = 0$,

$$8 = 60e^{-C} \implies C = \ln\frac{15}{2} \approx 2.0149.$$

Since $y = 15$ when $t = 3$,

$$15 = 60e^{-2.0149e^{-3k}}$$

$$\frac{1}{4} + e^{-2.0149e^{-3k}}$$

$$\ln\frac{1}{4} = -2.0149e^{-3k}$$

$$k = -\frac{1}{3}\ln\left(\frac{\ln 1/4}{-2.0149}\right) \approx 0.1246.$$

Thus,

$$y = 60e^{-2.0149e^{-0.1246t}}$$

When $t = 10$, $y \approx 34$ beavers.

25. Following Example 4, the differential equation is

$$\frac{dy}{dt} = ky(1 - y)(2 - y)$$

and its general solution is

$$\frac{y(2 - y)}{(1 - y)^2} = Ce^{2kt}$$

$$y = \frac{1}{2} \text{ when } t = 0 \implies \frac{(1/2)(3/2)}{(1/2)^2} = C \implies C = 3$$

$$y = 0.75 = \frac{3}{4} \text{ when } t = 4 \implies \frac{(3/4)(5/4)}{(1/4)^2} = 15 = 3e^{2k(4)}$$

$$\implies k = \frac{1}{8} \ln 5 \approx 0.2012.$$

Hence, the particular solution is

$$\frac{y(2 - y)}{(1 - y)^2} = 3e^{0.4024t}.$$

Using a symbolic algebra utility or graphing utility, you find that when $t = 10$,

$$\frac{y(2 - y)}{(1 - y)^2} = 3e^{0.4024(10)}$$

and $y \approx 0.92$.

27. (a) $\dfrac{dQ}{dt} = -\dfrac{Q}{20}$

$$\int \frac{dQ}{Q} = \int -\frac{1}{20} \, dt$$

$$\ln|Q| = -\frac{1}{20}t + C_1$$

$$Q = e^{-(1/20)t + C_1} = Ce^{-(1/20)t}$$

Since $Q = 25$ when $t = 0$, we have $25 = C$, thus, the particular solution is $Q = 25e^{-(1/20)t}$.

(b) When $Q = 15$, we have $15 = 25e^{-(1/20)t}$.

$$\frac{3}{5} = e^{-(1/20)t}$$

$$\ln\left(\frac{3}{5}\right) = -\frac{1}{20}t$$

$$-20 \ln\left(\frac{3}{5}\right) = t$$

$$t \approx 10.217 \text{ minutes}$$

29. $\dfrac{ds}{dh} = \dfrac{k}{h}$

$\displaystyle\int ds = \int \dfrac{k}{h}\, dh$

$s = k \ln h + C_1 = k \ln Ch$

Since $s = 25$ when $h = 2$ and $s = 12$ when $h = 6$, it follows that $25 = k \ln 2C$ and $12 = k \ln 6C$, which implies

$$C = \dfrac{1}{2} e^{-(25/13)\ln 3} \approx 0.0605 \qquad \text{and} \qquad k = \dfrac{25}{\ln 2C} = \dfrac{-13}{\ln 3} \approx -11.8331.$$

Therefore, s is given by the following.

$$s = -\dfrac{13}{\ln 3} \ln\!\left[\dfrac{h}{2} e^{-(25/13)\ln 3}\right]$$

$$= -\dfrac{13}{\ln 3}\!\left[\ln \dfrac{h}{2} - \dfrac{25}{13}\ln 3\right] = -\dfrac{1}{\ln 3}\!\left[13 \ln \dfrac{h}{2} - 25 \ln 3\right]$$

$$= 25 - \dfrac{13 \ln(h/2)}{\ln 3},$$

$$2 \le h \le 15$$

31. $\displaystyle\int \left(\dfrac{1}{y}\dfrac{dy}{dt}\right) dt = \int \left(\dfrac{1}{x}\dfrac{dx}{dt}\right) dt$

$\displaystyle\int \dfrac{1}{y}\, dy = \int \dfrac{1}{x}\, dx$

$\ln|y| = \ln|x| + C_1 = \ln|Cx|$

$y = Cx$

33. $\displaystyle\int \dfrac{1}{rA + P}\, dA = \int dt$

$\dfrac{1}{r}\ln|rA + P| = t + C_1$

$rA + P = Ce^{rt}$

$A = \dfrac{1}{r}(Ce^{rt} - P)$

Since $A = 0$ when $t = 0$, it follows that $C = P$. Therefore, we have

$$A = \dfrac{P}{r}(e^{rt} - 1).$$

35. Since $A = 120{,}000{,}000$ when $t = 8$ and $r = 0.1625$, we have

$$P = \dfrac{(0.1625)(120{,}000{,}000)}{e^{(0.1625)(8)} - 1} \approx \$7{,}305{,}295.15.$$

37. (a) $\displaystyle\int \frac{dC}{C} = \int -\frac{R}{V}\, dt$

$\ln|C| = -\dfrac{R}{V}t + K_1$

$C = Ke^{-Rt/V}$

Since $C = C_0$ when $t = 0$, it follows that $K = C_0$ and the function is $C = C_0 e^{-Rt/V}$.

(b) Finally, as $t \to \infty$, we have $\displaystyle\lim_{t\to\infty} C = \lim_{t\to\infty} C_0 e^{-Rt/V} = 0$.

39. (a) $\displaystyle\int \frac{1}{Q - RC}\, dC = \int \frac{1}{V}\, dt$

$-\dfrac{1}{R} \ln|Q - RC| = \dfrac{t}{V} + K_1$

$Q - RC = e^{-R[(t/V) + K_1]}$

$C = \dfrac{1}{R}\left(Q - e^{-R[(t/V) + K_1]}\right) = \dfrac{1}{R}\left(Q - Ke^{-Rt/V}\right)$

Since $C = 0$ when $t = 0$, it follows that $K = Q$ and we have $C = \dfrac{Q}{R}\left(1 - e^{-Rt/V}\right)$.

(b) As $t \to \infty$, the limit of C is Q/R.

Practice Test Solutions for Chapter 0

1. Rational   (Sec. 0.1)

2. (Sec. 0.1)
   (a) Satisfies   (b) Does not satisfy
   (c) Satisfies   (d) Satisfies

3. $x \geq 3$ (Sec. 0.1)

4. $-1 < x < 7$ (Sec. 0.1)

5. $\sqrt{19} > \frac{13}{3}$ (Sec. 0.1)

6. (Sec. 0.2)
   (a) $d = 10$
   (b) Midpoint: 2

7. $-\frac{11}{3} \leq x \leq 3$ (Sec. 0.2)

8. $x < -5$ or $x > \frac{33}{5}$ (Sec. 0.2)

9. $-\frac{25}{2} < x < \frac{55}{2}$ (Sec. 0.2)

10. $|x - 1| \leq 4$ (Sec. 0.2)

11. $3x^5$ (Sec. 0.3)

12. 1   (Sec. 0.3)

13. $2xy\sqrt[3]{4x}$ (Sec. 0.3)

14. $\frac{1}{4}(x + 1)^{-1/3}(x + 7)$ (Sec. 0.3)

15. $x < 5$ (Sec. 0.3)

16. $(3x + 2)(x - 7)$ (Sec. 0.4)

17. $(5x + 9)(5x - 9)$ (Sec. 0.4)

18. $(x + 2)(x^2 - 2x + 4)$
    (Sec. 0.4)

19. $-3 \pm \sqrt{11}$ (Sec. 0.4)

20. $-1, 2, 3$ (Sec. 0.4)

21. $\dfrac{-3}{(x - 1)(x + 3)}$ (Sec. 0.5)

22. $\dfrac{x + 13}{2\sqrt{x + 5}}$ (Sec. 0.5)

23. $\dfrac{1}{\sqrt{x}(x + 2)^{3/2}}$ (Sec. 0.5)

24. $\dfrac{3y\sqrt{y^2 + 9}}{y^2 + 9}$ (Sec. 0.5)

25. $-\dfrac{1}{2(\sqrt{x} - \sqrt{x + 7})}$ (Sec. 0.5)

26. $-1, 2, 4$ (Sec. 0.4)

Practice Test Solutions for Chapter 1

1. $d = \sqrt{82}$ (Sec. 1.1)

2. Midpoint: $(1, 3)$
   (Sec. 1.1)

3. Collinear   (Sec. 1.1)

4. $x = \pm 3\sqrt{5}$
   (Sec. 1.1)

5. x-intercepts: $(\pm 2, 0)$ (Sec. 1.2)
   y-intercept: $(0, 4)$

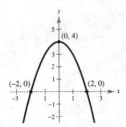

6. x-intercepts: $(2, 0)$ (Sec. 1.2)
   No y-intercept

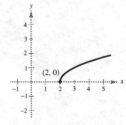

7. x-intercept: $(3, 0)$ (Sec. 1.2)

  y-intercept: $(0, 3)$

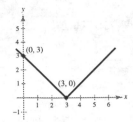

8. $(x - 4)^2 + (y + 1)^2 = 9$ (Sec. 1.2)

  Center: $(4, -1)$

  Radius: 3

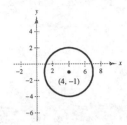

9. $(0, -5)$ and $(4, -3)$ (Sec. 1.2)

10. $6x - y - 38 = 0$ (Sec. 1.3)

11. $2x - 3y + 1 = 0$ (Sec. 1.3)

12. $x - 6 = 0$ (Sec. 1.3)

13. $5x + 2y - 6 = 0$ (Sec. 1.3)

14. (a) 4   (Sec. 1.4)

  (b) 31

  (c) $x^2 - 10 + 20$

  (d) $x^2 + 2x(\Delta x) + (\Delta x)^2 - 5$

15. Domain: $(-\infty, 3]$ (Sec. 1.4)

  Range: $[0, \infty)$

16. (a) $2x^2 + 1$ (Sec. 1.4)

  (b) $4(x + 1)(x + 2)$

17. $f^{-1}(x) = \sqrt[3]{x - 6}$ (Sec. 1.4)

18. 22   (Sec. 1.5)

19. 12   (Sec. 1.5)

20. Does not exist   (Sec. 1.5)

21. $\dfrac{\sqrt{5}}{10}$ (Sec. 1.5)

22. 5   (Sec. 1.5)

23. Discontinuities: $x = \pm 8$ (Sec. 1.6)

  $x = 8$ is removable.

24. $x = 3$ is a nonremovable discontinuity.

  (Sec. 1.6)

25. (Sec. 1.6)

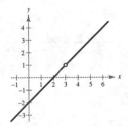

26. $y = \pm \sqrt{-x^2 - 6x - 5}$ (Sec. 1.2)

  Domain: $[-5, -1]$

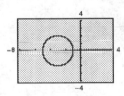

27. The graph does **not** show that the function does not exist at $x = 3$ on many graphing utilities.

  $\lim\limits_{x \to 3} f(x) = 6$ (Sec. 1.5)

Practice Test Solutions for Chapter 2

1. $\lim\limits_{\Delta x \to 0} \dfrac{f(x + \Delta x) - f(x)}{\Delta x} = \lim\limits_{\Delta x \to 0} (4x + 2\Delta x + 3)$

$$= 4x + 3$$

(Sec. 2.1)

2. $\lim\limits_{\Delta x \to 0} \dfrac{f(x + \Delta x) - f(x)}{\Delta x} = \lim\limits_{\Delta x \to 0} \dfrac{-1}{(x + \Delta x + 4)(x + 4)}$

$$= -\dfrac{1}{(x - 4)^2}$$

(Sec. 2.1)

3. $x - 4y + 2 = 0$ (Sec. 2.1)

4. $15x^2 - 12x + 15$ (Sec. 2.2)

5. $\dfrac{4x - 2}{x^3}$ (Sec. 2.2)

6. $\dfrac{2}{3\sqrt[3]{x}} + \dfrac{3}{5\sqrt[5]{x^2}}$ (Sec. 2.2)

7. (Sec. 2.3)

Average rate of change: 4

Instantaneous rates of change:

$$f'(0) = 0, \quad f'(2) = 12$$

8. (Sec. 2.3)

Marginal cost: $4.31 - 0.0002x$

9. $5x^4 + 28x^3 - 39x^2 - 56x + 36$ (Sec. 2.4)

10. $-\dfrac{x^2 + 14x + 8}{(x^2 - 8)^2}$ (Sec. 2.4)

11. $\dfrac{3x^4 + 14x^3 - 45x^2}{(x + 5)^2}$ (Sec. 2.4)

12. $-\dfrac{3x^2 + 4x + 1}{2\sqrt{x}(x^2 + 4x - 1)^2}$

(Sec. 2.4)

13. $72(6x - 5)^{11}$ (Sec. 2.5)

14. $-\dfrac{12}{\sqrt{4 - 3x}}$ (Sec. 2.5)

15. $\dfrac{18x}{(x^2 + 1)^4}$ (Sec. 2.5)

16. $\dfrac{\sqrt{10x}}{x(x + 2)^{3/2}}$ (Sec. 2.5)

17. $24x - 54$ (Sec. 2.6)

18. $-\dfrac{15}{16(3 - x)^{7/2}}$ (Sec. 2.6)

19. $-\dfrac{x^4}{y^4}$ (Sec. 2.7)

20. $-\dfrac{2(xy^3 + 1)}{3(x^2y^2 - 1)}$ (Sec. 2.7)

21. $\dfrac{8\sqrt{xy + 4} + y}{10\sqrt{xy + 4} - x} = \dfrac{41y - 32x}{50y - 41x}$

(Sec. 2.7)

22. $-\dfrac{8x^2}{y^2(x^3 - 4)^2}$ (Sec. 2.7)

23. $\dfrac{5}{12}$ (Sec. 2.8)

24. $\dfrac{dA}{dt} = 2\pi r \dfrac{dr}{dt}$ (Sec. 2.8)

$$\dfrac{dr}{dt} = \dfrac{5}{4\pi}$$

25. (Sec. 2.8)

$$V = \dfrac{4}{3}\pi h^3$$

$$\dfrac{dV}{dt} = 4\pi h^2 \dfrac{dh}{dt}$$

$$\dfrac{dh}{dt} = \dfrac{1}{8\pi}$$

26. (Sec. 2.4)

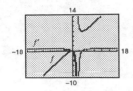

Horizontal Tangents at $x = 0$ and $x = 4$.
$f'(0) = f'(4) = 0$

27. (Sec. 2.7)

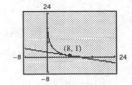

Tangent line: $y = -\dfrac{1}{4}x + 3$

Practice Test Solutions for Chapter 3

1. Increasing: $(-\infty, 0), (4, \infty)$

Decreasing: $(0, 4)$

(Sec. 3.1)

2. Increasing: $\left(-\infty, \frac{2}{3}\right)$

Decreasing: $\left(\frac{2}{3}, 1\right)$

(Sec. 3.1)

3. Relative minimum: $(2, -45)$

(Sec. 3.2)

4. Relative minimum: $(-3, 0)$

(Sec. 3.2)

5. Maximum: $(5, 0)$ (Sec. 3.2)

Minimum: $(2, -9)$

6. No inflection points   (Sec. 3.6)

7. Points of inflection: (Sec. 3.6)

$$\left(-\frac{1}{\sqrt{3}}, \frac{1}{4}\right), \left(\frac{1}{\sqrt{3}}, \frac{1}{4}\right)$$

8. $S = x + \dfrac{600}{x}$ (Sec. 3.4)

First number: $10\sqrt{6}$

Second number: $\dfrac{10\sqrt{6}}{3}$

9. $A = 3xy = 3x\left(\dfrac{3000 - 6x}{4}\right)$

$3x = 750$ feet, $y = 375$ feet

(Sec. 3.4)

10. $x \approx 13{,}333$ units   (Sec. 3.5)

11. $p = \$14{,}088$ (Sec. 3.5)

12. -1 (Sec. 3.5)

13. $-\infty$ (Scc. 3.6)

14. -2 (Sec. 3.6)

15. (Sec. 3.7)

Intercept: $(0, 0)$

Vertical asymptotes: $x = \pm 3$

Horizontal asymptote: $y = 1$

Relative maximum: $(0, 0)$

No inflection points

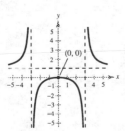

16. (Sec. 3.7)

Intercepts: $(-2, 0), \left(0, \frac{2}{5}\right)$

Horizontal asymptote: $y = 0$

Relative maximum: $\left(1, \frac{1}{2}\right)$

Relative minimum: $\left(-5, -\frac{1}{10}\right)$

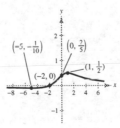

17. Intercept: $(0, -1)$ (Sec. 3.7)

No relative extrema

Inflection point: $(-1, -2)$

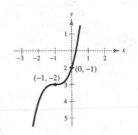

18. (Sec. 3.7)

Intercepts: $(0, 4), (2, 0)$

Relative minimum: $(2, 0)$

No inflection points

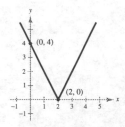

19. (Sec. 3.7)

Intercepts: $(2, 0), \left(0, \sqrt[3]{4}\right)$

Relative minimum: $(2, 0)$

No inflection points

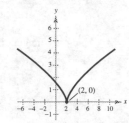

20. $\sqrt[3]{65} \approx 4.0208$ (Sec. 3.8)

21.

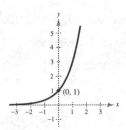

(Sec. 3.7)

Horizontal asymptotes at $y = \pm 5$.

No relative extrema.

22.

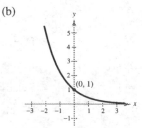

Yes, the graph crosses the horizontal asymptote $y = 2$.

(Sec. 3.7)

Practice Test Solutions for Chapter 4

1. (a) 81   (Sec. 4.1)

(b) $\frac{1}{32}$

(c) 1

2. (a) $x = 2$ (Sec. 4.1)

(b) $x = 32$

(c) $x = 5$

3. (a)

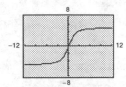

(b)

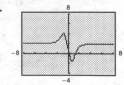

(Sec. 4.1)

4. (Sec. 4.1)

(a) $A \approx \$3540.28$

(b) $A \approx \$3618.46$

(c) $A \approx \$3626.06$

5. $6xe^{3x^2}$ (Sec. 4.2)

6. $\dfrac{e\sqrt[3]{x}}{3\sqrt[3]{x^2}}$ (Sec. 4.2)

7. $\dfrac{e^x - e^{-x}}{2\sqrt{e^x + e^{-x}}}$ (Sec. 4.2)

8. $x^2 e^{2x}(2x + 3)$ (Sec. 4.2)

9. $\dfrac{xe^x - e^x - 3}{4x^2}$ (Sec. 4.2)

10. $e^{1.6094\ldots} = 5$ (Sec. 4.3)

11. (a)

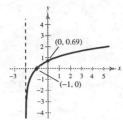

(0, 0.69)

(−1, 0)

(Sec. 4.3)

(b)

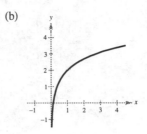

12. (a) $\ln\left(\dfrac{3x+1}{2x-5}\right)$ (Sec. 4.3)

   (b) $\ln\left(\dfrac{x^4}{y^3\sqrt{z}}\right)$

13. (a) $x = e^{17}$ (Sec. 4.3)

   (b) $x = \dfrac{\ln 2}{3\ln 5}$

14. $\dfrac{6}{6x-7}$ (Sec. 4.4)

15. $\dfrac{4x+15}{x(2x+5)}$ (Sec. 4.4)

16. $\dfrac{1}{x(x+3)}$ (Sec. 4.4)

17. $x^3(1 + 4\ln x)$ (Sec. 4.4)

18. $\dfrac{1}{2x\sqrt{\ln x + 1}}$ (Sec. 4.4)

19. (a) $y = 7e^{-0.7611t}$ (Sec. 4.5)

   (b) $y = 0.1501e^{0.4970t}$

20. $t \approx 5.776$ years   (Sec. 4.5)

21.

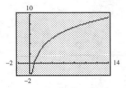

(Sec. 4.5)

The graphs are the same.

22.

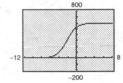

(Sec. 4.1)

$\lim\limits_{t\to\infty} f(t) = 600$

$\lim\limits_{x\to-\infty} f(t) = 0$

Practice Test Solutions for Chapter 5

1. $x^3 - 4x^2 + 5x + C$ (Sec. 5.1)

2. (Sec. 5.1)

$\dfrac{x^4}{4} + \dfrac{7x^3}{3} - 2x^2 - 28x + C$

3. $\dfrac{x^2}{2} - 9x - \dfrac{1}{x} + C$ (Sec. 5.1)

4. $-\dfrac{1}{5}(1 - x^4)^{5/4} + C$ (Sec. 5.2)

5. $\dfrac{9}{14}(7x)^{2/3} + C$ (Sec. 5.2)

6. $-\dfrac{2}{33}(6 - 11x)^{3/2} + C$ (Sec. 5.2)

7. $\dfrac{4}{5}x^{5/4} + \dfrac{6}{7}x^{7/6} + C$ (Sec. 5.1)

8. $-\dfrac{1}{3x^3} + \dfrac{1}{4x^4} + C$ (Sec. 5.1)

9. $x - x^3 + \dfrac{3}{5}x^5 - \dfrac{1}{7}x^7 + C$

   (Sec. 5.1)

10. $-\dfrac{5}{12(1 + 3x^2)^2} + C$

   (Sec. 5.2)

11. $\left(\dfrac{1}{7}\right)e^{7x} + C$ (Sec. 5.3)

12. $\left(\dfrac{1}{8}\right)e^{4x^2} + C$ (Sec. 5.3)

13. $\left(\frac{1}{16}\right)(1 + 4e^x)^4 + C$

(Sec. 5.3)

14. $\left(\frac{1}{2}\right)e^{2x} + 4e^x + 4x + C$

(Sec. 5.3)

15. $\left(\frac{1}{2}\right)e^{2x} - 4x - e^{-x} + C$

(Sec. 5.3)

16. $\ln|x + 6| + C$ (Sec. 5.3)

17. $-\left(\frac{1}{3}\right)\ln|8 - x^3| + C$

(Sec. 5.3)

18. $\frac{1}{3}\ln(1 + 3e^x) + C$ (Sec. 5.3)

19. $\frac{(\ln x)^7}{7} + C$ (Sec. 5.3)

20. $\frac{x^2}{2} + x + 6\ln|x - 1| + C$ (Sec. 5.3)

(Use long division first)

21. -3 (Sec. 5.4)

22. $\frac{381}{7}$ (Sec. 5.4)

23. 2 (Sec. 5.4)

24. $A = 36$ (Sec. 5.5)

25. $A = \frac{1}{2}$ (Sec. 5.5)

26. $A = \frac{2}{3}$ (Sec. 5.5)

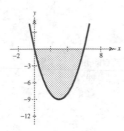

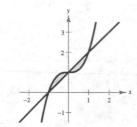

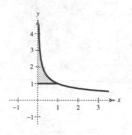

27. 1.4949   (Sec. 5.6)

28. 0.1472   (Sec. 5.6)

29. 3π (Sec. 5.7)

30. $\frac{5000\pi}{3}$ (Sec. 5.7)

31. $n = 50$: 22.442278   (Sec. 5.6)

$n = 100$: 22.443875

32. (c) Actual area is 4.5

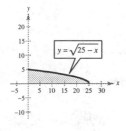

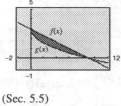

(Sec. 5.5)

Practice Test Solutions for Chapter 6

1. $\frac{2}{5}(x + 3)^{3/2}(x - 2) + C$

(Sec. 6.1)

2. $-\frac{x - 1}{(x - 2)^2} + C$ (Sec. 6.1)

3. $\frac{2}{3}\ln|3\sqrt{x} + 1| + C$ (Sec. 6.1)

4. $\frac{(\ln 7x)^2}{2} + C$ (Sec. 6.1)

5. $\frac{1}{4}e^{2x}(2x - 1) + C$ (Sec. 6.2)

6. $\frac{x^4}{16}[4(\ln x) - 1] + C$ (Sec. 6.2)

7. $\frac{2}{35}(x - 6)^{3/2}(5x^2 + 24x + 96) + C$

(Sec. 6.2)

8. $\frac{1}{32}e^{4x}(8x^2 - 4x + 1) + C$

(Sec. 6.2)

9. $\ln\left|\frac{x + 3}{x - 2}\right| + C$ (Sec. 6.3)

10. $\ln \left| \dfrac{x^3}{(x+4)^2} \right| + C$ (Sec. 6.3)   **11.** $5 \ln |x+2| + \dfrac{7}{x+2} + C$ **12.** $\dfrac{3}{2}x^2 + \ln \dfrac{|x|}{(x+2)^2} + C$

(Sec. 6.3)   (Sec. 6.3)

13. $-\dfrac{\sqrt{16-x^2}}{16x} + C$ (Sec. 6.4)   **14.** $x[(\ln x)^3 - 3(\ln x)^2 + 6(\ln x) - 6] + C$ (Sec. 6.4)

15. $1200x - 20,000 \ln(1 + e^{0.06x}) + C$ (Sec. 6.4)   **16.** (a) 15.567   (Sec. 6.5)

(b) 15.505

17. (a) 1.191   (Sec. 6.5)   **18.** Convergent; 6   (Sec. 6.6)

(b) 1.196

19. Divergent   (Sec. 6.6)   **20.** Divergent   (Sec. 6.6)

21. $n = 50$:   1.652674   (Sec. 6.5)   **22.** $n = 100$:   8.935335   (Sec. 6.5 and 6.6)

$n = 100$:   1.652674 $n = 1000$:   2.288003

$n = 10,000$:   1.636421

Converges $\left(\text{Actual answer is } \dfrac{\pi}{2} \right)$

Practice Test Solutions for Chapter 7

1. (a) $d = 14\sqrt{2}$ (Sec. 7.1)   **2.** $(x-1)^2 + (y+3)^2 + z^2 = 5$ **3.** Center: $(2, -1, -4)$ (Sec. 7.1)

(b) Midpoint: $(4, 2, -2)$ (Sec. 7.1)   Radius: $\sqrt{21}$

4. (Sec. 7.2)

(a) x-intercept: $(8, 0, 0)$ (b) $y = 2$

y-intercept: $(0, 3, 0)$ Parallel to xz-plane

z-intercept: $(0, 0, 4)$

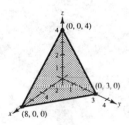

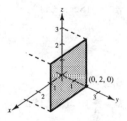

5. (Sec. 7.2)   **6.** (a) Domain: $x + y < 3$ (Sec. 7.3)

(a) Hyperboloid of one sheet   (b) Domain: all points in the xy-plane except the origin

(b) Elliptic paraboloid

7. $f_x(x, y) = 6x + 9y^2 - 3$ (Sec. 7.4)   **8.** $f_x(x, y) = \dfrac{2x}{x^2 + y^2 + 5}$ (Sec. 7.4)

$f_y(x, y) = 18xy + 12y^2 - 6$

$f_y(x, y) = \dfrac{2y}{x^2 + y^2 + 5}$

9. $\dfrac{\partial w}{\partial x} = 2xy^3 \sqrt{z}$ (Sec. 7.4)

$\dfrac{\partial w}{\partial y} = 3x^2 y^2 \sqrt{z}$

$\dfrac{\partial w}{\partial z} = \dfrac{x^2 y^3}{2\sqrt{z}}$

10. $\dfrac{\partial^2 z}{\partial x^2} = 2x\left(\dfrac{x^2 - 3y^2}{(x^2 + y^2)^3}\right)$ (Sec. 7.4)

$\dfrac{\partial^2 z}{\partial y \partial x} = 2y\left(\dfrac{3x^2 - y^2}{(x^2 + y^2)^3}\right)$

$\dfrac{\partial^2 z}{\partial x \partial y} = 2y\left(\dfrac{3x^2 - y^2}{(x^2 + y^2)^3}\right)$

$\dfrac{\partial^2 z}{\partial y^2} = 2x\left(\dfrac{3y^2 - x^2}{(x^2 + y^2)^3}\right)$

11. Relative minimum: $(1, -2, -23)$ (Sec. 7.5)

12. Saddle point: $(0, 0, 0)$ (Sec. 7.5)

Relative maxima: $(1, 1, 2)$, $(-1, -1, 2)$

13. $f(2, -8) = -16$ (Sec. 7.6)

14. $f(4, 0) = -36$ (Sec. 7.6)

15. $y = \frac{1}{65}(-51x + 355)$ (Sec. 7.7)

16. $y = \frac{1}{6}x^2 - \frac{7}{26}x + \frac{7}{3}$ (Sec. 7.7)

17. $\frac{81}{16}$ (Sec. 7.8)

18. $-\frac{135}{4}$ (Sec. 7.8)

19. (a) $A = \displaystyle\int_{-2}^{2}\int_{3}^{7-x^2} dy\, dx = \int_{3}^{7}\int_{-\sqrt{7-y}}^{\sqrt{7-y}} dx\, dy$ (Sec. 7.8)

(b) $A = \displaystyle\int_{0}^{1}\int_{x^2+2}^{x+2} dy\, dx = \int_{2}^{3}\int_{y-2}^{\sqrt{y-2}} dx\, dy$

20. $V = \displaystyle\int_{0}^{4}\int_{0}^{4-x}(4 - x - y)\, dy\, dx = \dfrac{32}{3}$ (Sec. 7.9)

21. $y \approx 0.832t + 20.432$ (Sec. 7.7)

$r \approx 0.983$

22. 1.028531×10^{17} (Sec. 7.8)

Practice Test Solutions for Chapter 8

1. $\frac{11}{16}$ (Sec. 8.1)

2. $\frac{5}{13}$ (Sec. 8.1)

3. $E(x) = \frac{1}{2}$ (Sec. 8.1)

$V(x) = 4.05$

$\sigma \approx 2.012$

4. $k = \dfrac{1}{4}$ (Sec. 8.2)

5. (a) $\dfrac{25}{64}$ (Sec. 8.2)

(b) $\dfrac{63}{64}$

6. (a) 4   (Sec. 8.3)

(b) $\dfrac{4\sqrt{5}}{5}$

(c) 4

7. (a) $6\ln\left(\dfrac{3}{2}\right) \approx 2.433$ (Sec. 8.3)

(b) $\sqrt{6 - 36\left(\ln\dfrac{3}{2}\right)^2} \approx 0.286$

(c) $\dfrac{12}{5}$

8. $\mu = \dfrac{1}{7}$ (Sec. 8.3)

Median: $\dfrac{\ln 2}{7}$

$\sigma = \dfrac{1}{7}$

9. 0.0469   (Sec. 8.3)

10. $P(19 < x < 24) = 0.4401$ (Sec. 8.3)

Practice Test Solutions for Chapter 9

1. $y' = 3x^2 - 4 - \dfrac{C}{x^2}$ (Sec. 9.1)

$xy' + xy = x\left(3x^2 - 4 - \dfrac{C}{x^2}\right) + \left(x^3 - 4x + \dfrac{C}{x}\right)$

$\qquad = 3x^3 - 4x - \dfrac{C}{x} + x^3 - 4x + \dfrac{C}{x}$

$\qquad = 4x^3 - 8x = 4x(x^2 - 2)$

2. $y' = -5Ce^{-5x}$ (Sec. 9.1)

$y'' = 25Ce^{-5x}$

$y''' = -125Ce^{-5x}$

$y''' + 125y = -125Ce^{-5x} + 125Ce^{-5x} = 0$

3. $y^4 = 2x^2 + 8x + C$ (Sec. 9.2)

4. $y = C(x - 1) - 4$ (Sec. 9.2)

5. $y(\ln y - 1) - e^x(x - 1) + C$

(Sec. 9.2)

6. $y = \left(\dfrac{1}{2}\right)e^{-2x} + Ce^{-4x}$

(Sec. 9.3)

7. $y = -\dfrac{e^{1/x^2}}{2x^2} + Ce^{1/x^2}$

(Sec. 9.3)

8. $y = -3x^2 + \dfrac{1}{4} + Cx^4$

(Sec. 9.3)

9. $y = 30 - 26e^{-0.0523t}$ (Sec. 9.4)

10. $y = 1000e^{-2.9957e^{-0.1553t}}$ (Sec. 9.4)

11. $y = Ce^{x^3}$ (Sec. 9.2)

12. $t^2 + Q^2 = C$ (Sec. 9.2)

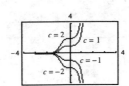

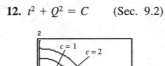

Practice Test Solutions for Chapter 10

1. $a_n = \dfrac{n}{n^2 + 1}$ (Sec. 10.1)

2. $a_n = (-1)^{n-1}(2n + 3)$

(Sec. 10.1)

3. Converges to $\dfrac{1}{3}$ (Sec. 10.1)

4. Converges to 4   (Sec. 10.1)

5. $\frac{1}{12}$ (Sec. 10.2)

6. Converges by the Ratio Test

(Sec. 10.3)

7. Converges since it is a p-series with $p = \frac{4}{3} > 1$.

(Sec. 10.3)

8. Diverges by the nth-Term Test   (Sec. 10.2)

9. Diverges since it is a geometric series with

$|r| = \left|-\dfrac{6}{5}\right| = \dfrac{6}{5} > 1.$

(Sec. 10.2)

10. Diverges since it is a p-series with $p = \frac{1}{6} < 1.$

(Sec. 10.3)

11. Diverges by the Ratio Test   (Sec. 10.3)

12. Converges since it is a geometric series with

$|r| = |0.27| = 0.27 < 1.$

(Sec. 10.2)

13. Diverges by the *n*th-Term Test
    (Sec. 10.2)

14. $R = 1$ (Sec. 10.4)

15. $R = \lim_{n \to \infty} (n + 2) = \infty$
    (Sec. 10.4)

16. $e^{-4x} = \sum_{n=0}^{\infty} \frac{(-4x)^n}{n!}$ (Sec. 10.5)

17. $\frac{1}{\sqrt[3]{x}} = 1 + \sum_{n=1}^{\infty} \frac{(-1)^n 1 \cdot 4 \cdot 7 \cdots (3n - 2)(x - 1)^n}{3^n n!}$
    (Sec. 10.5)

18. 0.214   (Sec. 10.5)

19. $x \approx 1.213$ (Sec. 10.6)

20. $\sqrt[4]{10} \approx 1.778$ (Sec. 10.6)

21. (Sec. 10.2)
    (a) 7.9961
    (b) 138,699

22.

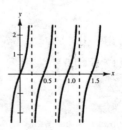

(Sec. 10.5)

23. -1.2090 and 0.9021
    (Sec. 10.6)

Practice Test Solutions for Appendix D

1. (a) 93.913°   (Sec. D.1)
    (b) $\frac{7\pi}{12}$

2. (a) 140°, $-580°$ (Sec. D.1)
    (b) $\frac{25\pi}{9}$, $-\frac{11\pi}{9}$

3. $\sin \theta = \frac{y}{r} = -\frac{5}{13}$ $\csc \theta = \frac{r}{y} = -\frac{13}{5}$

   $\cos \theta = \frac{x}{r} = \frac{12}{13}$ $\sec \theta = \frac{r}{x} = \frac{13}{12}$

   $\tan \theta = \frac{y}{x} = -\frac{5}{12}$ $\cot \theta = \frac{x}{y} = -\frac{12}{5}$

   (Sec. D.2)

4. $\theta = 0, \frac{\pi}{2}, \frac{3\pi}{2}$ (Sec. D.2)

5. (a) Period: 8π (Sec. D.3)
        Amplitude: 3

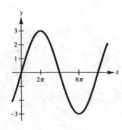

(b) Period: $\frac{1}{2}$

6. $3(1 + \sin x)$ (Sec. D.4)

7. $x(x \sec^2 x + 2 \tan x)$
    (Sec. D.4)

8. $3 \sin^2 x \cos x$ (Sec. D.4)

9. $\frac{\sec x(x \tan x - 2)}{x^3}$ (Sec. D.4)

10. $5 \cos 10x$ (Sec. D.4)

11. $-\frac{1}{2}\sqrt{\csc x} \cot x$ (Sec. D.4)

12. $\sec x$ (Sec. D.4)

13. $-2e^{2x}\csc^2 e^{2x}$ (Sec. D.4)

14. $3\sec(x^2 + y) - 2x$ (Sec. D.4)

15. $-\dfrac{1}{3}\sin^2 3y\sec^2 x$ (Sec. D.4)

16. $\dfrac{1}{4}\sin 4x + C$ (Sec. D.5)

17. $-8\cot\dfrac{x}{8} + C$ (Sec. D.5)

18. $-\dfrac{1}{2}\ln|\cos x^2| + C$

(Sec. D.5)

19. $\dfrac{\sin^6 x}{6} + C$ (Sec. D.5)

20. $\ln|\csc x - \cot x| + \cos x + C$

(Sec. D.5)

21. $e^{\tan x} + C$ (Sec. D.5)

22. $-\ln|1 + \cos x| + C$

(Sec. D.5)

23. $2\tan x - 2\sec x - x + C$

(Sec. D.5)

24. $x\sin x + \cos x + C$

(Sec. D.5)

25. $\dfrac{\pi^2 - 8\sqrt{2}}{16}$ (Sec. D.5)

26. ∞ (Sec. D.6)

27. $\dfrac{7}{3}$ (Sec. D.6)

28. $\dfrac{5}{9}$ (Sec. D.6)

29.

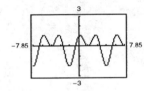

Minimum: -2

Maximum: 1.125

(Sec. D.3)

30.

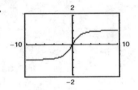

The limit is 1. L'Hôpital's Rule fails.

(Sec. D.6)